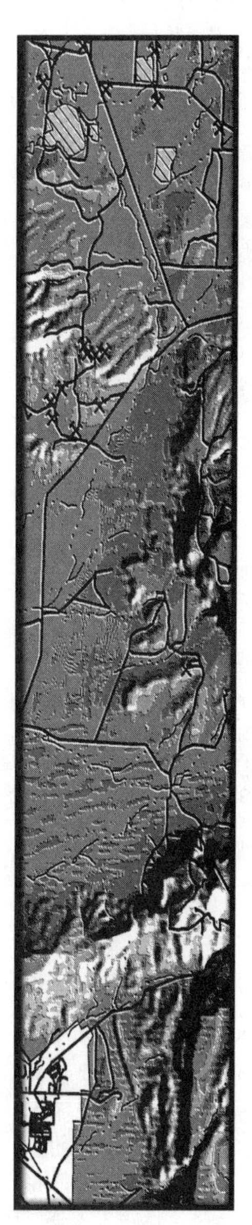

Raster Imagery in Geographic Information Systems

Stan Morain & Shirley López Baros, Editors

Raster Imagery in Geographic Information Systems

Stan Morain and Shirley López Baros, Eds.

Published by:
OnWord Press
2530 Camino Entrada
Santa Fe, NM 87505-4835 USA

First Edition, 1996
SAN 694-0269

10 9 8 7 6 5 4 3 2 1

Printed in the United States of America

Library of Congress Cataloging-in-Publication Data

Raster imagery in geographic information systems / [edited by] Stan Morain and Shirley V. López Baros.

p. cm.
Includes index.
ISBN 1-56690-097-2

1. Geographic information systems. 2. Computer graphics.
I. Morain, Stanley A., 1941- II. López Baros, Shirley V., 1962-
G70.212.R37 1996
910'.285--dc20

96-14278
CIP

Trademarks

Warning and Disclaimer

About the Editors

Stan Morain is a professor of geography and director of the Earth Data Analysis Center at the University of New Mexico (UNM-Albuquerque). During his tenure at UNM, he has pursued a career in geography, remote sensing, and spatial analysis through teaching, research, and application projects. Since 1973, Stan has consulted with the United Nations Food and Agriculture Organization (FAO), the UN Development Programme (UNDP), the U.S. Agency for International Development (USAID), and many private corporations on projects using modern spectral and spatial analysis techniques in developing countries. He has assisted in program design, implementation, and evaluation in over 20 countries of Asia, Africa, and Central America, usually in the role of project team leader, chief-of-party, or chief technical adviser.

Professionally, Stan has served the American Society for Photogrammetry and Remote Sensing in several roles, most notably as the 1992/93 national president. He is currently serving on the ASPRS Committees on Ethics and Professional Conduct, Nominations, and Data Preservation and Archives. He also serves on the Academic Advisory Committees for SPOT Image Corporation and is a member of the NAS/NRC Space Studies Board, Committee on Earth Studies. He serves on the editorial boards of three professional journals and has authored over 90 professional articles, books, and technical papers. He presently coauthors a bimonthly remote sensing column for *GIS World* with Amelia Budge.

Shirley López Baros has been using geographic information systems (GIS) technology since 1987 and global positioning systems (GPS) technology since 1992. As manager of GIS/GPS Services at the Earth Data Analysis Center, University of New Mexico, her experience includes managing large multitask and multidisciplinary contracts and grants. She provides database design, project management, training, and technical support for many diverse GIS/GPS activities. Shirley is proficient in ARC/INFO software. Her project experience includes integration of remote sensing, GIS, GPS, and relational database management system (RDBMS) technologies as they are applied to planning, natural and cultural resources, and environmental issues. Prior to EDAC, she worked for the City of Albuquerque in geographic information systems. A graduate of the University of New Mexico, she has taught GIS courses in the Department of Geography.

Acknowledgments

The editors are indeed grateful to all who provided their stories and illustrations for this effort. On behalf of the editors and contributors, we wish to thank also the following people: Kevin Corbley (Corbley Communications, Denver, CO) for writing and editing several of the case studies; Kerri Mich (Earth Data Analysis Center, University of New Mexico, Albuquerque NM) for coordinating and preparing editorial copy for 51 separate case study contributions to ensure their timely delivery; Amelia Budge, EDAC, for mastering electronic art on short notice to produce the illustrations in Chapter 1; and Jeanette Albany, EDAC, for transforming a plethora of electronic contributions into a common format and presentation for publication. The winter of 1996 will be remembered as the time of government shutdowns coupled with severe snow storms, floods, and other impediments to labor. Through it all, this team worked selflessly—early, long, and late hours—for three months to meet a stringent production schedule. They realized up front that similar commitments could not be expected, nor would they be needed from all contributors. Thus, the team's challenge became one of extracting exactly what was needed, when it was needed.

Their principal reward comes in proving that it could be done. The stimulus in this case was to produce a book for which industry leaders believe there is an eager and expanding market. Everyone familiar with the technology could cite at least a few interesting raster applications. No one could have guessed the wide variety of models being developed or the breadth of hardware, software, and data sets being employed. Industry leaders at the Environmental Systems Research Institute (ESRI) and SPOT IMAGE believed that a book celebrating achievements in raster GIS applications and that exercised the variety and breadth of approaches would promote wider use of these techniques. It is thanks to the foresight of ESRI and SPOT IMAGE that this project was even undertaken; and it is to them the book is dedicated on behalf of the entire GIS community.

Stan Morain and Shirley López Baros
Earth Data Analysis Center
April 15, 1996

OnWord Press...

OnWord Press is dedicated to the fine art of professional documentation.

In addition to the editors and contributors who developed the material for this book, other members of the OnWord Press team contributed their skills to make the book a reality. Thanks to the following people and other members of the OnWord Press team who contributed to the production and distribution of this book.

Dan Raker, President
Janet Leigh Dick, Associate Publisher
Gary Lange, Vice-President of Contracts & Administration
David Talbott, Director of Acquisitions
Barbara Kohl, Senior Editor
Carol Leyba, Senior Production Manager
Kristie Reilly, Production Editor
Andy Lowenthal, Production Manager
Kristie Reilly, Assistant Editor
Beverly Nabours, Production Artist
Lynne Egensteiner, Cover designer
Kate Bemis, Indexer

See the special offer for SPOTView image samples on the following page.

Contents

Introduction . . . . **xiii**

Book Structure xv

Chapter 1: Image Formation and Raster Characteristics **1**

by Stan Morain, Jack Estes, Timothy Foresman, and Joseph Separr

Background 1

Historical Notes 3

Data and Information 5

The Electromagnetic Spectrum 7

Popular Passive Instruments 11

Raster (Pixel) Characteristics 14

Chapter 2: Image Display and Processing for GIS **29**

by Shirley López Baros, Paul R. H. Neville, and Joe Messina

Image Display. 30

Cell-based GIS Processing 35

Image Processing 37

Preprocessing 50

Chapter 3: Using Scanned Aerial Photographs **55**

by Roy Welch and Thomas Jordan

Conversion of Aerial Photographs to Raster Format 56

Orthophotographs 66

Summary 68

Chapter 4: Data Collection Systems, Formats, and Products 71

by Amelia Budge and Stan Morain

Primary Satellite Sensor Systems 72

Satellite Pour l'Observation de la Terre (SPOT)............ 72

Landsat 77

Television and Infrared Observation Satellite (TIROS) (aka NOAA Polar Orbiting Environmental Satellite - POES series) 82

European Resource Satellite (ERS-1) 88

Indian Remote Sensing Satellite (IRS) 92

RADARSAT 97

Standard Satellite Digital Data Products................. 101

Chapter 5: Modeling Techniques 111

Economic Applications.. 112

Evaluating Franchise Locations 112
by Bradley Cullen, Richard Curtiss, Paul Neville and Teri Bennett

Cellular Phone Transceiver Site Selection 117
by John R. Jensen, Xueqiao Huang, Derek Graves, and Richard Hanning

Finding a Least-cost Path for Pipeline Siting 125
by Sandra C. Feldman, Ramona E. Pelletier, Ed Walser, James C. Smoot, and Douglas Ahl

Visualizing the Albury Bypass 132
by Michael Byrne

Choosing Efficient, Cost-effective Transportation Routes . . . 141
by Dana Nuñez Brown

Assessing Tourism Potential: From Words to Numbers..... 149
by João Ribeiro da Costa

Georgia Tax Pilot Study 154
by D. Kasouf

Diverse Modeling Scenarios **159**

Editing and Updating Vector GIS Layers Using Remotely Sensed Data 159
by Douglas Stow, Dong Mei Chen, Robert Parrott and Sue Carnevale

Modeling Vegetation Distribution in Mountainous Terrain 165
by Diego Fabián Lozano García and René G. González Murguía

An Urban Mask Raster Image for Vector Street Files. 172
by Jeffrey T. Morisette, Heather Cheshire, Casson Stallings, and Siamak Khorram

Combining Polygon and Raster Data in Cartography 178
by Anthony D. Renaud

ASSESS: A System for Selecting Suitable Sites 182
by Simon M. Veitch and Julie K. Bowyer

Shaded Relief Images: Visualizing Terrain Without Contour Lines 192
by Jan Benson and Bob Greene

Building Attribute Tables for Raster GIS Files with ARC/INFO 198
by Zhenkui Ma and Roland L. Redmond

Chapter 6: Water, Crops and Weather **203**

Hydrology. **205**

Targeting Wetlands Restoration Areas. 205
by Richard G. Kempka, Frederic A. Reid, Scott Flint and Kari Lewis

Water Resources Management Applications 211
by Thomas H. C. Lo

Mapping Glaciers with SPOT Imagery and GIS 219
by Andrew G. Klein and Bryan L. Isacks

Identifying Efficient and Accurate Methods for Conducting the Irrigated Water Use Inventory, Estancia Basin, New Mexico 226
by Douglas J. Paulson

Agriculture 231

Polygon Mode Filter for Remotely Sensed Agricultural Land Cover Data 231
by Casson Stallings, Siamak Khorram, and Rodney L. Huffman

SPOT, ARC/INFO, and GRID: A New Concept in Agro-environmental Monitoring 238
by Paul de Fraipont, Stephen Clandillon, Dominique Esnault, Didier Georgieff, Jean-Daniel Hennemann, and Alain Lefeuvre

Vegetative Index for Characterizing Drought Patterns 247
by James Rowland, Andrew Nadeau, John Brock, Robert Klaver, Donald Moore, and John E. Lewis

CropWatch: Monitoring Irrigation Water Use in the Desert 254
by Tom Elder

Mapping Long-term Change in Irrigated Land 259
by Anthony Morse, William J. Kramber, David Palmer, and Dewayne McAndrew

Meteorology 264

Lightning Analysis at Bonneville Power Administration 264
by Robert White

Integrating Raster and Vector GIS with Climate Data for Winter Road Maintenance 267
by Dan Cornford

Spatial Distribution of Global Atmospheric Water Vapor Storage 272
by Julie Driver

Chapter 7: Land Use and Planning 277

Landscape Analysis 279

Landscape Structure as Input to Ecological Planning in Northland, New Zealand 279
by Russell L. Watkins

Mapping the Wilderness Continuum Using Raster GIS..... 283
by Steve Carver

Defining Biophysical Land Units (BLU)................ 289
by Crista S. Carroll

The Definition of Topographic Divisions for Italy 293
by Fausto Guzzetti and Paola Reichenbach

Regional Forestry and Biodiversity in Tierra del Fuego, Chile................................... 301
by George R. Carlson, Robert J. Henry, and James A. Pugh

Land Use/Land Cover 307

Enhancement, Identification, and Quantification of Land Cover Change 307
by Douglas Stow, Dong Mei Chen, and Robert Parrott

Updating Vector Land Use Inventories Using Multi-date Satellite Imagery 313
by Robert Parrott, Sue Carnevale, and Douglas Stow

The U.S.-Mexico Border GIS: The Tijuana River Watershed Project................................. 322
by Richard D. Wright, John F. O'Leary, and Douglas A. Stow

Development of a Land Use/Land Cover Map 328
by Jerry L. Whistler, Stephen L. Egbert, Edward A. Martinko, David Baumgartner, Re-Yang Lee, and Mark E. Jakubauskas

A Structural Vegetation Database for the Murray Darling Basin 334
by Kim Ritman

Urban and Regional Planning 343

Measuring and Modeling Urban Growth in Kathmandu 343
by Barry Haack, David Craven, and Susan McDonald Jampoler

Merged Raster and Vector Data for Regional Environmental and Land Use Planning: Tamaulipas, Mexico 350
by Mindy Roberts, Keith E. Miller, and Philip R. Chernin

Mapping Urban Growth in Metropolitan Beirut 358
by Yousef Nizam

Chapter 8: Environment and Mineral Exploration 365

Detecting Contamination from Uranium Mines........... 366
by Paul R.H. Neville

Mapping Biodiversity 373
by Teri Brotman Bennett and Esteban Muldavin

Remote Sensing and GIS for Monitoring Grassland Environments 378
by Mark Chopping

Change Detection of Pacific Coast Estuaries 386
by Ruth E. Spell, Richard G. Kempka, Jon K. Graves, and Patrick T. Cagney

A Method for Identifying Wetlands Mitigation Sites 397
by Floyd Stayner

Monitoring and Mapping the Belchatow Mining Complex in Poland.................................. 408
by Stanislaw Mularz

Vector and Raster Analysis of the Solitario Dome and Terlingua Uplift, West Texas..................... 413
by Michael Clark

Chapter 9: Forest Management 419

Developing a Multi-scale Forests Database for Australia ... 420
by Philip Tickle

Mapping Needleleaf Forests in Great Smoky Mountains National Park.. 431
by John B. Rehder

Information Technologies to Support the Licensing of Forestry Activities.. 437
by G. W. Turner, R. M. C. Ruffio, and M. W. Roberts

Timber Harvest Scheduling with Adjacency Constraints ... 441
by Bruce Carroll, Vaughan Landrum, and Lisa Pious

Fire!™ Using GIS to Predict Fire Behavior............... 448
by Kass Green, Mark Finney, Jeff Campbell, David Weinstein, and Vaughan Landrum

Analyzing the Cumulative Effects of Forest Practices 454
by Kass Green

Using GIS to Produce Fire Control Maps................ 462
by Sonny Parafina

Appendix A: List of Contributors 469

Appendix B: Contact Points 475

Index.. 487

Introduction

GIS applications have entered the Information Age by relying more and more on data sets prepared by others and made available on the Internet, CD-ROM, and other digital formats and media. Among the most sweeping of these trends is greater use of raster image data from aircraft and satellite platforms. Such data sets are available now from a relatively few suppliers but will become much more accessible via the information superhighway by the end of the millennium. Raster data that were once difficult to obtain or expensive to use are rapidly becoming core requirements for geographical information system (GIS) problem solutions. Unfortunately, for many GIS developers and operators trained specifically in spatial data analysis, the spectral properties of raster data are not well understood or appreciated. Conversely, remote sensing specialists trained in spectral analysis sometimes overlook the role spatial context plays in their applications. It is possible, of course, to create and apply GISs without using raster image data, and there are image processing applications that survive without a GIS.

This book focuses on the vast middle ground where raster image data (i.e., spectral data) enhance vector GIS applications. In order for the GIS community to better appreciate how spectral attributes promote their applications in a cost-effective manner, the introductory chapters are intended as a brief tutorial on how the data are collected, the kinds of satellite systems and sensors currently operating, some of the primary physical trade-offs between spectral and spatial resolution, and some of the principal data structures and file formats.

Most of the book is devoted to case studies that have been edited only to achieve a common format and style; otherwise, they are presented as submitted. The fact that the case studies cluster naturally into recognizable topic areas is reassuring. Some rather important topic areas, however, are not as well populated as desired. Applications in arms control, hazardous waste management, and in business and finance are not represented in true proportion to the current amount of work in these areas.

The aim of everyone's effort was to produce an introductory book on raster imagery for vector GIS applications. This is not a new concept, but there is room for growth. The not so subliminal messages are that (1) raster data, especially multispectral data, have found numerous uses in vector-based GIS; (2) these raster data contain unique information about the landscapes they portray; (3) the information content can be updated frequently and economically; and (4) the hardware, software, and modeling techniques for raster data are already available for use by the GIS community. Commercial and government applications are exploding onto the GIS scene, but only a few, relatively speaking, are accessing or employing satellite and/or aerial raster data in their modeling solutions.

Book Structure

Chapter 1 is an overview of remote sensing concepts presented as system design trade-offs. The author-editors felt this approach would not only link remote sensing technical terms in the reader's mind, but would illustrate that no single sensor can produce rasters having all the properties users want. Chapter 2 discusses how raster data are processed, and Chapter 3 is focused on desktop mapping practices and the use of scanned aerial photographs. Chapter 4 is a summary of raster formats, media, and products currently available from six of the leading satellite systems.

The remaining Chapters (5 through 9) are collections of case studies. Where possible, the editors have linked these studies to relevant paragraphs in earlier chapters, especially Chapters 2 and 3.

If the book serves no other purpose than to put would-be GIS developers and users into contact with those who have contributed, at least one aim of the book will have been achieved. To promote this exchange, Appendices A and B provide complete addresses and contact points for each chapter and case study. Our greatest hope is to influence a greater fusion of the technologies to stimulate even further growth.

Finally, a detailed index is provided at the end of the book.

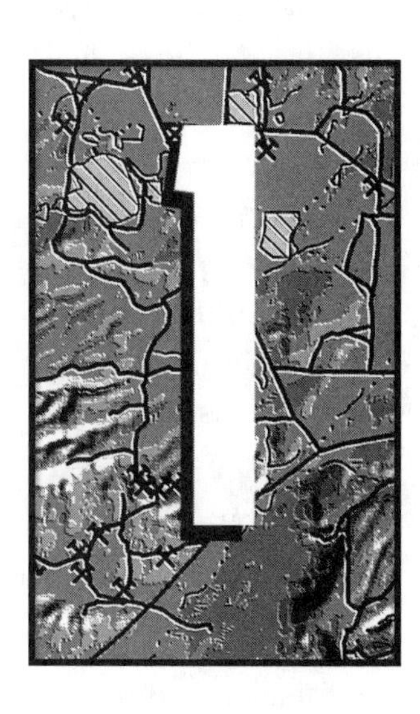

Image Formation and Raster Characteristics

Stan Morain, Earth Data Analysis Center, University of New Mexico
Jack Estes, Department of Geography, University of California, Santa Barbara
Timothy Foresman, Department of Geography, University of Maryland
Joseph Separr, Remote Sensing Research Unit, University of California, Santa Barbara

Background

Throughout human history, technology has been a key facilitator for change. Today's technologies are creating change on spatial and temporal scales never before possible. Application of a number of these same technologies enables investigations into the character and rate of change, and is leading to better understanding of global resources at scales from local to global. Remote sensing and geographic information systems (GIS) are two of the leading technologies answering questions about spatial and temporal dimensions of landscapes and resources. Processed and analyzed images, combined with other data and visualization techniques, are being employed in GIS by a wide variety of resource planners and managers, private sector entrepreneurs, scientists, and public policy decision makers.

Geographic information systems are designed for assembling, integrating, and analyzing spatial information in a decision-making context. These systems help in modeling complex geospatial phenomena, and their output serves to better manage global resources. GIS evolved for handling diverse data sets for specific geographic areas by using coordinates as the basis for an information system. Key enablers for GIS include hardware and software for manipulating digital data, the rise of information science for creating and extracting information from complex databases, and developments in digital cartography. Other advances that make these tools even more powerful include desktop workstations with increased data processing speed and data storage capabilities; inexpensive plotters, lowering the cost and improving the quality of maps, charts and tables; development of user friendly GIS software and image processing software; and improved graphics and visualization tools.

GIS is such a compelling technology that traditional lines separating developers and users are getting blurred. In today's complex work environment, public, commercial, and research applications are so numerous that users are not only developing their own data sets, but learning to model their own problem solutions. Data fusions and the relations between data sets are the heart of any GIS, but this heart is growing to include both image and attribute data that more fully utilize both raster and vector data structures. The aim of this book is to provide basic understanding of how raster data (primarily digital images) can be used to augment or enhance traditional vector GISs, and to illustrate how raster data alone are useful as contributions to investigations in areas such as land management, natural resource assessment, and economic, cultural, and environmental solutions.

This chapter provides a tutorial on (1) how raster data from images are created; (2) raster data property description and identification, and (3) the primary sensors and satellite systems currently being employed. The terminology used is primarily derived from remote sensing, but the chapter is not intended as a primer on that subject. Key raster concepts and attributes are described in non-rigorous ways so as to introduce remote sensing language to the vector GIS community.

Historical Notes

When employed in a GIS, remote sensing data facilitate mapping and monitoring tasks. Basically, the two technologies are linked at the most fundamental levels of measurement, mapping, monitoring, modeling, and management.

Transforming data into information.

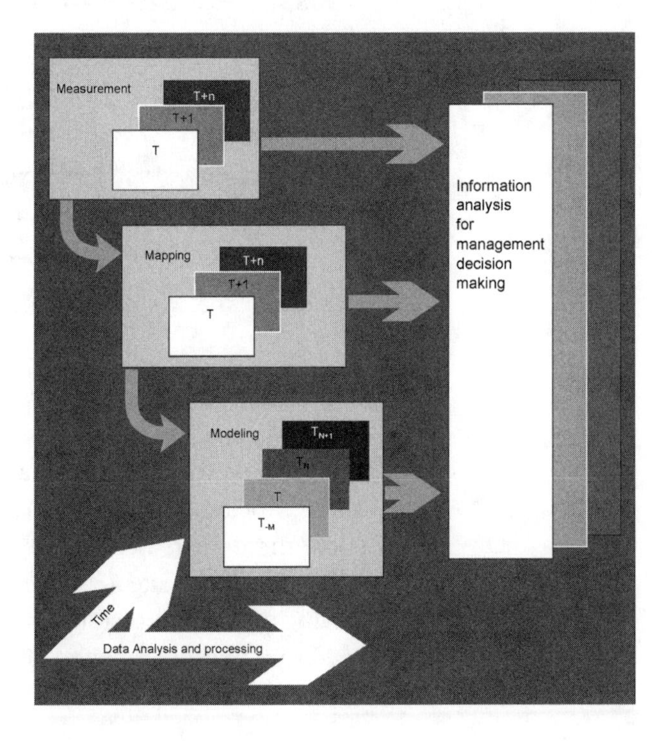

Airborne remote sensors have traditionally acquired data for environmental applications. Since the early days of the space program, however, there has been a growing awareness of space-derived imagery to capture data about Planet Earth. While remote sensing technology precedes the space flight era by a century, there is probably no more familiar reminder of it than the TV weather report. From data obtained from tethered balloons in the 1850s to those obtained by modern satellites, the quantity and types of data that are routinely acquired have increased astronomically. The following figure illustrates this growing complexity.

Increasing complexity of platforms, sensors, and tasks.

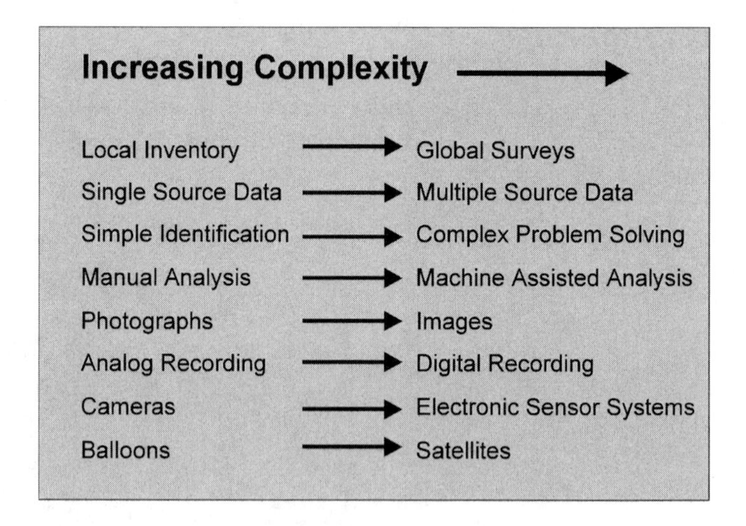

Balloons were first employed as a camera platform to acquire aerial photographs. The first *photogrammetric application* was used to map German forests in the late 1870s. Rocketry was used to acquire aerial photos in 1906, some three years before the first aircraft photography was acquired. In the late 1800s, experiments to extend the sensitivity of film emulsions into non-visible regions of the spectrum were conducted. Thermal and active microwave sensors came into civilian use in the 1960s. This is about the time (1958) that the term "remote sensing" was coined. Coincidently, the term "geographic information system" was coined in the early 1960s. The first weather satellites (TIROS) were launched in 1960, and the first civilian Earth satellite in 1972. These satellites carry electronic sensor systems to record images of the Earth in digital form.

In the past, a single type of photography or imagery for a given application was typically used. While this is still true, there are more sophisticated applications that require multiple images from different regions of the EM spectrum and/or different dates. More importantly, applications have evolved that require both raster and vector data sets, and from local studies to those using data at a variety of scales.

GIS and remote sensing are thus linked both historically and functionally. In a historic context, some of the early work leading to the development of GIS revolved around methods to better access aerial photographic coverage of specific areas.

Data and Information

The only goal of information systems is to convert data into information. Data are considered to be "input" to a process where information is created. For example, daily rainfall recorded over long periods is used to produce a monthly mean average for a given location. Measurements are also employed to model processes. The model's output then provides information for a management decision which might in turn generate other information types (see following illustration). Converting measurements and observations into information requires great care to ensure the quality and validity of the process. It is not uncommon that the cost for verifying the output of a data set (e.g., a map, statistical summary, or resource inventory) is greater than the cost of the original data.

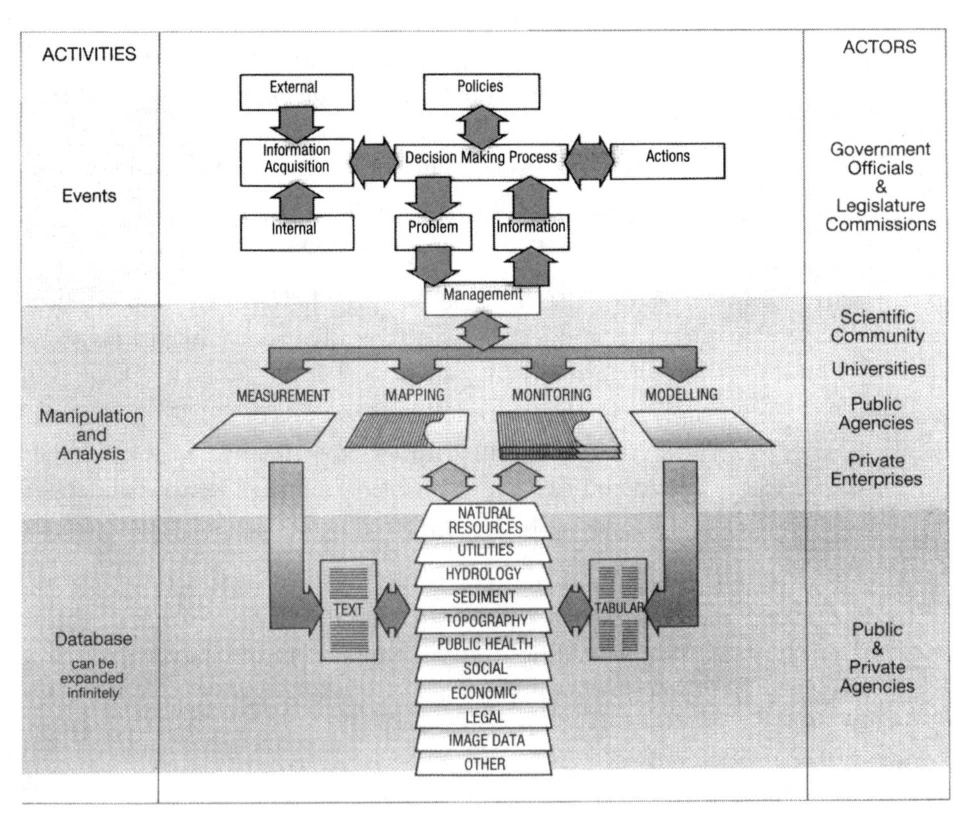

GIS protocols for decision making.

In addition to producing high quality data and information, there is the need to distinguish between *accuracy* and *precision*, not only in raw data, but also in derived information. Accuracy is associated with fidelity to a given standard and a lack of bias. Precision, in contrast, involves an ability to make fine distinctions. For example, one could measure annual rainfall to the nearest 10cm in a simple rain gauge. This process would provide highly accurate (unbiased) data that are rather imprecise. Such imprecision, inappropriately extrapolated over space and time, or across scales, might invalidate the data and lead to incorrect decisions. On the other hand, use of precise but biased data (e.g., a rain gauge precise to 0.1mm, but with a pin hole in the gauge bottom) would absolutely invalidate the results.

In the growing sophistication of applications, GIS developers need to consider both spatial and non-spatial data and information. Because of the complexity of data sets—some raster, some vector, some non-spatial, some quite qualitative, some quite precise—quality control becomes an important issue. From this issue yet another class of data and information called "metadata" has arisen. Metadata provide end users with the characteristics of a given data set. While metadata are often described as data about data or about data sets, in reality, they are not data but bits of information about data. Metadata are static and non-manipulable. They cannot be analyzed, only interpreted. For example, the date when a data set was verified for accuracy and measurement scale (e.g., elevations were recorded in meters, rather than feet) is a form of metadata. Metadata are key elements of catalogs and are being developed to improve user access to spatial data and information.

The Electromagnetic Spectrum

Human eyes are systems detecting the radiant energy environment. The senses of smell and taste are keyed to chemical surroundings, hearing to sound vibrations, and touch to direct contact. Considering the radiant environment alone, eyes detect only a narrow region in the electromagnetic (EM) spectrum. Even though energy from gamma rays to radio waves swirl around us, the human eye responds to light from only a small region from blue to red (Plate 1). To improve the perception of the energy environment, a variety of specialized instruments has been constructed to record information and present it in image format. While a photograph is an image, the term *image* more commonly refers to any picture created from sensor data.

Remote sensing is a technology used to gather data from a distance, with sensors recording reflected or emitted energy (Plate 2). Vegetation types, soils, and other natural and artificial features emit and reflect energy differently throughout the EM spectrum. These differences make it possible to measure, map, and employ imagery collected through time, or monitor the status of environmental features, or observe the effects of phenomena acting through time (e.g., urban sprawl). While aircraft still collect most aerial photography, satellite sensors collect the bulk of operational remote sensing data. Most of these data are available in two formats: electronic (i.e., digital) and analog (i.e., image). While analog data in the form of hardcopy images are still popular, aerial photographs are now commonly scanned into digital format. Data acquired in digital format or scanned into digital raster format must be processed by computer before they can be used in an application (see Chapter 4).

For Earth observations, the electromagnetic (EM) spectrum typically spans wavelengths from ultraviolet (UV) to microwave (MW). Ultraviolet wavelengths (0.25 micrometers—μm, or 10^{-6}m to about 0.39μm) are shorter than visible blue and are mainly used in fluorescence studies for mineral exploration, oil spill detection, lichenology, or for other applications where electron shifts result in photons of radiated light.

The visible region (VIS) extends from about 0.40μm (blue) to about 0.70μm (red), and records pigments resulting from particular chemical and molecular structures. Reflective infrared radiation (RIR) occupies the region between 0.70μm and roughly 1.5μm, and is used mainly as a means for distinguishing vigorous and stressed vegetation. At wavelengths between 1.5μm and about 3.0μm, the energy recorded by detectors is a mixture of reflected and emitted radiation. It is called the short wavelength infrared (SWIR) region. These wavelengths are most useful in atmospheric studies and to distinguish snow from clouds.

Beyond 3.0μm to about 5μm, very hot objects such as forest fires and volcanoes can be detected. The peak power radiation of most Earth objects is between 8.0μm and 14μm. The objects can be detected on the basis of molecular vibrations which can be measured by sensors as ambient temperatures. The 3-5μm window is called the mid-wavelength infrared (MWIR) region, and the 8-14μm window, the long wavelengths infrared (LWIR) region. In the millimeter wavelengths active microwave (radar) and passive microwave (radiometry) sensing are performed. Both are sensitive to dielectric properties of materials and moisture contents. Radar is also sensitive to object geometry.

Remote sensor systems can be broadly categorized into active and passive systems. A camera is an example of a passive system as its film records light energy illuminated by the Sun. Add a flash attachment to the camera and it becomes an active system, since it now provides its own source of illumination. In a similar way, active microwave systems provide their own illumination by transmitting a signal and receiving an echo from the ground.

Most satellite remote sensing data are currently acquired by multispectral scanners and linear array devices. These are passive systems which record solar radiation reflected from the Earth's surface. Data derived from multispectral scanners can provide information on (among other things) vegetation types, distribution, and condition; geomorphology; soils; surface waters; and river networks.

SWIR sensors record emitted energy from surfaces and have been particularly useful for monitoring fires and for studying areas of volcanic and geothermal activity. LWIR sensors have been popular for mapping ocean temperatures related to the dynamics of coastal waters and currents. On land, plant water stress induces changes in canopy temperatures that are detectable. Thermal maps are also popular for urban areas, industrial sites, manufacturing centers, and agricultural scenes.

Active systems operating in the visible spectrum are using laser technologies (LIDARS) primarily for oceanographic and forestry applications. Regardless of the wavelengths they employ, active systems do not depend on reflected energy from the Sun for image formation. They are able to acquire data during the day or night. In addition, because longer wavelength microwave radiation is not distorted by the atmosphere to as great a degree as shorter wavelength energy, radar systems can collect data through cloud layers and some precipitating clouds.

Other remote sensing systems are employed to detect the Earth's magnetic and gravitational fields. These tools are used extensively in oil and mineral exploration. Depicted in the following illustration are examples of black-and-white photographs of the Santa Barbara area in California. Analyses of such multispectral data can, if properly designed and carried out, increase both the quality and quantity of information for given applications.

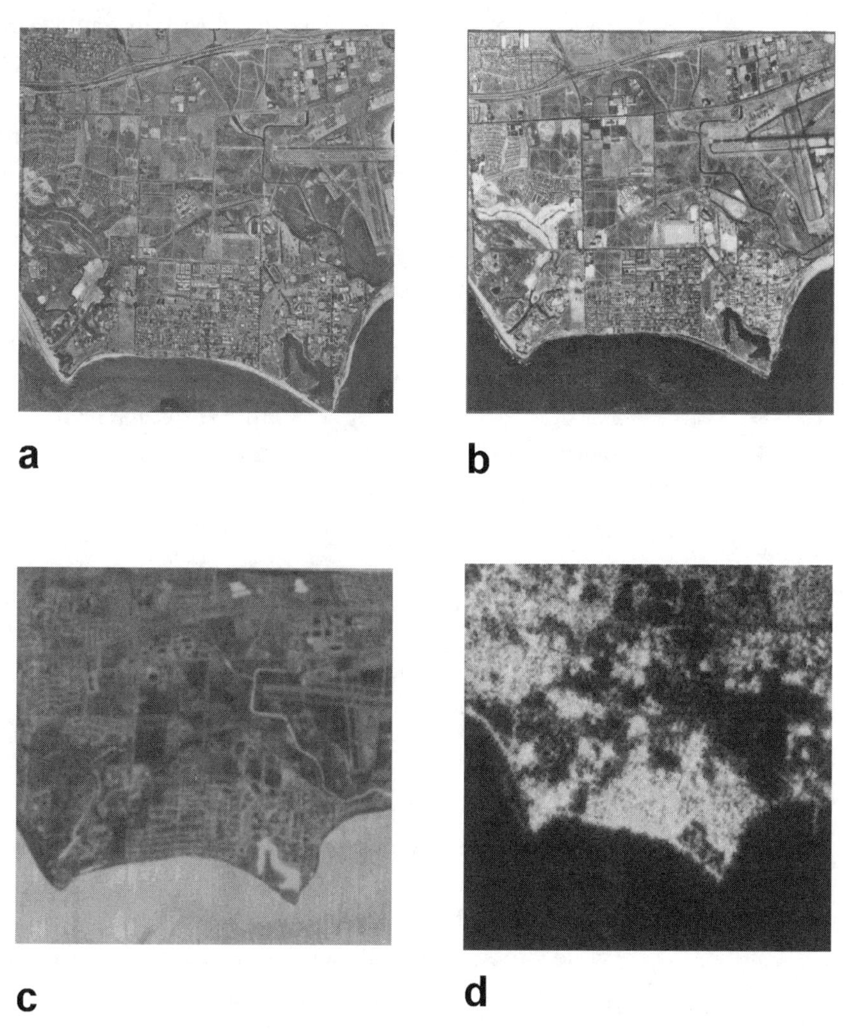

Views of the Goleta (University of California, Santa Barbara) area: (a) rasterized black-and-white photograph; (b) rasterized black-and-white infrared photograph; and (c) shuttle imaging radar (SIR) image.

Popular Passive Instruments

Cameras with appropriately filtered films are able to image scenes in the visible near infrared (VNIR) region. Such images can be obtained as continuous grayscale products, as normal color photographs, or color infrared photographs. If image data are required for wavelengths shorter than blue or longer than reflective infrared, then devices other than cameras are required to collect the reflected or emitted radiation. These devices are commonly called radiometers and scatterometers. Although such devices are usually employed to measure energy in the non-visible wavelengths, they can also serve to measure visible wavelengths.

Scanners and imaging spectrometers are standard instruments for gathering reflected energy in the VNIR. Since continuously operating, unmanned satellites are not useful platforms for carrying heavy cameras with bulky films and filters, these instruments represent the satellite data gathering equivalent of airborne cameras. They supply VNIR and data in digital raster form, and the properties and attributes of data in such form enhance vector GISs. Of course, it is also possible to electronically scan photographs to convert them to digital raster format, as discussed in more detail in Chapter 2.

Satellite and airborne instruments use the forward motion of the platform to create an image of the ground track beneath the sensor. More sophisticated sensors are being designed that can acquire overlapping fore and aft images along the flight path, known as along-track stereo imaging, while others are designed to view terrain off-nadir to the side(s) of the platform. Off-nadir viewing can be used either to create cross-track stereo imagery or to decrease the time lag between viewing opportunities.

Electromechanical Scanners

Imaging systems that produce routine satellite data for use in today's GISs acquire data line by line, much like a TV picture is generated. Electromechanical scanners like the one in Plate 3 (far left) use an oscillating mirror backed by a telescope to sweep perpendicularly across the forward motion of the platform. Called "whisk-broom scanning," reflected energy from the ground is collected on the forward motion of the mirror scan, and, after passing through the telescope, is directed to a fiber optic array where discrete detectors sense the amount of energy in each spectral channel. Because the mirror is moved mechanically and is relatively slow by comparison to the speed of the satellite (and consequently, the amount of ground being covered), the usual design is to sweep several scan lines per mirror sweep. The detector materials are silicon photodiodes.

Push-broom Scanners

"Push-broom scanners" (Plate 3, second from left) are similar to whisk-broom instruments except that they use a linear array of silicon photodiode detectors to image the entire line across the swath at once. This design obviates the need for an oscillating mirror and reduces other elements of distortion caused by time delays introduced into the line by the mirror motion. While a whisk-broom mirror is collecting data, the platform is moving forward. This means that each scan line has a slight curvature to it which must eventually be corrected. By collecting the entire line at once, these problems are largely overcome.

Electromechanical Imaging Spectrometers

Electromechanical imaging spectrometers (Plate 3, second from right) use an oscillating mirror backed by a telescope and coupled to a light disperser. The disperser partitions surface reflectance into its spectral components and projects them onto a linear detector array. A light-projecting subsystem comprised of a pin-hole aperture, an object lens to focus the image point onto the aperture, and a collimating lens for projecting the aperture image onto the disperser ensures that only one point of light from the ground is cast onto the disperser at a given time.

Solid-state Imaging Spectrometers

Solid-state imaging spectrometers are similar to their electromechanical cousins, but employ a two-dimensional detector array instead of a linear array (Plate 3, far right). After energy passes through the objective lens, it passes through a slit aperture and collimator lens before reaching the light disperser. Each line of detectors in the two-dimensional array records a separate spectral channel.

Raster (Pixel) Characteristics

The digital raster data received at ground stations comes in a continuous stream as long as the satellite is within line-of-sight of the receiver. This procedure is more fully discussed in the next section under "Radiometric Resolution." Since the data will be in binary form for satellite-to-ground transmission, the amount of data to be telemetered cannot exceed data transmission technology and the bandwidth over which the data are carried. Telemetry rates and carrier wave bandwidths have increased significantly in recent years from kilobytes per second (kbs) to megabytes per second (mbs). This means that 6-bit partitioning of the y axis (64 levels of gray) in older systems can now be designed for 8-bit accuracy, or higher. Nevertheless, it is clear that higher Nyquist sampling rates and increased bit rates combine to create exponential growth rates in data transmission requirements. This data stream must be processed and formatted by data suppliers into usable image sets before they become available for GIS. Chapter 4 describes in more detail how the data are formatted. The rest of this chapter describes some of the attributes of the picture elements (pixels, or rasters).

Formation of the Raster

Incoming reflected radiation is recorded in analog form as a voltage measured in microwatts per square centimeter per steradian (μwatt/cm^2/sr). This analog reflectance curve must be converted into its digital equivalent. Plate 4 illustrates analog reflectance data for an 8-channel scanner. These curves are sampled on the x axis at an interval that accounts for the scaling factor between the ground and the focal plane of the sensor to achieve a desired ground resolution. The Nyquist Theorem states that a continuous signal can be reconstructed from a set of discrete samples if those samples are obtained at twice the frequency of interest. The illustration below shows an example of how this is done for a whisk-broom scanner. In the case of a whisk-broom scanner, the time it takes to sweep a line of data becomes the sampled x axis. If the mirror collects data for a period of a few milliseconds, the line can be sampled every few microseconds to produce thousands of samples or rasters, each representing a quantifiable distance on the ground. Having partitioned the x axis into ground distances, the next step is to divide the magnitude of reflectance (y axis) into a digital grayscale.

Spectral Channels

The four types of instruments in Plate 3 are generic designs for scanners and imaging spectrometers Specific sensors may differ in additional design features. Depending on a satellite's orbital parameters, such as altitude and orbit period (which determine platform velocity), other parameters can be engineered to optimize sensing goals. For example, if a single detector and oscillating mirror are employed, the only way to achieve an image without data gaps on the ground is to move the mirror very fast. This means that either the length of the scan line (or field of view, FOV) must be relatively narrow, or that the viewing time for any given spot on the ground (instantaneous field of view, IFOV) must be short.

Unfortunately, if the viewing time is short, the amount of energy recorded for a spot on the ground must be comparatively high to be recorded. The best way to enhance this dwell time is to either "look at," or sample, relatively large areas (ground sampling distance, GSD), or employ spectral channels that are sufficiently wideband that they can collect enough reflected energy to activate the detectors. The longer the mirror can look at a given spot, the longer it will have to collect photons of reflected energy and the narrower the bandwidths of spectral channels can be. With longer dwell times, systems can be designed for smaller IFOVs and GSDs, as well as more spectral channels.

The earliest orbiting multispectral scanners had only four channels and relatively coarse GSD. Nevertheless, they provided the first synoptic views of the Earth and guided much of the discussion and engineering modifications for later scanners. Later systems were able to reduce the size of the GSD (e.g., SPOT Multispectral and Panchromatic) or increase the number of spectral channels (e.g., Landsat Thematic Mapper) through design improvements. In the case of SPOT, the High Resolution Visible Sensor (HRV) achieved a significant GSD improvement over the Landsat Multispectral Scanner by adopting the linear array "pushbroom" technology and decreasing the swath width. By further adding a wideband (panchromatic) channel and doubling the number of detectors per line, this pushbroom approach is able to optimize both the IFOV and energy sensitivity constraints to produce the best available GSD currently available from general purpose Earth observing satellites. The Landsat Thematic Mapper (TM) instrument was able to increase the number of spectral channels from four to seven by designing the oscillating mirror to collect data on both the forward and backward parts of the scan.

Bandwidth

Although the systems described in the next section are called *multispectral* scanners, their designs are such that they collect data only in a few broad spectral regions similar to the green, red, and reflective infrared dye layers used for color infrared photographs. The digital data these scanners collect are spectrally discontinuous in the sense that if the spectral values are plotted for each pixel, only three (HRV), four (MSS), or seven (TM) points are available to characterize the spectral curves of each ground sample. These broad bandwidths have been necessary, however, because the sensing goals have been to obtain synoptic global coverage with moderated GDS on a frequent revisit schedule. These goals equate to a wide FOV, high sampling rate (IFOV), and short dwell time. The bandwidths must be wide (100nm-plus) to collect enough photons to activate the detectors. For GIS applications, these bandwidths are adequate to identify broad categories of land cover/land use, and this is essentially how the data have been used thus far. (See Plate 5.)

Ideally, GIS applications could address many more problems assuming access to spectral data from tens, hundreds, or even thousands of channels having narrower bandwidths. When plotted, these would closely resemble the spectral signature of terrain phenomena. Generally speaking, sensors equipped with tens of channels are considered to be *multispectral*; hundreds of channels are called *hyperspectral*; and thousands of channels are called *ultraspectral* (Plate 6).

Recalling that the narrower the bandwidth, the longer the sensor must "stare" at a given point on the ground, it does not at first seem possible to simply add more spectral channels and at the same time maintain fine ground sampling distances. In fact, experimental hyperspectral sensors on aircraft platforms have been around for several years (e.g., *Advanced Visible/Infrared Imaging Spectrometer, AVIRIS*), and there are even a few that are commercially operated (e.g., *Compact Airborne Spectrographic Imager, CASI*). These two sensors have 224 and 288 channels, respectively, with bandwidths of 10 nanometers (nm, or 10^{-9}m) and 3nm.

In order to achieve sensing goals, the experimental hyperspectral sensors have adopted narrow swath widths (FOVs) and higher radiometric resolutions (number of digital gray values). From space altitudes, these engineering trade-offs result in GSDs on the order of meters, but it may take hundreds of orbits and long revisit schedules to obtain synoptic, global data. The first hyperspectral instrument to be launched will likely be the *Hyperspectral Imager (HSI)* on board the Lewis spacecraft. It is expected to have 256 channels across a 7.7km swath to achieve 30m GSD. These data are expensive to collect and do not exist for most areas of the world. Nevertheless, site-specific GISs might make future use of hyperspectral data.

Ultraspectral instruments are not available in civilian GIS applications, and probably should not be expected any time soon. They are experimental at best, and are so massive in terms of data storage and manipulation demands that current computer systems would be swamped. When the day comes that ultraspectral data are available, they will most likely be used in air and water quality applications, and other highly demanding detection requirements.

Resolution

Ground Sampling Distance

"Resolution" is a complex function involving several design trade-offs. In aerial photography, ground resolution is measured as "line pairs per millimeter," but there are over two dozen parameters that influence this measurement. For practical purposes, resolution panels are often placed on the ground so photogrammetrists can directly measure the minimum observable area on the ground.

For non-photographic sensors (scanners, digital cameras, radiometers, and imaging spectrometers) there are several kinds of resolution that influence each other. Whether spatial or GSD, the sensor is not divorced from radiometric and spectral resolution. In its simplest form, ground resolution is the smallest area on the ground from which reflectance measurements can be made by a given sensor at a given altitude having a specified swath width and scan time. This GSD is usually measured in meters, but some sensors, like the Advanced Very High Resolution Radiometer (AVHRR), have GSDs measured in kilometers. All other factors being equal, a one-meter GSD system should be able to record objects the size of card tables or man-hole covers; 10-meter GSD systems should "see" items the size of tennis courts or residential buildings; and, at 100 meters, an interpreter should see objects the size of a football stadium or golf course (but not the detail within those features). Plate 7 illustrates typical pixel sizes against the backdrop of a U.S. football field.

In reality, when the spectral contrast, or color, between adjacent items is lower than the detectivity of the scanning system (NE$\Delta\rho$), or is less than the bandwidth of the spectral channel, the ground resolvable distance (GRD) may be larger or smaller than the calculated GSD. Basically, if two adjacent pixels have virtually the same color, they cannot be separated. However, if their reflectance differs by more than the minimum spectral resolution, they should be distinguishable on the ground as having different colors. Adjacent rasters with high spectral contrast may reveal ground features much smaller than the GSD. This is why roads, canals, and other linear features can sometimes be seen on images of rather coarse GSD.

Plate 8 is a sample SPOT XS Image for part of the city of Albuquerque, New Mexico. The top view shows the full data set, and the bottom shows full magnification. Some adjacent raster elements have recognizably different colors, but others are so similar they are not distinguishable. When the differences exceed a threshold value, they appear as a visibly different color. The threshold value is directly related to quantization, or radiometric resolution.

Radiometric Resolution

Data collected by scanners are designed to be sensitive to a range of reflectance intensities that are considered more or less normal for Earth features, and this is expressed as a voltage range. Extremely bright objects (as from a distress signal from a mirror), and very low reflecting surfaces (a newly paved parking lot or a cloud shadow) are recorded either as maximum voltage (saturation) or as "0" voltage (no response). The question then becomes one of how many values between zero and maximum voltage should be recorded. This parameter is called *radiometric resolution*, and is measured in bits. A two-bit system (no pun intended) would capture only zeros and ones, and result in an essentially black-and-white (high contrast) image. A six-bit system would convert the incoming signal intensities from each raster into 256 levels of gray, whereas a 12-bit system would convert the incoming signals to any one of 4,096 levels of gray. The higher the number of gray levels, the "smoother" and less "contrasty" the image. Obviously, the higher the bit number, the better the radiometric resolution.

Radiometric resolution refers essentially to how finely the spectral response of each GSD is divided. The voltage representing the range of responses from zero to saturation can be scaled to provide coarse or fine detail, but the finer the detail, the larger the amount of telemetered or recorded raw data. In practice, radiometric resolution is often referred to as quantization because digital data are telemetered as binary strings of zeros and ones. Six- and eight-bit quantization (26 and 28) are common word lengths for SPOT and Landsat data, meaning that each raster in each channel is characterized by combinations of either six or eight zeros and ones. Values (digital numbers, or DNs) from zero to 63 are possible with 6-bit radiometric resolution, or zero to 256 with 8-bit resolution. The following table contains examples of these strings. Many other current systems, both airborne and spaceborne, have 10-, 11-, or 12-bit quantization. While it is clear that such partitioning of spectral responses into longer word lengths provides ever finer detail, the total number of bits to be processed grows exponentially. This fine level of radiometric detail, coupled with sensors designed for larger numbers of channels and narrower bandwidths, could translate into data overload for GIS users.

Examples of 6-bit and 8-bit binary words and DN equivalents

6-bit binary words			
DN=0 000000	DN=1 000001	DN=2 000010	DN=3 000011
DN=15 001111	DN=32 100000	DN=45 101101	DN=63 111111
8-bit binary words			
DN=0 00000000	DN=1 00000001	DN=2 00000010	DN=3 00000011
DN=15 00001111	DN=32 00100000	DN=45 00101101	DN=63 0011111
DN=126 01111110	DN=197 11000111	DN=236 11101110	DN=256 11111111

In short, the "effective spatial resolution" of non-photographic scanners operating in the VNIR is the smallest area on the ground that can be detected by the sensor, recorded by the analog-to-digital converter, and whose color exceeds NE$\Delta\rho$.

Spectral Resolution

Spectral resolution, like ground resolution, depends on interrelated design trade-offs. The most important task is to select the spectral channels required to detect desired ground phenomena. Most satellite scanners sense the VNIR region because the interaction of light with land cover results in reflections that allow interpreters to distinguish broad categories of water, vegetation, bare ground, clouds, urban areas, and agricultural cropping patterns. Other sensors operating in the SWIR, LWIR, passive microwave, and radar frequencies are "blind" to distinctions based on color, but are very sensitive to other attributes of surface materials that may be desired coverages in a GIS. Environmental studies, for example, may require mapping surface ambient temperatures, locating "hot spots," or determining broad-scale soil moisture status. Clearly, it is necessary to understand the spectral properties of surface materials and processes well enough to select the most appropriate spectral channels. Some GIS applications might well require data sets collected by different sensors from across the EM spectrum to model the problem solution.

Having located where data are required in the spectrum, the next task is to narrow the spectral bandwidth to detect specific attributes of interest. For broad categories of surface features, wider bandwidths permit a smaller GSD so that greater spatial detail is available. If very fine distinctions are required (e.g., between wheat and oats), it may be necessary to use much narrower bandwidths in certain channels and to dispense with spatial detail.

It would seem from this discussion that an important part of a raster GIS would be a spectral library of cover types and surface features. Such libraries do, in fact, exist as attribute features in more advanced image processing software. Unfortunately, these libraries are currently limited. They tend to be best in the areas of clay minerals and rock types, and not very well developed for biological or manufactured materials. One of the hopes for hyperspectral imaging is that spectral libraries for known surface materials under a variety of environmental conditions may be developed for broader use in vector GIS.

Temporal Resolution

Many GISs are being developed that require time-sensitive coverages of features that are either short-lived or cyclical (diurnal, seasonal, annual, or decadal). In any case, GIS applications recognize the dynamic attributes of problem solutions. Because of cost, it is unrealistic to believe that aerial photography can be collected on demand for these time-sensitive phenomena. Therefore, an important feature of satellite sensing is the possibility of frequent revisits of an area. For the reasons given above, it is now clear that such revisits may become less and less frequent as more channels and finer GSDs are desired. Both trade-offs can be accommodated only if swath width is decreased, thereby increasing the number of orbits required to image the entire Earth in a reasonable amount of time. The "Primary Satellite Systems" section in Chapter 4 describes the characteristics of SPOT and Landsat satellites. Because of its narrower ground swath, the SPOT satellite revisits a given ground track every 26 days, compared to Landsat's 18-day revisit. However, the SPOT system has a viewing capability up to 27° off-nadir, so that it can view the same area up to 11 times per 26-day cycle.

Ultimately, design trade-offs to optimize data collection are constrained by the amount of data that can be reasonably collected and analyzed for GIS applications. The equation for calculating the rate of data collection appears below. Designs that lead to very high data collection rates either require massive through-put capabilities, data storage for later transmission, or short duty cycles.

$$R = \frac{SQVB}{g^2}$$

R = million bits per second (mbps)

S = swath in km

Q = quantization

V = ground velocity in km/sec

B = number of spectral channels

g = GSD

SPOT		Landsat MSS		Hypothetical	
If	S=60km Q=2^8	If	S=185km Q=2^6	If	S=5km Q=2^8
	V=2.5km/sec		V=2.34km/sec		V=3km/sec
	B=3 and g^2=400m		B=4 and g^2=6400m		B=300 and g^2=900m
then	R=288mbps	then	R=17.3mbps	then	R=1280mbps

Calibration and Registration

Many kinds of calibration are required for spectral image data. Detector sensitivity is measured before launch and checked periodically while in operation to account for detector drift. *Pixel-to-pixel registration* between spectral channels along a scan line and between images of the same area collected on different dates are other important measurements. Finally, the *pointing accuracy* of the telescopes and mirrors must be considered.

For VNIR sensors, detector sensitivity is often measured as a signal-to-noise (SNR) ratio, or the minimum amount of reflected energy that can be recorded above the system noise. This is a measure of the detector material itself. Another measurement records the minimum amount of energy required to distinguish between adjacent pixels having nearly the same reflectance, and is called the noise-equivalent change in reflectance (NE$\Delta\rho$). Common NE$\Delta\rho$s for spacecraft instruments are on the order of 0.5 percent. For LWIR channels, the noise-equivalent change in temperature (NEΔT) is the corresponding measurement. In civilian systems, the NEΔT is on the order of 0.1°C.

Pixel-to-pixel registration between spectral channels for GSDs along a scan line is usually measured as a percentage of the IFOV. Obviously, the smaller the percentage, the better the registration. A value of 0.1 would indicate that for a 30m pixel size, a band-to-band registration of 3m would be expected. This number is important because it provides a measure of how well the spectral values obtained between channels actually represent the same items on the ground.

Image-to-image registration from one observation at t_1 to the next at t_2 is a common problem for GIS developers. For image data, the problem is somewhat more complicated because platform stability in roll, pitch, and yaw directions, and sensor pointing accuracy are both variables that influence the beginning and end points of a given scan line. It is unreasonable to expect that a given pixel at t_1 will be exactly superimposed at t_2. This means that digital spectral values do not represent the exact same spots on the ground from one time to the next, especially pixels collected at the edges of features like agricultural fields, roadsides, or lake shores; or pixels representing complex scenes like residential neighborhoods, urban areas, or other mixed cover types. Registration of multitemporal imagery is, therefore, a matter for GIS specialists to perform as part of data and metadata file preparation. Registration methods include (1) find control points on the ground that are fixed as to their location and which are also observable on each image being used; or (2) identify the geolocations of objects in the scene by GPS techniques.

The selection of ground control points usually includes stable features such as mountain tops or major cross-roads, but not ephemeral features such as lake shores or bridges over rivers (since water levels rise and fall). The aim of selecting these control points is to achieve image-to-image registrations on the order of one pixel through a process of electronically "rubbersheeting" one image to fit the other. GPS locations are preferred because their coordinates can be measured to sub-meter accuracies, and the points selected can be included in vector GISs as permanent attributes. With GPS it is no longer a matter of fitting one image to another in a relative sense, but of fitting multiple images to a known coordinate system.

Platform stability is usually achieved through 3-axis momentum or reaction wheels, or by gyroscopes. SPOT's 3-axis pointing accuracy is 0.1° maintained by three momentum wheels. The Landsat series of satellites were stabilized to ±1° in each plane. Landsat -4,-5 were stabilized by gyroscopes. Landsat Follow-on (L-7) is being designed for 3-axis stabilization within 0.05°.

GIS users routinely require timely data for input in order to optimize their systems for analysis and decision making. In many instances, raster data are preferred. Processing and analysis of these data are improved when an analyst has access to collateral information on the area covered by the image. In addition, raster data from satellite sensors have inherent information content beyond their use as an image backdrop. A broad variety of image processing and statistical analyses are utilized to extract information from raster data. Chapters 2, 3, and 4 describe some of the more important procedures and their implementation.

Image Display and Processing for GIS

Shirley López Baros and Paul R. H. Neville, Earth Data Analysis Center, University of New Mexico
Joe Messina, SPOT Image Corporation

In a world filled with points, lines, and polygons, the incorporation and eventual utilization of satellite imagery within GIS systems may seem incredibly complex and confusing. In most cases, the traditional methods of display and processing are not nearly as relevant in deriving information from satellite imagery as are specific "image processing" models. Fortunately, there are two very different yet effective routes to complete data incorporation. Current image processing software provides a vast suite of tools to permit fast and simple data manipulation. Alternatively, commercial raster data are usually already corrected before sale, and can be used directly as a backdrop image map or for "heads-up" screen digitizing. In most cases, additional on-site processing is performed to enhance the data for more advanced applications. The data may be processed into themes so that they can be treated as separately attributed polygons in a GIS. Filtering can also be used to visually enhance the data, and thus make it easier to interpret.

For those who are new to cell-based GIS processing tools, a brief synopsis of ARC/INFO's GRID module is presented to aid in understanding case studies presented in later chapters. This chapter provides brief descriptions of image processing techniques and approaches. While one of the primary aims of this book is to promote wider integration of raster and vector GIS models and applications, image processing is too large a subject area to thoroughly cover here. The editors encourage interested readers to refer to any of the numerous textbooks available on this topic.

Image Display

Raster image data structure is simple. Most commercial data are processed to a standard format, often a simple flat raster binary file. Furthermore, virtually all GIS and desktop mapping software systems can automatically load the image data. The emergence of *GeoTIFF* as an industry-wide standard for image transfer format will likely make the concern over formats a thing of the past.

Image data are presented as a raster array of individual squares called *pixels*. Each pixel is characterized by a DN (digital number) value, one for each multispectral channel. The higher the response received by a channel detector, the higher the DN value. Pixels of a single data channel displayed on the screen are known as a *grayscale display*. DN values of 0 are displayed as black, increasing DN values as shades of gray between black and white, and the highest DN values as white.

The ability of the sensor to distinguish between different amounts of reflected or emitted energy within a single channel (shades of gray) is referred to as *radiometric resolution*. Most imagery is collected as 8-bit per channel data, or 256 shades of gray, with 0 being black and 255, white. The DN values represent a gradient of response, or continuous data, from lowest to highest values as opposed to another data format (discussed later in the chapter) known as *thematic data* where each DN value represents a separate feature or theme.

Display systems use VRAM to compile, dither, and display the data. The ability of a system to display data can be a limiting factor in its interpretability. A system incapable of displaying more than 256 colors may have trouble with a single channel 8-bit data set. The 8-bit data, 256 potential values may be displayed as 6-bit, 64 data values. Dithering the data to 64 shades of gray may reduce the apparent spatial and radiometric resolution of the data. Most hardware systems allow the user to define the maximum number of colors displayed. Within the performance limitations of the system, it is always best to choose the maximum available.

Alternatively, the image's DN values may not have the full dynamic range of the output display. Contrast stretching improves image contrast by expanding the range of DNs to utilize the full dynamic range of the display. If the data have a normal distribution, a linear stretch from the data set's minimum DN value to its maximum DN value is often sufficient.

A non-linear stretch is required if the data are skewed or have a multi-modal distribution. Most software packages contain a variety of stretching procedures. The simplest and most common tools are brightness and contrast slider bars. These tools along with more advanced interactive non-linear adjustment tools permit the user to manipulate the appearance of the data to best meet interpretability requirements.

Every time a continuous data image is displayed, a look-up table (LUT) is applied. For non-stretched data, the output LUT values are the same as the DN values. Stretching the data does not change the DN values, but does change the output LUT values to the new stretched values. Not only can new grayscale values be assigned in a LUT, but specific colors or shades of gray can be assigned to each DN value. This is known as *density slicing*. The technique is used to separate the DN range of an image into a relatively few discrete colors for easier interpretation. Thus, a single data channel may be "colorized" to enhance regions of interest. The divisions can be combined, taken out, or reassigned to reveal desired patterns.

For examples of how the above display techniques have been used, see case studies by Benson and Greene, and Renaud, Chapter 5 (Plate 37); Guzzetti and Reichenbach, Chapter 7; and Turner et al., Chapter 9 (Plates 74 and 75).

RGB Color Composite

Three data channels can be viewed as a composite by displaying one channel in each of the display monitor's three color guns (red, green, and blue). The resulting RGB composite will appear as a color image in which the colors are in direct proportion to the grayscale ranges of each channel. Mixing various proportions of RGB (additive primaries) results in the full spectrum of colors. Plate 18 shows that red and green light combine to form yellow; blue combined with green forms cyan; and red with blue forms magenta. Equal proportions of light from all three primaries produces a shade of gray. Roughly 16.7

million colors can be created if the monitor has an 8-bit display in each of the color guns.

If the data are displayed as the colors they represent (i.e., visible red wavelength as red, visible green as green, and visible blue as blue), the composite is essentially a natural color image (e.g., Landsat TM bands 3,2,1 as R,G,B). All other composites are called *false color composites* (FCCs), the best known of which is created by displaying data from the green channel through the blue color gun, data from the red channel through the green gun, and data from the reflective infrared channel through the red color gun (e.g., SPOT XS bands 3,2,1 as R,G,B). (See Plate 19.) The high reflectance of healthy vegetation in the near infrared is revealed in shades of red in this particular false color composite combination. Other combinations tend to highlight other features, but most require special interpretation because what can be accomplished in practice does not always have meaning in terms of applications. Case studies by Renaud in Chapter 5 and Lo in Chapter 6 (Plate 40) are especially useful adjuncts to this discussion.

HSV Transformation

Another way of displaying raster data is to look at three channels of data that have been transformed into *hue-saturation-value* (HSV) space (Plate 20). This is also described in the literature as *intensity-hue-saturation* or IHS transformation. The IHS domain is a complement to the RGB domain. An image's intensity is a function of the overall brightness of all three channels, and the hue is a function of the dominant wavelength of color as it lies along the circumference of the color sphere. Saturation is the purity of the color with the least pure color being a shade of gray. The IHS transformation can enhance subtle differences in color not readily observable in the RGB domain, but it can only be applied simultaneously to three channels.

Registration and Rectification

The next step after simple display is the registration or rectification of the image data within an existing GIS database or as the foundation for a new one. The decision to purchase rectified imagery depends upon the ability of the end-user's GIS software to accurately correct the data, the time required to perform the corrections, and of course budgetary limitations. *Registration* refers to aligning two data sets, image to image or image to GPS points, and so forth, so that corresponding points on the ground appear at the same location on each image. For simple image-to-image rectification, the image being used as a base does not necessarily need to be map rectified. For GIS applications, however, all data sets must be registered to a map base. Most commercial GIS and image processing software packages have this capability. Most image processing systems allow for image-to-map, image-to-vector, image-to-image, and image-to-keyboard or ASCII file input using GPS surveyed points. (See Plate 17.)

Rectification is the correction of imagery for X and Y displacement. During rectification the image is projected onto a horizontal reference plane, such as a map projection base, so each pixel in the image is at its correct map location. This involves correlating a number of pixel coordinates on an image to corresponding ground control points (GCPs) on the map base. *Orthorectification* requires using a digital elevation model (DEM) to further correct for Z displacement. (See Plate 22.) Base level imagery usually contains geometrical distortions that can hamper their use in spatial applications requiring map accuracy. Consequently, it is always important to consider the unique correction and incorporation processes of imagery. An example of a significant issue involves using imagery with GPS points or GPS derived vectors.

Most GPS data are collected using a spherical coordinate system like latitude/longitude WGS84. While these collected points are entirely valid, many users fail to project the coordinates when displaying or printing the data. This error is not immediately apparent with point, line, and polygon topology, but is very apparent with imagery, especially high resolution imagery at high latitudes. Projections were developed to categorize and control these distortions. Image data are best used within a grid coordinate system like UTM WGS85 or State Plane NAD 27 rather than a spherical coordinate system. Rectification case studies by Jensen et al. in Chapter 9 and Klein and Isacks in Chapter 6 are good examples of image rectification procedures. Selection of GCPs is currently the most accurate method of geometric correction, but future satellite systems are being designed to include pixel-by-pixel GPS fixes.

Cell-based GIS Processing

Whereas image data are stored in a raster format, GIS data can be stored in vector or grid-based systems. Surface data are stored differently in raster and vector formats.

Vector systems define an object or surface and its characteristics and attributes, such as the x,y coordinate location. Grid-based systems divide the surface into cells, such as a square meter. Within a grid data structure, the location is not defined as a characteristic but is inherent to the storage structure. In other words, each cell is assigned a value that corresponds to the characteristic located at or descriptive of a site, such as a watershed, crop type, or land use classification. Vector systems maintain the integrity of the object and are less concerned with its location. In contrast, a grid system's primary concern is the location of the object.

Points, lines, and polygon surfaces are treated as cells in a grid. The grid environment allows for simple and fast mathematics when generating an analysis between grids. Some GISs allow for both vector and grid data management. In fact, the environment for updating vector GIS layers is evolving into one that is based on raster imagery. The direction of this evolution is due in part to increasing availability of remote sensing data and integration of such data into GIS through image processing technology.

A common approach to revising vector data is the display of classified pixel data and addition or revision of vectors through a "heads-up" digitizing process. (See Plate 23.) Heads-up digitizing (also known as on-screen digitizing) can be executed in the vector environment by using ARC/INFO IMAGE INTEGRATOR, or in the raster environment by using the ERDAS Vector Module. Raster data are often used simply as a backdrop in a vector display for the visual properties that pictures provide. GIS thematic layers may be exploited to facilitate interpretation of imagery in a raster system.

The case study in Chapter 7 by Stow et al. was written specifically to provide information on these techniques. Other case studies that describe aspects of the procedures are Paulson in Chapter 6, and Wright et al., and Parrott et al., in Chapter 7. In addition to ARC/INFO's vector capabilities, the ARC/INFO GRID module contains data input, management, and multivariate statistical tools such as PCA, cluster analysis, and maximum likelihood classification for raster processing and analysis. There are dispersion tools for predictive modeling, DEM interpolation, and raster editing and display. While GRID is not an image processing module, it has the ability to perform many of the procedures outlined in later sections. GRID is the primary tool for raster or cell-based geoprocessing within the ARC/INFO environment.

Advanced image processing can be performed using an external image processing solution, such as ERDAS. ARC/INFO's IMAGE INTEGRATOR supports numerous raster storage formats, including TIFF, SUNRASTER, ERDAS, BIL, BIP, and GRASS. Data in any raster format supported by the IMAGE INTEGRATOR can be imported into a grid by using the IMAGEGRID command. At the same time, some image processing software packages have begun to handle both vector and raster data. Several analytical tools can be used to manipulate the data, including neighborhood analysis, contiguity analysis, overlaying, recoding, masking, and indexing. In addition, the ERDAS "live link" to ARC/INFO allows GIS developers the full vector editing capability of ARC/INFO with the complete functionality of an image processing software package.

Authors of over half the case studies in Chapters 5 through 9 have used GRID or similar capabilities to process and incorporate raster data into respective problem solutions.

Image Processing

Image processing is achieved through a variety of mathematical operations on the DN values to improve feature detectability and interpretability. The image processing techniques used depend on the desired results. These techniques include spatial enhancement, spectral enhancement, theme extraction (classification), data fusion, and change detection.

Commercially available image processing software typically contains functions that accomplish the above procedures with a few keystrokes or mouse clicks. GIS developers and modelers should be familiar with the mathematical and statistical operations being applied to their raster data sets, but the details of the procedures are embedded in commercial packages. For everyday processing, it is not necessary to be expert in the theoretical foundations of the techniques.

Spatial Enhancement

Spatial enhancements through filtering are transformations that reduce noise or enhance desired characteristics of an image by emphasizing or deemphasizing data of various spatial frequencies. The results of these transformations are *apparent* improvements in the spatial resolution of the data. The algorithms improve the interpretability of the data by enhancing spatially related features. Several filtering strategies can be employed depending on the desired results. A few software packages will allow filtering of the image display only; otherwise, filtering will cause a permanent change in the image data. Therefore, filtering should be used as an end process or for applications where the original integrity of the data is not an issue. High pass filters emphasize fine detail and edges, while low pass filters emphasize gradual change. A range of different filtering techniques can be executed using image processing software packages. Choices are typically more limited in desktop mapping software.

Smoothing Filters

A smoothing filter is used to remove speckle and high frequency noise. Smoothing filters are also known as low pass filters because they pass over the low frequency data and operate on the high frequency data. These random, but sometimes regular, effects are usually caused by limitations in sensing operations, signal digitization, or the data recording process, and have not been removed by standard preprocessing functions.

Smoothing filters are routinely used to improve the appearance of RADAR data. One of these types of filters finds the average DN value within a user-specified kernel, typically 3x3 or 5x5 pixels, and places the value in the center pixel before it moves one column over and begins the process again. Another example of this filter follows the same process, but replaces the middle pixel with the median value for the kernel. Both filters have the disadvantage of smoothing the overall contrast in the image, although the effects of the median filter are not as dramatic. When the user wants to remove speckle from a thematic layer, only a median or majority filter should be used.

Edge Enhancement Filters

Edge enhancement emphasizes the edges (or boundaries) of features by forcing edge pixels to a higher or lower value depending on whether they are on the bright or dark side of an edge. If no change occurs within the filter kernel, the image remains the same. The result gives the image a sharper look, almost as if the GSD or spatial resolution has been improved. In fact, only the contrast has been enhanced. These filters are sometimes known as high pass filters because they pass over high frequency data and operate on low frequency data. This technique is useful when the image is to be used for hardcopy display or where contrast must be improved for on-screen digitizing. Both image processing and GIS software packages contain edge enhancement filters. There is no reason to use this type of filter on a thematic image.

An edge detection filter is similar to an edge enhancement filter except that output responses occur only where there is an edge. By highlighting only the edges, this filter makes interpretation relatively easy for cultural features—such as roads, fence lines, the effects of adjacent land management practices, and in some cases even international borders—or natural edge features such as fault lines. Edge enhancement filters come in linear and non-linear as well as directional and non-directional flavors. Such a filter can be used as a layer draped over other layers or in FCCs; vectorized in GIS software packages; or digitized into a vector coverage.

Case studies using filtering techniques include Morisette et al., Chapter 5; Stallings et al., Chapter 6; and Wright et al., and Parrott et al., Chapter 7.

Spectral Enhancement

Spectral resolution measures the numbers and dimension of the electromagnetic spectrum recorded by the sensor. Objectives of spectral enhancement techniques include using the spectral information found in the multiple data bands to create a band for enhancing a feature of interest, and reducing data size and redundancy. Image enhancement techniques result in specific channels representing relative intensities of a feature of interest. After enhancement, these channels can be displayed individually as a grayscale image, a density sliced image based on user-defined thresholds, an FCC with other data, or a combination with other channels in a classification.

Indices

Indices are simple algebraic operations applied to raster DNs to enhance features of interest. Strategies employed in creating indices are based on the spectral properties of the feature of interest. The most commonly used index is a two-channel ratio in which the high response of a feature in one channel is divided by a low response in another channel. In addition to enhancing certain features, *channel ratios* (also known as *band ratios*) also tend to suppress differential illumination caused by varying relief in the image.

Most image processing software packages contain several indices as standard modules. Indices are generally not found in GIS software, but can be created with any software that allows the user to create algorithms.

A potential problem when using indices is that they may unintentionally enhance other features. Consequently, results must be evaluated for confusion errors. In addition, band ratios tend to enhance residual noise in the data.

The following table presents selected frequently used ratios for enhancing vegetation and surface soil properties. Indices are described in case studies by Rowland et al. and Elder in Chapter 6; Chopping, Chapter 8; and Parafina, Chapter 9.

Commonly Used Indices

Name	Equation
Vegetation Difference Index	RIR-Red
Vegetation Index	RIR/Red
Normalized Difference Vegetation Index	(RIR-Red)/(RIR+Red)
Iron Oxide Ratio	Red/Blue
Clay Ratio	MWIR1/MWIR2
Where: RIR = 0.7-1.5μm; Red = 0.6-0.7μm; Blue = 0.4-0.5μm; $MWIR_1$ = 1.6μm; and $MWIR_2$ = 2.2μm.	

Principal Components Analysis

Principal components analysis (PCA) is a technique for transforming multivariate data from different spectral channels into a series of statistically uncorrelated axes. The data are transformed to describe the same total variance with the same number of variables, but in such a way that the first axis accounts for as much of the total variance as possible. The second axis accounts for as much of the remaining variance as possible (while being uncorrelated with the first axis), and so on. Normally, this results in a few axes that account for most of the total variance and a larger number of smaller axes that, when combined, account for very little of the total variance.

One use of PCA is to reduce the number of channels to the three components that represent most of the data variance and can be easily displayed as a false color composite. Another use is to filter noise from an image by removing the last (noise) component, inverting the eigenvectors, and retransforming the image back to its original dimensionality. Specific features can be enhanced by selecting the channels used in the PCA and carefully interpreting the eigenvectors. Examples of PCA appear in case studies by Cullen et al. in Chapter 5, and Neville, Chapter 8.

Tasseled Cap Transformation

The tasseled cap transformation uses a pre-set linear combination of coefficients which reduces the dimensionality of the image data to bands enhancing soil brightness, greenness, and yellowness in the case of Landsat MSS, or soil brightness, greenness, wetness, and haze for Landsat TM. Many image processing software packages contain this transformation as a pre-set program, but the transformation coefficients can be used in any software that allows for user-created algorithms. For an example, see the case study by Parafina, Chapter 9.

Theme Extraction (Classification)

The object of theme extraction strategies is to classify an image into useful categories for a given application. For some applications, like surface water mapping, a single theme may be all that is desired. In others, like land use mapping, the entire scene may require classification. Classification is a pattern-based process that assigns individual pixels to categories based on spectral properties. The resulting thematic image is different in many ways from the previously discussed continuous data images. Each DN value is a class representing separate user-designated surface feature categories. Every class carries a list of attributes and is generally color-coded for quick visual identification.

Thematic layers can be directly operated on by a cell-based GIS system or vectorized into a polygon coverage. The two main steps in theme extraction are clustering and gathering class statistics, and classification of the continuous multispectral image data into a single thematic layer. The clustering process can be further divided into two different methods: unsupervised and supervised. The information gathered in the clustering process is used to classify the image. Several classification rules are used to create the final thematic layer. In the end, the classified image may require combining some classes to better meet application needs, filtering to remove speckle in the data, and an accuracy check.

Unsupervised Clustering

Unsupervised clustering is a way of creating groups of similar surface features without a priori knowledge of the items to be classified. The total spectral response of all pixels is inspected for statistical relationships and clustered into similar groupings. It is then the interpreter's responsibility to assign names to the groupings.

A standard clustering technique is *iterative self-organizing data analysis* (ISODATA). The ISODATA clustering technique initially divides the image data into equal areas of data variance based on the output number of classes designated by the user. Each pixel is then evaluated on how close its DNs are to each cluster mean and assigned to the nearest cluster. Each pixel's spectral distance is reevaluated after adjusting the cluster means to better fit the image variance. This process repeats itself until a user-defined percentage of pixels which remain unchanged in each cluster is reached. The resulting classes in the signature file are evaluated as to which feature they best represent. The evaluation is based on user interpretation of the derived image or spectral libraries embedded in the software.

Some classes may be eliminated because they represent too many different features, or classes may need to be merged because they represent redundant information. Occasionally, classes are further divided into more classes. Several evaluation tools are available in most image processing software packages.

Unsupervised clustering techniques have the following attributes: (1) No prior knowledge of the site is required. (2) User bias is minimized, but output classes may not have anything to do with what the user wants (e.g., a desired result may be a vegetation map, but in an arid location the results will be driven by soil response patterns). (3) Interpretation of some output classes may be difficult. (4) Because the number of unique groups in an image is unknown, some output classes may require further splitting or merging.

Supervised Clustering

Supervised clustering allows the user to define which groups are based on a priori knowledge of the area. One of several methods is to draw polygons on the screen defining known features. Another method is to statistically seed similar areas based on a known point. Thematic images derived from other classifications or GIS layers can be used to create clusters.

Because a principal assumption is that the training area is typical of the whole category, great care must be taken to select the training area. Sometimes several conditions or states of the feature exist simultaneously in the data set. Each requires a separate training site. As with unsupervised clustering, supervised clustering produces a signature file that should be evaluated and possibly modified or refined.

The main attributes of supervised classification follow: (1) Prior knowledge (usually by field observation) is required for the area to be classified. (2) The areas chosen for the training sites may not be entirely representative of a given class description and, therefore, the final map output could be inaccurate. (3) Because the meaning of the output classes is known, the user does not need to spend time trying to interpret each class. (4) Features in the image that should have been separated but were not specifically trained for will be misclassified as errors of commission. (5) The classification process is directed to classify features according to a user-defined theme.

Decision Rules

The results of clustering processes are used to classify image rasters. The classification process applies a decision rule to the image using the information found in the signature file. The most frequently used decision rules are parallelepiped, minimum distance, Mahalanobis distance, and maximum likelihood. The *parallelepiped*, the simplest decision rule, is premised on creating spectral boxes based on the minimum and maximum DN values for the clusters and evaluating each pixel as to which box it best fits. One advantage to this method is that it works best for features that are not normally distributed, that is, the boxes do not overlap. In cases where boxes overlap, however, this procedure is not recommended.

The *minimum distance* rule classifies pixels based on proximity to a cluster. The output from an ISODATA cluster is essentially a minimum distance classification. The *Mahalanobis distance* rule uses the covariance matrix as well as the spectral distance found in the signature file to classify an image. The *maximum likelihood* rule uses the same variables and incorporates a probability function to determine which pixel belongs to a particular class. Both the Mahalanobis distance and the maximum likelihood rules are generally the most accurate, but they are highly parametric and require that the data are normally distributed.

Classified and *probability images* are then written for display and analysis. Probability images can be used in image analysis to segregate areas of correct assignment from areas that have been incorrectly assigned or require more training. The final classified image can then be evaluated by comparing the resulting classes with a random sampling of known points. Many case studies in subsequent chapters highlight classified images and the methods of their creation. Collectively, these case studies demonstrate that the DNs characterizing individual rasters are a valuable form of data for vector GIS. The only way these DNs can be exploited for their information content is to merge raster and vector capabilities.

Among the case studies using classified imagery are Nuñez et al. in Chapter 5; Carlson et al. (Plate 51), Guzzetti and Reichenbach, Whistler et al. (Plate 57), Haack et al., Roberts et al., and Nizam, Chapter 7; Spell et al., Chopping, Bennett and Muldavin (Plate 63), and Neville, Chapter 8; and Rehder, Chapter 9.

Change Detection

In addition to creating layers of information representing a specific moment in time, image data can also be used to create data sets representing different moments in time. This process is known as *change detection*. Depending on the objective of the application, several methods are used to detect change. (See Plate 21.) Three of the most common methods are described below.

Change detection uses two or more images acquired at different times, called *reference images*, to create yet another image that pinpoints areas of change. This is a useful application for reviewing large areas for recent activity, or identifying changes in activity levels between two dates. The two temporal images are registered to each other and displayed in RGB such that a particular channel from one date is one color, and the same channel from a different date in another color. Equal mixtures of the two colors indicate where there is no change, whereas emphasis of one color over another indicates where a particular date has a stronger response in that channel. While the change image will highlight a condition of change, it will not indicate the significance of the change. Knowledge of the spectral properties of surface features in the channel showing change are important for interpretation.

A variation on the reference image method is to use a spectral enhancement, such as a vegetation index, that points to a specific feature of interest for both dates and then compares the two outputs in RGB. If the two images have also been radiometrically corrected and at least equilibrated by a linear regression model, still other techniques can be used. Under the *image differencing* technique, one image is subtracted from another. The positive or negative results indicate the direction of change for each band with no change indicated by values close to zero. Similarly, the two images can be ratioed. Values greater than 1 or less than 1 indicate the direction of change for each channel. Values equal to 1 indicate no change. The first component indicates where the images are the same, and the second component indicates where they are different with the loadings of the eigenvalues showing which type of change is enhanced in the high and low values.

When comparing more than two dates, PCA can be used on all images. Interpretation of the eigenvalues indicates changes being highlighted by each component. The PCA method has the advantage of highlighting complex temporal patterns, but may be difficult to interpret.

The above techniques work well if continuous data are a suitable output. The results can be used as grayscale or RGB displays, as density sliced images, or as input to "heads-up" digitizing. However, if a thematic layer is needed, classification techniques are required. Each temporal image can be classified separately and overlaid to determine areas of change and areas of constancy. All multidate images can be combined into a single file for classification and the resulting change classes interpreted by the signature file results. The above listed change indices and PCA images can also be classified. Case studies by Kasouf in Chapter 5; Kempka et al., and Lo in Chapter 6; Watkins, Carroll, Parrott et al., Haack et al., and Nizam in Chapter 7; and Spell et al., and Mularz in Chapter 8 focus on change detection methods.

Data Fusion

Different types of image data may be merged in a process called *data fusion*. Data fusion is especially useful when the spectral qualities of a lower GSD data set can be enhanced by higher GSD data. Another use of data fusion is to combine satellite imagery with other kinds of data sets such as RADAR, digital elevation models (DEMs), radiometric surveys, or aero-magnetic data to aid visualization. Because the principal goal of data fusion is to preserve the original spectral data while incorporating other information, the results are hybridized values and should not be further spectrally enhanced or classified.

Three major techniques are used to merge different image data sets. The simplest method is to resample and convolve the lower resolution data to the same pixel size as the higher data. An edge enhancement filter is used on the higher resolution scene to help enhance spatial detail. The edge enhanced scene is then histogram-matched separately to each of the lower resolution channels. The results from each histogram match are multiplied through each of the lower GSD channels.

A more involved method is to execute a PCA on the resampled lower resolution scene. The first component, sometimes called the albedo component because it enhances overall surface response, is considered a close approximation of higher resolution panchromatic imagery. The higher resolution image is histogram-matched to the albedo image and then substituted for the albedo image. The eigenvalues are then inverted and this PCA-higher resolution image hybrid is transformed back into the original channels, but ingrained with the original data is new, higher frequency information.

Another method begins at the same place as the PCA method, but instead an IHS transform is carried out on the lower resolution data. In this case, the intensity image is also considered a close approximation to higher resolution panchromatic imagery. The higher resolution data are histogram-matched to the intensity image and then substituted for it. The IHS-higher resolution data hybrid is transformed back into the RGB domain with the higher resolution data ingrained. This technique may resolve more subtle differences in color contrast, but its major disadvantage is that it can only be executed using three channels of lower resolution data.

Examples of data fusion appear in Paulson in Chapter 6, and Cullen et al. in Chapter 9. In these case studies, image sharpening has been achieved by fusing a data set having a relatively coarse GSD with data for the same area collected by SPOT images having a finer GSD. Typically, a SPOT panchromatic image of 10m GSD is used to sharpen a Landsat 30m image to an apparent resolution of 12 to 13m.

Preprocessing

The optional use of raw or unprocessed data permits the user to define the initial parameters necessary to successfully complete a given application. Raw data may be used for classification and DEM generation, or to provide the initial input of highly accurate map bases. Every step or algorithm produces a data entropy effect that reduces the overall statistical accuracy of the data set. Because it is impossible to reverse engineer the models and expect mathematically valid results, raw data are often used. The use of raw or unprocessed image data requires the application of an entirely new set of pre-GIS use operations. This "preprocessing" refers to clean-up operations to a data set. Several unique imagery issues require attention. Detector corrections are required to adjust raster DN values so that they are comparable across the period of time it took to collect them. Correcting for the effects introduced by the Earth's curvature and rotation in the context of satellite orbital parameters is also important. Earth corrections may also be applied to remove the larger errors introduced by atmospheric noise (commonly known as *haze removal*).

Detector Corrections

Depending on the product, systematic errors in raster data are corrected by the vendor prior to sale. It is important to know what corrections have or have not been made, especially if further processing is desired for specific applications. Additional corrections, however, may be required by the GIS specialist before the data can be used for specific applications. For instance, *radiometric correction* is concerned with removing errors associated with sensor detectors. Errors associated with sensor detectors, such as *line dropout* caused by a detector failure during a scan; *banding or striping* caused by a detector out of adjustment; and *line start errors* caused by system failure to collect data at the beginning of a scan. Other corrections that should be considered are related to sensor signal deviations, namely *gain* and *offset*. Line dropouts and banding can sometimes be handled by filtering methods. Gain and offset corrections can be handled by any software that allows insertion of user-defined equations. This type of preprocessing is generally referred to as *Level 1A* processing.

Geometric Corrections

Many applications require geometric rectification of the imagery. All image processing systems and many GIS packages are able to rectify imagery. Because the level of rectification accuracy varies, a thorough understanding of the limitations of specific packages becomes necessary. The option to purchase the data with some gross geometric corrections applied should always be considered. The gross geometric corrections associated with *Level 1B* preprocessing are applied directly during the Level 0 through Level 1B process. The Level 1A steps discussed in the preceding section are also applied, but the data never technically become 1A data. This direct level of processing minimizes data entropy. Deconvolution and systematic distortion corrections associated with the Earth's rotation and incidence angle effects produce data with pixels in their proper spatial place with respect to every other pixel, and exhibit a constant GSD.

Atmospheric Corrections

Atmospheric effects represent alterations of DNs caused by scattering and absorption of solar energy in the atmosphere. These effects result in increased brightness of DNs because scattering increases DN value, or lowered brightness because absorption can reduce DN value. Atmospheric corrections can be complicated depending on how completely scattering and absorption effects must be removed. Time- and computer-intensive methods involve using an atmospheric modeling software package to derive input values to transform the image DN values to pure reflectance units. This level of processing is more important in scientific studies requiring the actual reflectance of a surface feature or in cases where a digital spectral library is used to classify the data.

Among the simpler but less accurate methods of atmospheric correction are *dark pixel subtraction* and *linear regression.* In the subtraction method, a haze component is determined by identifying DNs whose value should be zero (no response) and subtracting the amount necessary to make them zero. By subtracting an equal amount from all pixels, it is assumed that this form of atmospheric noise has been removed. Surfaces typically used to represent these "no response" features are cloud shadows, freshly paved parking lots, or large bodies of open water where shoreline sediment loads and algae are not contributing to the response. Limitations to the subtraction approach are fairly obvious. It does not work well under thin cirrus clouds or with data sets acquired during multiple thin cloud decks.

In cases where two scenes of different dates are to be used together, such as for change detection, *linear regression* is usually sufficient to correct for atmospheric distortion. In this method, the user records DN values of both images from each of the bands. The points where the DN values are recorded should represent areas of no change such as building tops, asphalt parking lots, or rock outcrops. Care should be taken not to include areas where a change in vegetation response or water level may occur, or where one of the images might have clouds or shadows. These numbers are then used in a statistical program that can calculate linear regression equations for each band. The values for the independent variable are from the image used as the baseline. The resulting equations are then applied to the bands from the other image.

Aside from haze removal, other more sophisticated adjustments to DN values account for solar elevation (seasonal effects in reflectance) and solar altitude (diurnal effects). Most satellites have orbital designs that reduce solar altitude problems by more or less constant equatorial crossing times. This procedure ensures that data are acquired at about the same time everywhere, thereby reducing Earth rotation problems. Seasonal effects, however, are related to Earth's revolution around the Sun, and must be dealt with separately, especially in cases where seasonality is a confusing element of DN values.

Preprocessing is based on the reasonable assumption that raster DNs can be statistically and mathematically processed for additional information content. Satellite imagery can be the most important addition to a GIS data base. The ability to see reality rather than cartographic generalizations permits the user to adjust for and manipulate the data in ways that automated systems cannot. The incorporation of satellite imagery can be straightforward and painless, but the responsibility to learn about the very different issues associated with the imager rests with the data user. This chapter is intended to provide guidelines on the issues and applications associated with satellite imagery rather than a cookbook ordering tool.

Using Scanned Aerial Photographs

Roy Welch and Thomas Jordan, Center for Remote Sensing and Mapping Science (CRMS), University of Georgia

Aerial photographs are a major data source in the construction of GIS databases. Typical uses include delineation of land use/cover information, generation of highly accurate planimetric and topographic data, and revision of existing base maps or vector GIS database files. In general, aerial photographs are of much higher resolution than satellite images. They must be used where information requirements dictate the inclusion of 3-D features positioned so as to meet the stringent accuracy requirements associated with large-scale mapping projects or detailed GIS databases.

With the introduction of computer-based, softcopy photogrammetric mapping and GIS systems that operate in a desktop computing environment and are able to manipulate both vector and raster data, the demand for images in raster (digital) format has increased sharply. While the intent of this chapter is to provide guidelines related to the use of aerial photographs in digital format, it must be stressed that aerial photographs in hardcopy format may prove to be a more economical solution for many mapping and GIS database construction tasks.

Conversion of Aerial Photographs to Raster Format

Aerial photographs recorded with standard mapping cameras must be scanned and converted to raster format before they can be used in digital image processing/mapping environments. Decisions made concerning the scanning process will determine in large part the geometric accuracy of the scanned photographs, the amount of information that can be extracted, and the speed with which the images can be processed.

The basic function of a scanner is to convert a hardcopy photograph to digital (i.e, softcopy) format. Scanners are designed to capture image data in reflection (paper print) and/or transmission (film transparency) modes. Because film transparencies tend to have higher spatial resolutions and a greater range of gray values than paper prints, they are the preferred source material when converting aerial photographs to digital images.

Low-cost, flatbed scanners are generally designed for desktop publishing applications. They offer adequate spatial and radiometric resolution for capturing detail from large-scale aerial photographs, but may introduce unwanted geometric distortions into the scanned image. Some flatbed scanners advertise high resolution, but are actually scanning at low resolution and resampling to higher resolutions through the use of software. These types of scanners are useful for documents and graphics applications, but are not optimal for scanning aerial photographs to be used in applications where both high spatial resolution and good geometric accuracy are essential.

Top quality scanners designed specifically for digital photogrammetry include those manufactured by Carl Zeiss, Inc., Helava Associates, Inc., Vexcel Imaging Corporation, and Wehrli and Associates, Inc. These scanners can meet all requirements for resolution and accuracy, but costs are generally higher than for flatbed scanners. Consequently, it may be appropriate to make use of available commercial scanning services offered by companies such as ImageScans (Denver, Colorado) and Precision Photo Laboratories (Dayton, Ohio). Photographs are scanned for a nominal fee and the data returned on CD-ROM or tape. In the following paragraphs, some of the considerations associated with scanning aerial photographs, rectifying the digital image files, and using orthophoto products are discussed.

Photo Scale, Scanning Resolution, Information Content and Data Volume

Aerial photographs are recorded with photogrammetric mapping cameras from altitudes of approximately 300m to over 20,000m. If a standard lens focal length of 15cm is assumed, photo scales will range from about 1:2,000 (very large scale) to approximately 1:135,000 (very small scale).

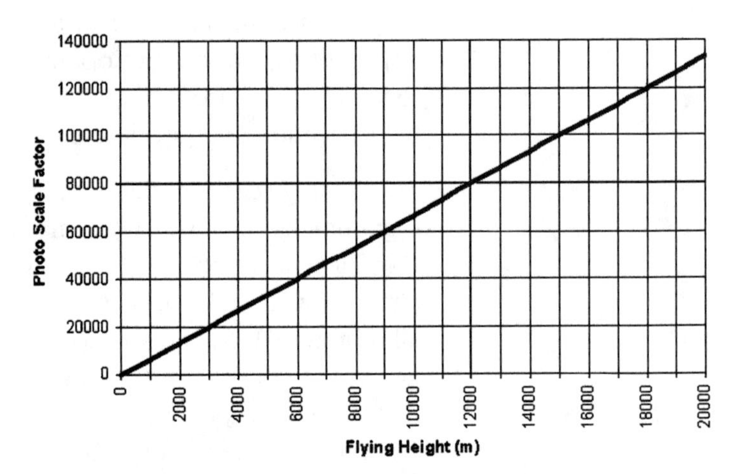

Photo scale factor plotted as a function of flying height for a typical photogrammetric camera equipped with a lens cone. The lens cone has a focal length of 15cm.

Typically, photographs are scanned as raster images at pixel resolutions ranging from about 250µm (100 dpi) to about 10µm (2,500 dpi). Each pixel or dot has a gray value ranging from white to black encoded as an 8-bit byte.

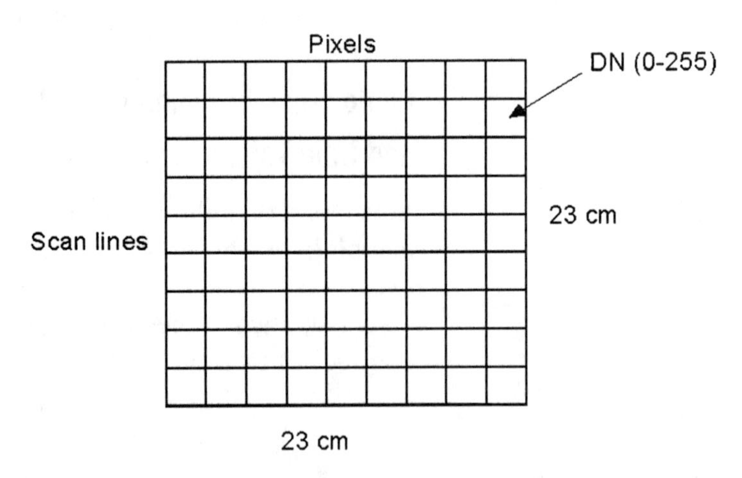

Raster format scanned aerial photograph in which pixel size is specified in micrometers or dots per inch. Each pixel is represented by a digital number (DN) on a scale of 0 (black) to 255 (white).

The relationship between scanning resolution in micrometers (µm) or dpi and pixel dimensions in meters on the ground is a critical factor. In the following figure, for example, these relationships are depicted for photographs of 1:5,000, 1:20,000 and 1:60,000 scale.

Pixel size versus ground dimension for photo scales of 1:5,000, 1:20,000, and 1:60,000.

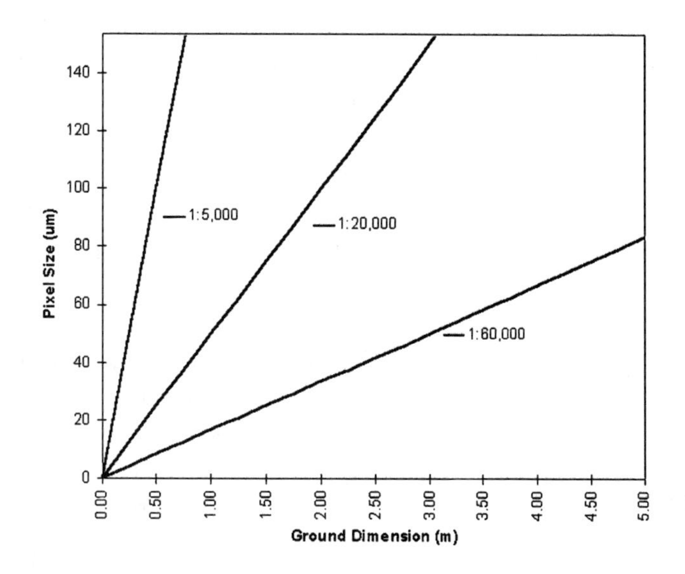

Data volume, which is normally computed in terms of the number of bytes per photograph, directly determines disk storage requirements and must be considered in relationship to the processing speed of the computer in which the images will be housed.

Data volume in megabytes increases sharply as scanning resolution is reduced below 25 μm.

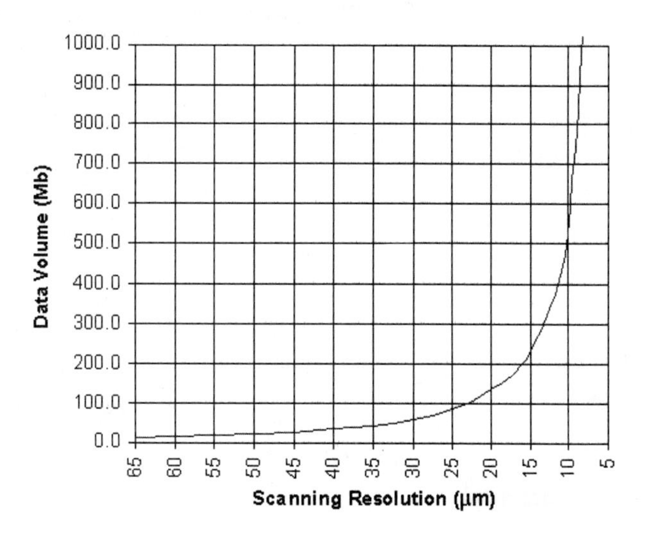

From the two previous figures, it is evident that scanning resolutions of 64μm (400 dpi) to 25μm (1000 dpi), will yield ground pixel dimensions of between 0.125m and 1.28m for photos at scales of 1:5,000 to 1:20,000 and with data volumes of approximately 13 to 85 MB per photo. These ground pixel dimensions are commensurate with many mapping requirements, and data volumes of this magnitude can be processed on 486 and Pentium PCs. Consequently, rather than attempting to retain all resolution inherent to the original photos, it may be appropriate to set the scanning resolution at a value needed for the mapping task and consistent with the data processing capabilities of the hardware. This argument is strengthened by the non-linear relationship between photo resolution as expressed in line pairs per millimetre (lpr/mm) and information content. In other words, increasing or decreasing the resolution of the photo by a factor of two does not improve or reduce interpretability by a corresponding amount as judged by the examination of fine detail. In such circumstances, the change in interpretability more closely approximates the square root of two. Therefore, setting the scanning resolution to the coarsest pixel size necessary to identify the smallest features to be extracted and mapped is an appropriate strategy that need not seriously degrade information content.

As a rule of thumb, at least two to four pixels are required to represent a feature in an image. Theoretical considerations also dictate that approximately two pixels are required to represent a line pair at the resolution limit of the photograph. Since most photographs recorded by photogrammetric cameras on mapping films will have resolutions (on the original negative) of between 20 and 40 lpr/mm for low contrast targets (1.6:1 contrast ratio) and perhaps 15 to 30 lpr/mm on second generation paper prints or film transparencies, a scanning aperture of 1/30 mm (33µm) to 1/60 mm (17µm) will retain the original photo resolution when converting the majority of analog photos to digital files. However, retaining the original photo resolution during the scanning process only becomes critical when the photos are small-scale.

Scanning resolution (µm) vs. photo resolution (lpr/mm). One pixel is used to represent one element (bar or space) of the line pair at the resolution limit.

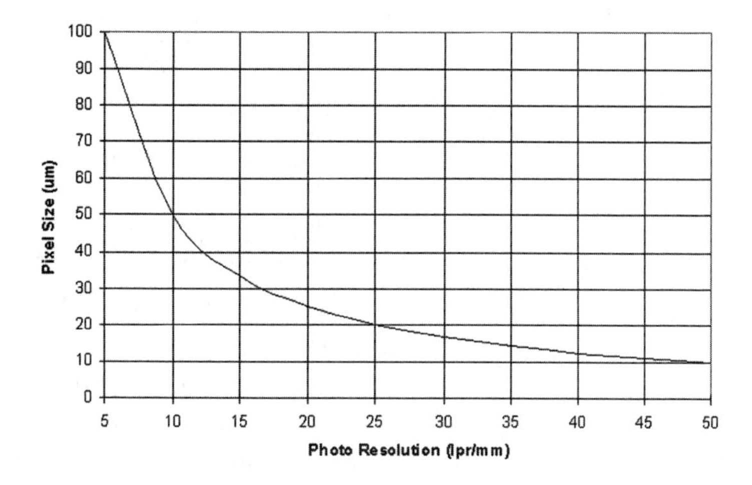

Relationships between scale and image feature size are illustrated in the next figure for symmetric objects such as buildings with a linear dimension of 10m, spaced at 10m intervals. In effect, as photo scale is reduced, these buildings represent increasingly higher frequencies in image space. Therefore, scanning resolutions must be adjusted accordingly. It is evident in the figure below and the preceding discussion that the image size of objects to be mapped and hence photo scale are extremely important considerations when determining scanning resolution.

The image size of features such as 10m buildings spaced at 10m intervals decreases as photo scale is reduced. At small photo scales, such as 1:40,000, a higher scanning resolution will be required to adequately represent the feature.

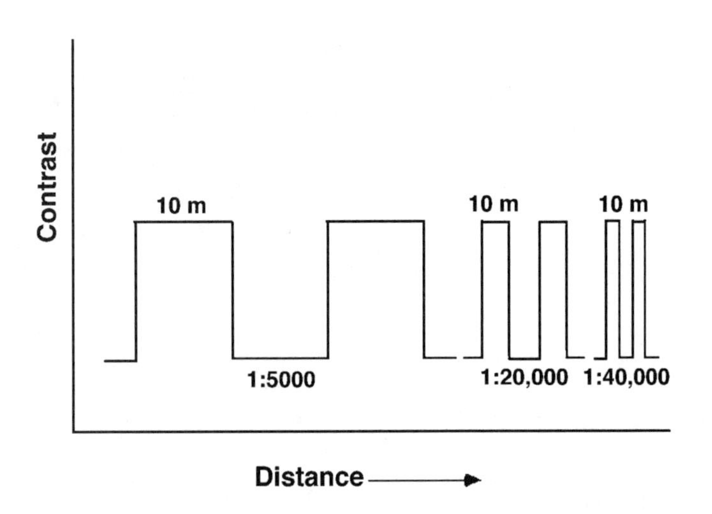

Rectification Methods for an Aerial Photograph

Aerial photographs must be geometrically corrected before the photos can be employed in a digital mapping system or GIS. For example, the displacements in the image due to lens distortion, earth curvature, refraction, camera tilt, and terrain relief must be removed (or at assessed and minimized). Correcting and removing these displacements are generally performed in two steps: preprocessing to correct the systematic displacements (i.e., lens distortion, earth curvature, refraction), and rectification to remove the effects of tilt and to establish scale. As part of the rectification process, the mathematical relationship between the map coordinate system and the scanned aerial photograph is determined and the digital image is resampled to create the rectified (or geocoded) image. Differential rectification adds a further correction for terrain relief through the use of a digital elevation model (DEM).

In many cases, it is not necessary to perform all of the corrections listed above. For example, scanning at a coarse pixel resolution may mask systematic displacements, leaving only camera tilt and terrain effects to be removed. In areas of low relief (i.e., where relief is less than 0.5% of the flying height), displacements due to terrain effects are likely to prove negligible and can be ignored for many planimetric mapping applications. The magnitude of displacements in mm and pixels at radial distances of 50mm and 100mm are provided in the following table.

	Relief displacement (de) at 50mm from photo center			Relief displacement (de) at 100mm from photo center		
Relief (% H)	mm	Pixels (25μm)	Pixels (50μm)	mm	Pixels (25μm)	Pixels (50μm)
0.5%	.25	10	5	.5	20	10
1%	.5	20	10	1.0	40	20
5%	2.5	100	50	5.0	200	100
10%	5.0	200	100	10.0	400	200

Relationships between photo and map coordinate systems are established by the use of ground control points (Δ). To obtain a satisfactory rectification, ground control points should be well-distributed throughout the photo as shown here.

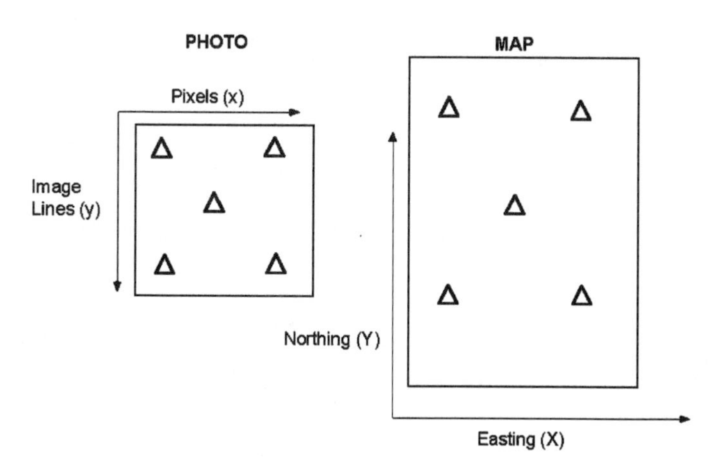

Planimetric displacements (de) caused by terrain relief are expressed in mm and the number of 25μm and 50μm pixels for radial distances of 50mm and 100mm from the center of a vertical aerial photograph. Differential rectification must be implemented when relief is greater than about 0.5% of the flying height (H).

Displacements (de) due to terrain relief increase radially from the center of the photograph.

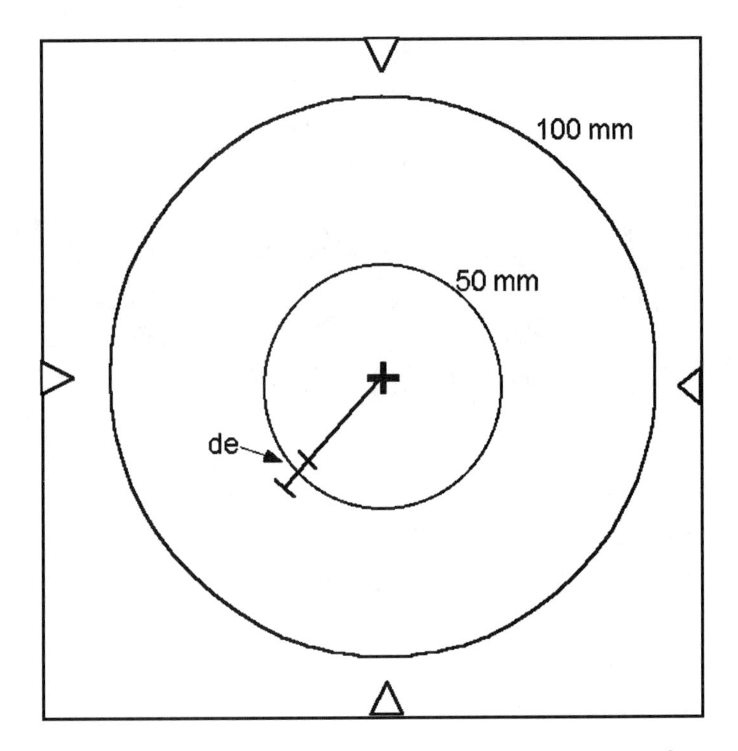

The following general steps are required to rectify an aerial photograph.

1. The aerial photograph in film format is scanned and converted to an 8-bit (256 gray levels) or 24-bit (color) raster image. The actual scanning aperture (resolution) depends on the original scale of the photograph and the desired output pixel size of the rectified photo. For example, scanning a 1:20,000 scale photograph at a resolution of 25μm will produce a digital photo in raster format having a 0.5m pixel resolution. This pixel dimension can be increased or decreased in the rectification/resampling process to achieve registration with other data sets and GIS layers.

2. A list of ground control points (GCPs) must be created and saved to disk. Ground control points are markers or features visible on the aerial photographs for which the x, y, and z terrain coordinates are known. In the U.S., Universal Transverse Mercator (UTM) and State Plane coordinate systems are typically employed. The ground coordinates of GCPs can be obtained by conventional ground surveys, from published maps, by global positioning system (GPS) surveys, or by aerotriangulation.

3. The image locations of the fiducial marks and GCPs must be located on the scanned photo and digitized to obtain x, y photo coordinates. This is accomplished by enlarging the scanned photograph on the computer display and measuring the pixel and line coordinates of the control point locations to a fraction of a pixel. These coordinates are also saved to a disk file.

4. The mathematical transformation parameters are computed to establish the relationship between the ground (x, y, z) and the image (x, y) coordinates. The algorithmic approach depends on the type of rectification method to be used (i.e., polynomial, projective, or differential rectification). A simple affine polynomial approach will require a minimum of three well-distributed GCPs, whereas more complex approaches will need at least five GCPs. Regardless of the approach, more than the minimum number of GCPs should be employed in order to obtain an optimum solution.

5. Finally, the digital photo is rectified to the ground coordinate system. During rectification the scanned photo is resampled and the pixel size modified as required to correspond with other raster data sets that may be part of the database.

6. Differential rectification requires an additional step and employs terrain height data from a DEM to correct for relief displacements in the image.

Orthophotographs

Orthophotographs are scaled aerial photographs from which displacements due to tilt and terrain relief have been removed and that contain the geometric characteristics of a map with all information content of the original photo. The procedure for producing digital orthophotos is well established and derived from techniques developed in the 1950s and 1960s for producing orthophotos in hard-copy format. Required inputs to the digital process include aerial photographs, ground control points, and a DEM. The DEM can be generated from the aerial photograph by photogrammetric techniques and automated stereocorrelation, derived from contours digitized from existing maps or, perhaps, acquired from government or commercial sources. Digital orthophotos provide an excellent coordinate or map reference as the foundation layer in a GIS database, thus facilitating the development and registration of coverages to the map coordinate system.

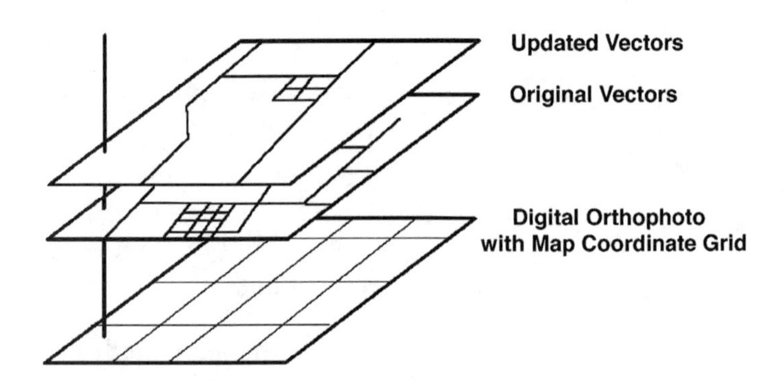

Integration of raster and vector data sets. The digital orthophoto base layer is ideal for updating vector data sets.

At present, the U.S. Geological Survey (USGS) is producing digital orthophoto quarter quads (DOQQs) for the land area of the United States. These DOQQs are derived from National Aerial Photography Program (NAPP) photos of 1:40,000 scale, and are reported to meet planimetric accuracy standards of ±3m on the ground.

The DOQQs and similar digital orthophoto products are ideal for GIS database construction and update using software packages such as ARC/INFO, Desktop Mapping System (DMS, by R-WEL, Inc.), EASI/PACE (PCI, Inc.), ERDAS IMAGINE, and MGE (Intergraph). However, it is important to carefully consider software capabilities when contemplating the use of aerial photographs for database construction and/or revision.

Map database compilation and revision from digital data sets necessitate overlay of existing vector files, enlargement or reduction of images, an ability to create vector files with attached attributes from raster image data and a method for editing the vector files in digital format. Some software packages such as ARC/INFO, AutoCAD (Autodesk, Inc.), MapInfo (MapInfo Corp.) and O'Map (R-WEL, Inc.) require rectified photos or orthophotos as the backdrop upon which to superimpose vector files already in a map coordinate system. Others, such as DMS, IMAGINE, and EASI/PACE, will allow rectification of uncorrected raster photo files, and also permit the user to generate vector files in registration with these data sets.

Emphasis is now being placed on registering vector files (e.g., USGS Digital Line Graph, Bureau of Census TIGER, ARC/INFO, and AutoCAD files) with raster images for update or analysis. For such applications, the ability to read different data formats is important, as is the ability to place the vector files in exact registration with the image data. Most mapping/GIS software packages provide for data format conversion; however, users should verify that conversions exist for the data formats that will be employed.

Example of a digital orthophoto with vector overlay.

Summary

Catalysts for rapid growth in the use of aerial photographs in digital (softcopy) format for mapping and GIS database construction applications include improvements in scanners, availability of commercial scanning services, and, most importantly, the release of reasonably priced software products that permit the use of raster images for the compilation and update of vector files. Growth in the use of scanned photographs will continue to be fueled as government agencies such as the USGS make DOQQ products available for large areas of the United States and commercial firms market black-and-white, color, and color infrared orthophoto products designed to meet customer demands for softcopy data products of sub-meter resolution. When linked with DEMs and map files in vector format, scanned photographs provide the basic ingredients for on-screen map and database compilation/update, as well as terrain visualization, environmental monitoring, and change detection.

Today, GPS, aerial photographs, and satellite images in both analog and digital formats are being used in conjunction with digital mapping and GIS techniques for environmental and natural resource studies. This integrated approach is much more powerful than any single technique. In addition, the utility of a traditional vector-based GIS can be greatly enhanced through the inclusion of raster data sets properly registered to the map coordinate system and tightly linked to the vectors and associated attributes.

For more detailed information on the topics of softcopy photogrammetry, digital orthophotos, and GIS database development, refer to publications such as *GIS World*, *Earth Observation Magazine (EOM)*, *Geodetical Info Magazine (GIM)*, and *Photogrammetric Engineering and Remote Sensing*. In addition, it should be noted that the American Society for Photogrammetry and Remote Sensing is about to publish a new book entitled *Digital Photogrammetry: An Addendum to the Manual of Photogrammetry*.

Data Collection Systems, Formats, and Products

Amelia Budge and Stan Morain, Earth Data Analysis Center, University of New Mexico

The path of multispectral data from collection to application in a GIS is not as straightforward as the path followed by conventional black and white photography. Conventional photography is a combination of wet chemistry and electronic manipulations, whereas multispectral processes are all digitized. Data must first be geometrically and radiometrically corrected. Whether the data were collected by a satellite or an airborne sensor, they must be sent to a ground processing site where they are recorded and preprocessed to radiometric and geometric accuracies, and placed into a standard format. At a workstation or PC, the raster data and various types of collateral information, such as digital terrain elevation data, are loaded into the system where they may then be manipulated using commercial image processing.

Primary Satellite Sensor Systems

There are several satellite sensor systems currently providing operational raster data to GIS developers and modelers. The characteristics of these systems vary according to the design goals and engineering trade-offs described in Chapter 1. In this chapter, we provide system descriptions for the primary platforms and sensors being used in GIS today. There is no attempt to be comprehensive because there are dozens of platforms and hundreds of sensors, both operational and planned. Readers interested in more complete listings should consult Earth Observing *Platforms and Sensors*, a searchable CD-ROM available from the American Society for Photogrammetry and Remote Sensing, or *Observation of the Earth and Its Environment: Survey of Missions and Sensors* available from Springer-Verlag.

Satellite Pour l'Observation de la Terre (SPOT)

Program Objectives

The SPOT program is committed to commercial remote sensing on an international scale and has established a global network of control centers, receiving stations, processing centers and data distributors. The SPOT satellite is owned and operated by the Centre National d'Etudes Spatiales (CNES), the French space agency. Worldwide commercial operations are anchored by the following private companies: SPOT Image Corporation in the U.S., SPOT IMAGE in France, SPOT Imaging Services in Australia and SPOT Asia in Singapore, and distributors in over 70 other countries. Central to SPOT's commercial approach is the commitment to launch a minimum of five satellites. This ensures a continuous source of image data into the 21st century. (See Plates 9 through 11.)

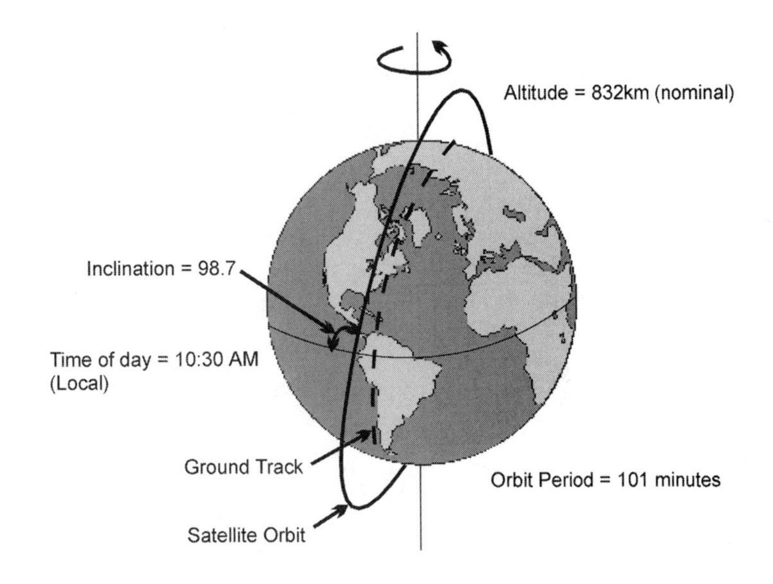

Orbital parameters for SPOT.

System Characteristics

Launch

- Dates: 1986, 1990, 1993, 1997, 2002
- Vehicle: Ariane V16; Ariane V34; Ariane
- Site: Kourou, French Guiana

Orbit

- Altitude: 832km
- Type: LEO, sunsynchronous
- Inclination: 98.7°
- Revisit: 26 days (satellite ground track); 1-3 days (off nadir mode)
- Coverage: Global
- Equatorial crossing: 10:30
- Period: 101 minutes

- Stabilization: 3-axis pointing accuracy 0.1° maintained by 3 momentum wheels

Dimensions

- Mass: 1750kg
- Size: 2m x 2m x 2.5m

Design lifetime

- 2 years (SPOT-1 through 3); 5 years (SPOT-4,-5)

Instruments

- HRV (2 each on SPOT-1, -2, -3); HRVIR (SPOT-4)

High Resolution Visible Sensor (HRV)

Dates of Operation

- 1986-present

Primary Mission

- Obtain Earth imagery for land use, agriculture, forestry, geology, cartography, regional planning, water resources, and GIS applications

Sensor Description

- The HRV sensor operates in two modes: multispectral and panchromatic. Each SPOT payload consists of two identical HRV imaging instruments that are pointable in the cross-track direction up to 27° from nadir (Plate 9). The viewing mirror is operated by a stepping motor that can tilt the mirror in 0.3° steps (= 0.6° change in the line-of-sight or about 8.7km on the ground). This enables the HRV to image a swath up to 475km off nadir (≈4.5 ground tracks). The telescope is a modified Schmidt with a focal length of 108.2cm and an aperture of 33cm.

Spectral Channels	Bandwidth	Signal-to-Noise Ratio	
50μm - .59μm	90nm	1	290
61μm - .68μm	70nm	2	380
79μm - .89μm	100nm	3	390
51μm - .73μm (panchromatic)	220nm	Pan	400

Detectors

On SPOT-1 each spectral band uses Fairchild's 122 DC detectors, and SPOT-2,-3 use four Thompson-CF TH 7801A 1728-element CCD arrays. All are push-broom linear CCD arrays. Each linear CCD consists of 6000 detectors.

Swath (Scan Angle)

FOV = ±2.08°; 117km with 3km overlap with up to 27° off-nadir viewing to create 970km swath via a tilting mirror that permits viewing on 7 successive passes at 0° latitude and 11 passes at 45° latitude.

Spatial Resolution (GSD)

- Multispectral mode = 20m (image area = 60km x 60km), 3000 pixels/line
- Panchromatic mode = 10m (image area = 60km x 60km), 6000 pixels/line

Radiometric Resolution

- 8-bit quantization (multispectral mode)
- 8-bit quantization (panchromatic mode) (See Plate 10.)

Calibrations

Calibration is performed weekly by taking each unit temporarily out of service. The procedure is performed by closing the mirror over the aperture during the nighttime ascending node. A tungsten-filament halogen lamp illuminates the detectors. The irradiance varies over the focal plane but the pattern is known and correctable. Other calibrations include solar, interband, absolute, and multidate calibrations.

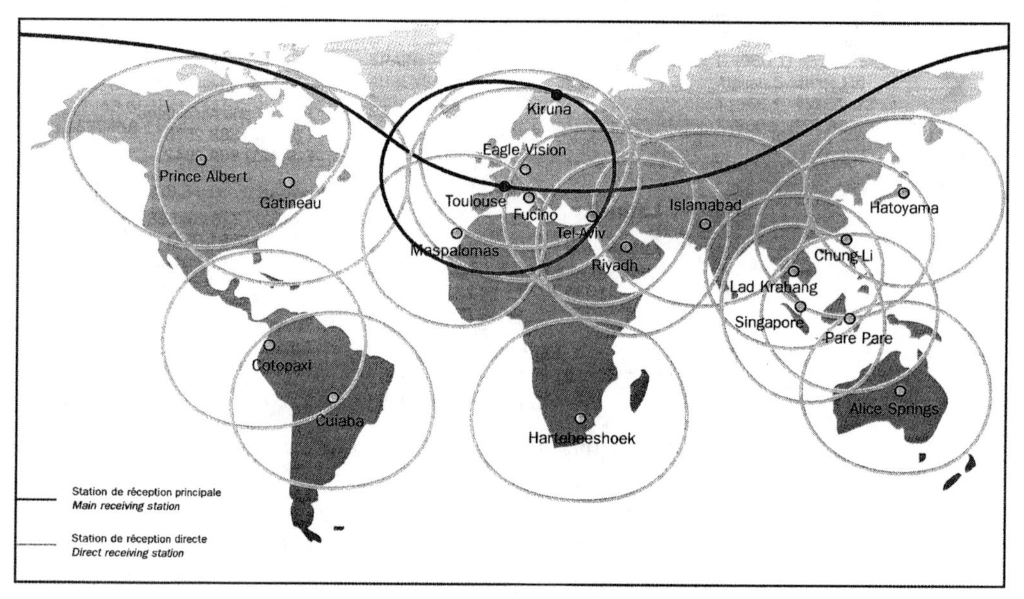

SPOT receiving stations.

Data Availability

Commercial data are distributed by SPOT Image SA, 5 Rue des Satellites, BP 4359, F-31030, Toulouse Cedex France, 33 6219 4040, fax 33 6219 4011; SPOT Image Corporation, 1897 Preston White Drive, Reston, VA 22091-4368 USA, (703) 620-2200, fax (703) 648-1813; SPOT Imaging Services, 165 Pacific Highway-St. Leonards, NSW 2065 Australia, (612) 906-1733, fax (612)906-5109; and SPOT Asia, 73 Amoy Street, Singapore 0106, (65)227-5582, (65) 227-6231. See also SPOT home pages: *http://www.spotimage.fr* (France) and *http://www.spot.com* (U.S.).

Landsat

Program Objectives

The Landsat program is a by-product of NASA's Earth Resources Survey Program with the collaboration and shared resources of other federal agencies. First known as the Earth Resources Technology Satellite (ERTS-1), Landsat-1 was the "proof-of-concept" that Earth-orbiting satellites could effectively monitor natural and cultural resources. Subsequent Landsats have continued to define spectral and spatial requirements for next generation sensors and stimulated research to determine optimum data processing and interpretation.

Soon after launch of Landsat-1 in 1972, quasi-operational uses for synoptic Earth data began to emerge. To accommodate this "instant" success, Congress moved the ground segment responsibility first to the USGS EROS Data Center in Sioux Falls, and shortly thereafter, to the Department of Commerce. In late 1985 Congress privatized both the space and ground segments by transferring responsibility to EOSAT Corporation, but in 1992 returned control to NASA and DOD. Landsat-6 failed to achieve orbit in 1993. Landsat-7 is a joint effort by NASA (provider), NOAA (operator), and USGS (archivist). Plate 12 shows the orbital swathing pattern for Landsat-4 and -5. A sample image from TM appears in Plate 13.

	Landsat-1, -2, -3	*Landsat-4, -5*	*Landsat-7*
Launch			
Dates	1972, 1975, 1978	1982, 1984	1998
Vehicle	Delta	Delta	Delta II
Site	Vandenberg	Vandenberg	Vandenberg
Orbit			
Altitude	L1=907km, L2=908km, L3=915km	705km	705km
Type	LEO, sun synchronous	LEO, sun synchronous	LEO, sun synchronous
Inclination	99°	98.2°	98°
Revisit	18 days (L2 & L3 were timed to provide < 12 day revisit)	16 days (233 orbits)	16 days (233 orbits)
Coverage	Global (81° N & S latitude)	Global	Global
Equatorial crossing	L1 = 850 am	0945 am	1000 am
	L2 = 908 am		
	L3 = 931 am		
Period	103 minutes (14 orbits/day)	99 minutes (14 orbits/day)	ND
Stabilization	±1.0° orbital plane (yaw)	Gyro Earth oriented	3 axis to within 0.05
Dimensions			
Mass	953kg	2200kg	2200kg
Size	3m x 1.5m with 4m solar paddles	2m x 4m	2.8m x 4.3m
Design lifetime	1 year minimum	3 years	5 years
Instruments	MSS	MSS, TM	ETM+

Multispectral Scanner (MSS)

Dates of Operation

- 1972-1995

Primary Mission

- Synoptic, multispectral images of Earth resources

Sensor Description

- MSS is a whisk-broom optical mechanical scanner that collects reflected radiation from the Earth via an oscillating mirror backed by a Ritchey Chretien telescope and silicon diode detectors. Scene energy is collected on the forward scan and directed to a 24-element fiber optic array. Six scan lines are collected per mirror sweep. It takes approximately 25 seconds to obtain a 185km image along-track.

Spectral Channels	Bandwidth	Signal-to-Noise Ratio and Detectivity
0.5μm - 0.6μm	100nm	125 NEΔρ = 0.5% (channels 1-4)
0.6μm - 0.7μm	100nm	108
0.7μm - 0.8μm	100nm	101
0.8μm - 1.1μm	300nm	108
10.4μm - 12.6μm	2.2μm (L-3)	NEΔT = 0.1°C (channel 5)

Detector Technology

- VNIR = 18 photomultiplier tubes and 6 silicon photodiode detectors per channel

Swath (Scan Angle)

- FOV = 11.56° (L-1, -2, -3), 15° (L-4, -5); 185km

Spatial Resolution (IFOV, GSD)

- IFOV = 0.086mrad, 80m (channels 1 to 4); 248m (channel 5)

Radiometric Resolution

- Landsat-1,-2,-3 = 6-bit quantization
- Landsat-4,-5 = 8-bit quantization

Calibrations

Every other mirror flyback period is used for calibration using an onboard lamp to illuminate the detectors through a rotating density wedge optical filter.

Data Availability

EOSAT Corporation, 4300 Forbes Boulevard., Lanham, MD 20706-9954, 1-800-344-9433, (301) 552-0642, fax (301) 552-3762; Customer Services, USGS EROS Data Center, Sioux Falls, SD 57198, (605) 594-6511, fax (605) 594-6589, URL: *http://sun1.cr.usgs.gov/landdaac/landdaac.html.*

NOTE: *EOSAT Corporation reports that it has relinquished rights to MSS data and no longer receives or processes these data.*

Thematic Mapper (TM)

Dates of Operation

- 1982-present

Primary Mission

- Monitor Earth resources

Sensor Description

- TM is a 7-channel scanning radiometer. Scene energy is collected in both directions cross-track while spacecraft forward motion provides the scan along-track. Light reflected from the scan mirror is directed to a 40.6cm clear aperture Ritchey-Chretien telescope with fl = 243cm. The variation in radiant flux passing through the field stop onto the detectors creates an electrical output that represents the radiant history of the line.

Detectors

- Channel 1-5: 16 detectors/channel = Si
- Channel 6: 4 detectors = HgCdTe
- Channel 7: 16 detectors = InSb

Swath (Scan Angle)

- FOV = 15.39°, 185km at 705km altitude

Spatial Resolution

- Channels 1-5, 7 = 30m
- Channel 6 = 120m
- Interband registration = 0.1 IFOV

Radiometric Resolution (IFOV, GSD)

- 8-bit quantization
- 4 calibrations

Three radiance controlled tungsten filament lamps are used for the reflective channels, and a blackbody, controllable to three known temperatures, is used for the thermal band. A black surface of known temperature provides the source for a dc-restoration of all bands. The calibrator is an oscillating shutter mechanism synchronized with the scan mirror to bring the calibration sources sequentially into view of the detectors during each scan mirror turnaround.

Data Availability

EOSAT, 4300 Forbes Boulevard, Lanham, MD 20706-9954, (301) 552-0537; USGS EROS Data Center, Sioux Falls, SD 57198, (605) 594-6099, DAAC URL: *http://sun1.cr.usgs.gov/landdaac/landdaac.html.*

Television and Infrared Observation Satellite (TIROS) (aka NOAA Polar Orbiting Environmental Satellite - POES series)

TIROS History

TIROS has a complicated history of name changes and akas, sensor packages, and other parameters. Between 1960 and 1965 ten TIROS satellites were launched. They were 18-sided cylinders covered on the sides and top by solar cells, with openings for two TV cameras on opposite sides. Each camera could acquire 16 images per orbit at 128-second intervals. Between 1966 and 1969 nine TIROS operational satellites (named TOS) were operated by the Environmental Science Services Administration (ESSA) and these were replaced in 1970 when TIROS-M was designated the Improved TOS (ITOS).

The TIROS program began as a joint NASA/DOD effort to develop meteorological satellites and demonstrate their value as orbital platforms for applications. By the time TIROS 10 was launched, operational systems were introduced (TOS) and NASA had inaugurated a more advanced platform called Nimbus. By the time TIROS-10 was terminated in 1965, more than 10,000 cloud cover pictures had been transmitted.

When TIROS operational satellites (aka ESSA series) were introduced in 1966, there were 400 receiving stations around the world providing weather services to 45 countries and 26 universities. The ITOS system added infrared sensors to the payload and was designated as a new NOAA series (the administration superceding ESSA). NOAA-1 (aka TIROS-M and ITOS-1) was thus launched in 1970 from Vandenberg AFB in a nearly circular orbit at 101° inclination. NOAA-3, -4, and -5 were placed in similar orbits in 1974 and 1976.

By 1978 a fourth generation of improved capabilities was inaugurated with TIROS-N and adopted as NOAA-6 and -7 in 1979 and 1981, respectively. This generation was replaced by Advanced TIROS-N (ATN) as NOAA-8 in 1983.

Case studies that utilize AVHRR data include Rowland et al. in Chapter 6, and Chopping in Chapter 8.

System Characteristics

Launch

- Date:

TIROS-1 thru -9 (NASA/DOD)	1960-1965
TIROS Operational Satellites (TOS) (aka ESSA-1 through -9)	1966-1969
Improved TOS (ITOS) (aka TIROS-M)	1970
NOAA-1 through -5 (aka TIROS-M)	1970-1976
NOAA-6, -7 (aka TIROS-N)	1978-1981
NOAA-8 through -14 (aka Advanced TIROS-N or -ATN)	1983-1994
NOAA-K through -N'	1996-2004
NOAA-O through -Q (first of a new generation of polar weather satellites)	2005-2011

- Vehicle: Atlas; Delta (TOS, ITOS)
- Site: Vandenberg

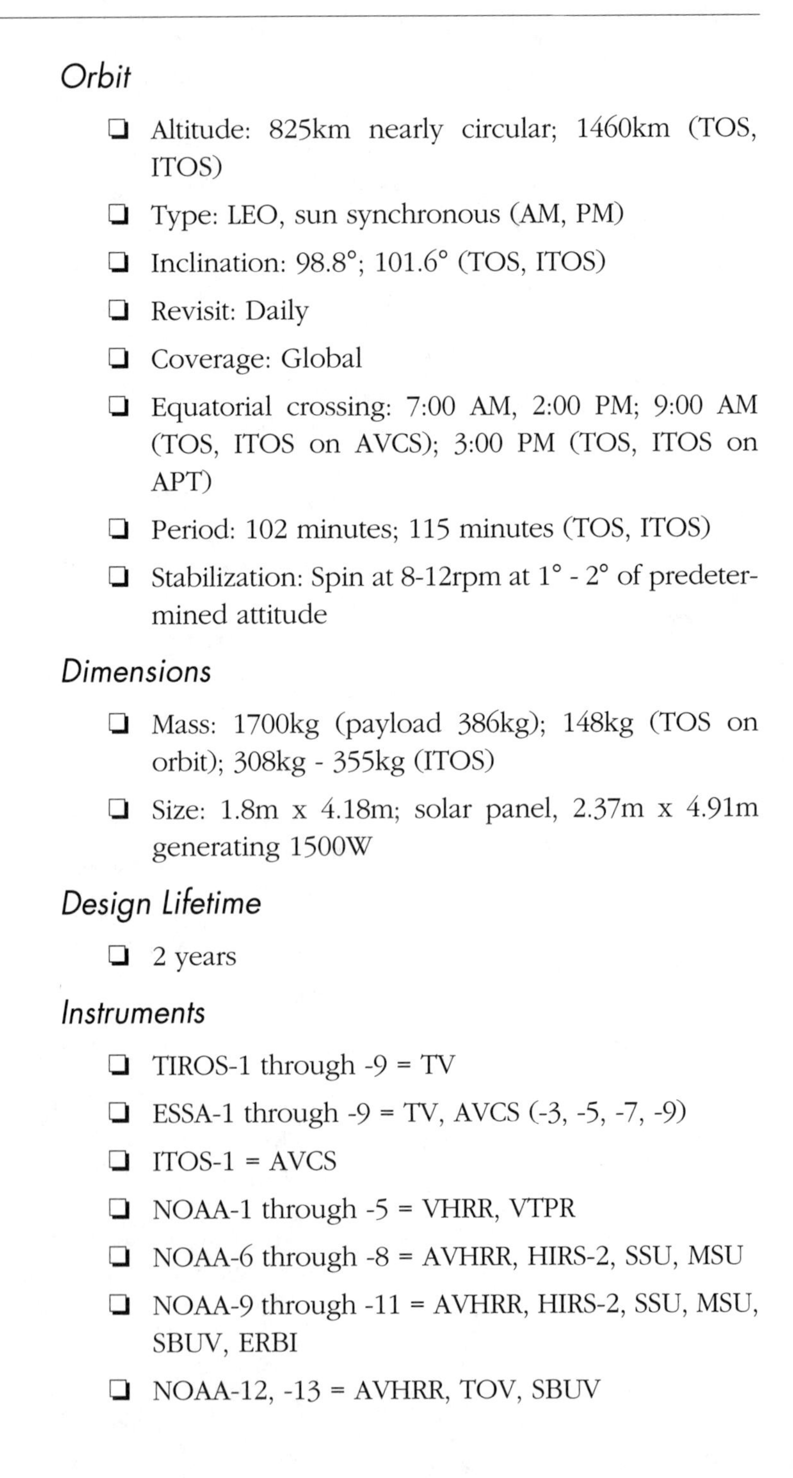

Orbit

- Altitude: 825km nearly circular; 1460km (TOS, ITOS)
- Type: LEO, sun synchronous (AM, PM)
- Inclination: 98.8°; 101.6° (TOS, ITOS)
- Revisit: Daily
- Coverage: Global
- Equatorial crossing: 7:00 AM, 2:00 PM; 9:00 AM (TOS, ITOS on AVCS); 3:00 PM (TOS, ITOS on APT)
- Period: 102 minutes; 115 minutes (TOS, ITOS)
- Stabilization: Spin at 8-12rpm at 1° - 2° of predetermined attitude

Dimensions

- Mass: 1700kg (payload 386kg); 148kg (TOS on orbit); 308kg - 355kg (ITOS)
- Size: 1.8m x 4.18m; solar panel, 2.37m x 4.91m generating 1500W

Design Lifetime

- 2 years

Instruments

- TIROS-1 through -9 = TV
- ESSA-1 through -9 = TV, AVCS (-3, -5, -7, -9)
- ITOS-1 = AVCS
- NOAA-1 through -5 = VHRR, VTPR
- NOAA-6 through -8 = AVHRR, HIRS-2, SSU, MSU
- NOAA-9 through -11 = AVHRR, HIRS-2, SSU, MSU, SBUV, ERBI
- NOAA-12, -13 = AVHRR, TOV, SBUV

- NOAA-K through -N' will introduce AMSU-A and AMSU-B

Transmission Modes

A continuous S-band broadcast is provided so that anybody can receive the data.

- HRPT = High Resolution Direct Readout (AVHRR)
- LAC and GAC = Recorded HRPT (1km) and reduced resolution (4km) AVHRR

Receiving Stations

Command and data acquisition stations: Wallops Island, VA; Fairbanks, AK (Gilmore Ck before 1984); Lannion, France (operated by CNES)

Ground stations (HRPT direct readout data within line-of-sight): Wallops Island, VA; Fairbanks, AK; Redwood City, CA

NOTE: *Wallops Island and Fairbanks receive 8-bit rather than 10-bit data and have 45-day, 90-day, and 120-day data pools, respectively.*

Advanced Very High Resolution Radiometer (AVHRR)

Dates of Operation

- October 1978-present

Primary Mission

- Measure cloud cover, day and nighttime sea surface temperature, cloud formations, vegetation index (see Plate 14).

Sensor Description

- AVHRR is neither "advanced," nor does it provide very high spatial resolution. It does, however, provide very high (10 bits/pixel) radiometric resolution. The scanning sequence starts when the mirror is pointing at 80° west of nadir. Calibration data are recorded for the first 29° of scan angle. The western horizon is at 69° from the nadir and data collection starts at 55°, collecting 2048 samples to 55° east of nadir (total data collection over 40°). The mirror rotates 360° at 6 rev/s. The cross-track sampling interval is 25μs.

Detector

- ❑ Telescope is an afocal confocal parabaloid with a 46.1cm aperture.
- ❑ Silicon (SI), Indium-antimonide (InSb), Mercury-cadmium-telluride (HdCdTe)

Detectivity

- ❑ Sea surface temperature with an accuracy of about 0.6°C if no clouds contaminate the FOV. Horizontal temperature gradients of about 0.2°C can be resolved. NEΔT = 0.12K at 300K.

Swath (Scan Angle)

- ❑ FOV = 55.4°, 2580km - 4000km

Spectral Channels

TIROS-N	NOAA-6,-8,-F	NOAA-7,-D,G to J	Bandwidth
.55μm-.90μm	.58μm-.68μm	.58μm-.68μm	350nm, 100nm
.73μm-1.1μm	.73μm-1.1μm	.73μm-1.1μm	375nm, 380nm
3.55μm-3.93μm	3.55μm-3.93μm	3.55μm-3.93μm	380nm, 380nm
10.5μm-11.5μm	10.5μm-11.5μm	10.5μm-11.3μm	1.0μm, 800nm
10.5μm-11.5μm	10.5μm-11.5μm	11.5μm-12.5μm	1.0μm, 1.0μm

Spatial Resolution (IFOV, GSD)	
IFOV (mrad)	**GSD**
1.39	1.1km local area coverage
1.41	4km global area coverage
1.51	
1.41	
1.3	

Radiometric Resolution

- ❏ 10-bit quantization

Calibrations

A two-point calibration of IR channels occurs every scan. As the line-of-sight passes zenith, the detectors are exposed to blackbodies of known temperature. This calibration source varies between 283K and 293K. The visible channels are uncalibrated and suitable for relative measurements only.

Data Availability

Atmospheric data from National Environmental Satellite and Data Information Center, FOB 4, Room 2069, Suitland, MD 20233, (301) 763-7190, fax (301) 763-4011, URL: *http://www.noaa.gov/nesdis/nesdis_intro .html.*

Land data from Request Coordination Office, National Space Science Data Center, Code 633, Goddard Space Flight Center, Greenbelt Rd., Greenbelt, MD 20771, (301) 286-6695, fax (301) 286-1771, email: *request@nssdca.gsfc.nasa.gov*, URL: *http://www.gsfc.nasa.gov/.*

Sea surface temperature data from Jet Propulsion Laboratory, 4800 Oak Grove Drive, Pasadena, CA 91109, (818) 354-4321, fax (818) 354-3437, URL: *http://seazar.jpl.*

nasa.gov or *http://sst-www.jpl.nasa.gov*. Global 1km coverage via ftp from Customer Services, EROS Data Center, Sioux Falls, SD 57198, (605) 594-6511, fax (605) 594-6589, URL: *http://sun1.cr.usgs.gov/landdaac/landdaac.html*.

European Resource Satellite (ERS-1)

Program Objectives

ERS-1 is intended for global measurements of sea wind and waves, ocean and ice monitoring, coastal studies, and limited land sensing. ERS-2 will augment land sensing capabilities by adding three visible channels.

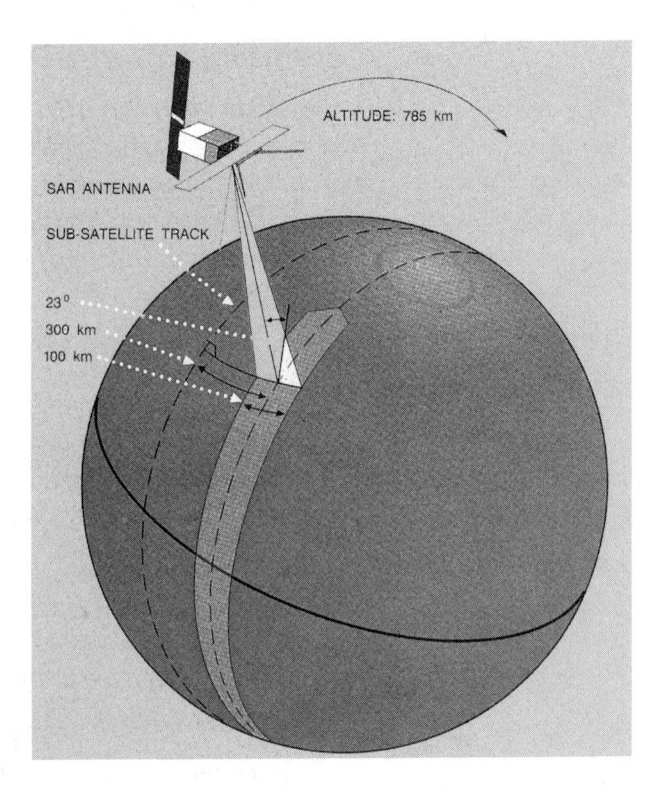

Orbital parameters for ERS-1.

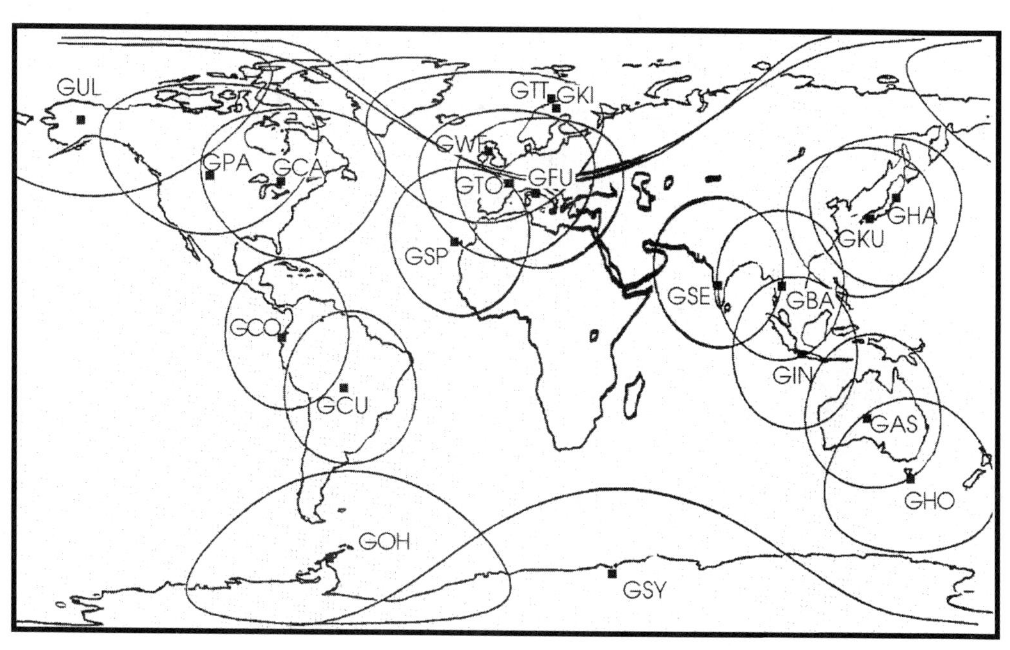

Receiving stations for ERS-1.

System Characteristics

Launch

- ❑ Date: 1991 (ERS-2 late 1994)
- ❑ Vehicle: Ariane V44
- ❑ Site: Kourou, French Guiana

Orbit

- ❑ Altitude: 777km
- ❑ Type: LEO, sun synchronous
- ❑ Inclination: 98.5° (13 orbits/day)

- Revisit: 35 days (501 revolutions) 16 days at ±60° latitude
- Coverage: Global
- Equatorial crossing: 10:15
- Period: 100 minutes
- Stabilization: 3 axis

Dimensions

- Mass: 2384kg
- Size: 11.8m x 11.7m x 2.4m; 2 solar paddles at 2.4m x 5.8m

Design Lifetime

- 3 years

Instruments

- AMI-SAR, RA

Data Availability

Worldwide cooperation in data reception and exchange exists. Data processing and archiving facilities are located in France, Italy, the UK, and Germany. In 1992 ESA authorized three data distributors: RADARSAT (Ottawa, Canada, for Canada and USA); EURIMAGE (Rome, Italy, for Europe, North Africa, and Middle East); and SPOT Image (Toulouse, France, for rest of world).

Active Microwave Imager-Synthetic Aperture Radar (AMI-SAR)

Dates of Operation

- 1991-present

Primary Mission

- ❑ Measure wind speed and direction; monitor ocean waves, land resources, and sea ice

Sensor Description

- ❑ AMI-SAR operates as a synthetic aperture radar and as a wind/wave scatterometer.

Antenna Description

- ❑ Wind scatterometer = 3 antennas (mid 0.35m x 2.3m; fore and aft, 0.25m x 3.6m)

Swath (Scan Angle)

- ❑ 80.4km - 99km (20.1°-25.9° incidence angle range) imaging and wave modes; 500km (scatterometer mode) FOR = 250km right side of nadir

Frequencies	Bandwidth
5.3GHz 5.66cm, C-band, VV polarization	15.55 MHZ

Spatial Resolution

- ❑ 8m - 200m (processing dependent); nominally 30m for image and wave modes and 50km for wind scatterometer; geolocation ≤ 1km

Radiometric Resolution

- ❑ 2.5dB at -18dB (scatterometer mode); 5-bit quantization (image mode)

Calibrations/Corrections

Wavelength accuracy of ±25% over 100m - 1000m and direction accuracy of ±20° for wave scatterometer. Wind scatterometer data can be analyzed with ±20° accuracy for wind vectors from 4m - 24m/sec and 0° - 360° at incidence angles between 27° and 58°.

Data Availability

Data are downlinked to Kiruna, Sweden; Gatineau, Canada; and Maspalomas, Canary Islands (Spain). Image data are normally collected only 7.5 minutes on each 90-minute orbit; 72 minutes are used for the "wind/wave" mode. The ERS-1 Help Desk, Frascati, Italy, serves as the central user service facility. North American AMI-SAR image products are available through RADARSAT, Richmond, B.C., and through the University of Alaska SAR Facility at Fairbanks, AK.

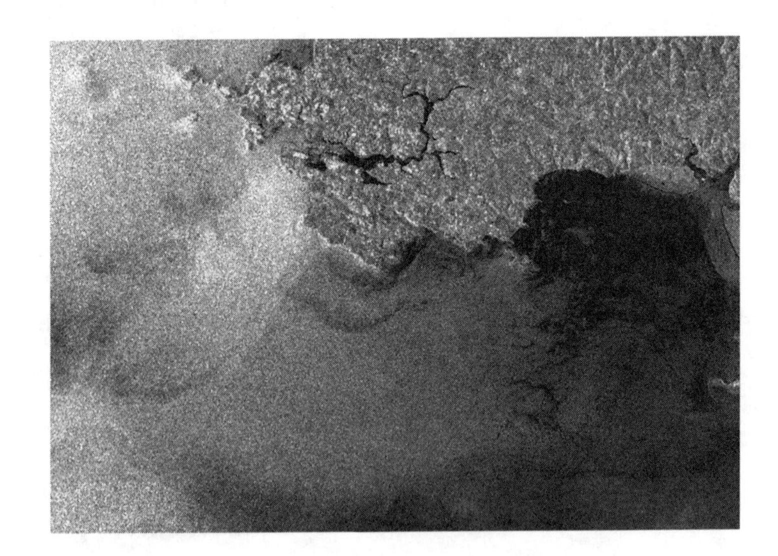

ERS-1 image of ocean and land areas, Wales, UK.

Indian Remote Sensing Satellite (IRS)

Program Objectives

IRS is India's dedicated Earth resources satellite system operated by ISRO and the National Remote Sensing Agency (NRSA). IRS-1A and -1B have been launched by the Soviet Vostok booster, and 1C is scheduled for launch in 1995 by the Russian Molniyam rocket. The primary objective of the IRS missions is to provide India's National Natural Resources Management System (NNRMS) with data derived from near state-of-the-art satellite sensors. (See Plate 15.)

IRS-1E (aka P1) was the first test of India's Polar Space Launch Vehicle (PSLV) in 1993. A sun synchronous orbit at 904km was intended with a 10:30 AM equatorial crossing and 22-day repeat cycle. IRS-P2 launched in 1994 and is in an 817km altitude at 98.7° inclination. It carries LISS-2 and the Multispectral Opto-electronic Scanner (MOS) developed by DLR, Germany.

System Characteristics

Launch

- Date: 1988, 1991, 1995 (C/D)
- Vehicle: SL-3 Vostok (1A, B, C/D)
- Site: Tyuratam (1A, B, C/D)

Orbit

- Altitude: 904km circular (1A, 1B); 817km (1C/D); 817km (IRS-P2)
- Type: LEO, sun synchronous
- Inclination: 99.03° (1A); 99.25° (1B); 98.69° (1C/D); 98.69° (IRS-P2)
- Revisit: 22 days (307 revolutions with 2872km ground track separation at equator); 24 days (1C/D)
- Coverage: Global
- Equatorial crossing: Descending node at 10:25 AM local time
- Period: 103 minutes
- Stabilization: 3-axis control with 0.3° pitch/roll and 0.5° yaw accuracy; 0-momentum reaction wheel using Earth/sun/star sensors and gyros (1A, B); 15° pitch/roll and 0.2° yaw accuracy for IRS 1 C/D

Dimensions

- Mass: 975kg (1A, B); 1350kg (1 C/D)
- Size: 1.6m x 1.56m x 1.1m with 2 sun-tracking solar arrays spanning 8.58m^2 and providing 709W (1A, B); 1.6m x 1.56m x 1.1m with 2 sun-tracking solar panels of 1.1m x 1.46m providing 830W

Design Lifetime

- 3 years

Instruments

- LISS-1, LISS-2, LISS-3, WIFS, PAN (MEOSS on 1E suffered launch failure)

Prime Contractors:

- Indian Space Research Organization (ISRO)

Sensor Package

- LISS-3: Similar to LISS-1,-2, but replaces one visible band with SWIR. Three 6000-element linear visible CCD imagers with spectral filters for bands 2, 3, and 4 (see LISS-1). These produce a 142km swath having 23.5m ground resolution and 7-bit radiometric resolution. Data rate = 35.70Mbps. The SWIR is a 2100 linear CCD of InGaAs cooled to -10°C. It operates in the 1.55μm - 1.70μm range producing a 148km swath with 70m ground resolution. Data rate = 2.02Mbps.
- PAN: Single channel (0.50μm - 0.75μm) imager with 3 linear CCDs with 23.9km swaths (70km total) producing better than 10m ground resolution with 6-bit radiometric resolution. Three off-axis mirrors permit swath steering of ±398km by a payload steering mechanism with ±26° in 0.2° steps. Data rate = 84.903Mbps. Five-day revisit cycle.

- WiFS: Wide field sensor similar to LISS-1. It uses 2048-element linear CCDs in 2 channels (0.62μm - 0.68μm and 0.77μm - 0.86μm). Each channel has a focal length of 56.420mm and FOV of 13.5° (±27° total), and provides a 774km swath with 188m ground resolution and 7-bit radiometric resolution. Five-day revisit cycle.

Linear Imaging Self-Scanning System (LISS)

Dates of Operation

- 1988-present

Primary Mission

- Image land, water, and coastal resources

Sensor Description

- The LISS is designed to take advantage of existing engineering/cost criteria. It uses a push-broom scanning strategy adapted to commercially available CCD line lengths enabling 7-bit quantization; multiple reflecting telescopes; multiple instruments (LISS-I, LISS-IIA, LISS-IIB) to achieve high spatial resolution; and an immobile calibration system. The telescopes have focal lengths of 16.2cm (LISS-I) and 32.4cm (LISS-II) with apertures of 3.6cm and 7.2cm, respectively. LISS-IIA and -IIB each image a different side of the swath, while LISS-I images the entire swath. Thus, there are 4 telescopes for LISS-I and 8 for LISS-II.

Detector Technology

- Linear CCD with 2048 elements. LISS-I = 4 CCDs, LISS-II = 8 CCDs

Detectivity

- NE = 0.5%
- SNR = 155 (LISS-1/IRS-1B, all channels)
- SNR = 142 (1), 152 (2), 155 (3), 147 (4) (LISS-2/IRS-1B)
- SNR > 128 (LISS-3/IRS-1C, all channels)
- SNR > 127 (LISS-2/IRS-P2, all channels)

Swath (Scan Angle)

- LISS-I FOV = ±9.4° either side of nadir, 148km
- LISS-II FOV = ±4.7°, 74km
- IRS-P2 = 4.7°, 67km at nadir

Spectral Channels		Bandwidth	
Liss-I, -II	Liss-III	-I, -II	-III
0.45μm - 0.52μm	0.52μm - 0.59μm	70nm	70nm
0.52μm - 0.59μm	0.62μm - 0.68μm	70nm	60nm
0.62μm - 0.68μm	0.77μm - 0.86μm	60nm	90nm
0.77μm - 0.86μm	1.55μm - 1.70μm	90nm	115nm

Spatial Resolution

- LISS-I = 72.5m
- LISS-II = 36.25m
- LISS-III= 23m (Ch 2,3,4), 50m (Ch 5)

Radiometric Resolution

- 7-bit quantization

Calibrations

- Light emitting diodes (LEDs) are used as the calibration source instead of tungsten filament lamps. There are two LEDs per CCD array mounted at 30° on either side of the telescope axis. Calibration is carried out at night about once a month.

Data Availability

Products are being sold nationally and internationally by the Indian Space Research Organization's National Remote Sensing Agency (NRSA) Data Center, Balanager, Hyderabad 500037, (842) 279572, fax (842) 278648. Data are also sold by EOSAT, 4300 Forbes Blvd., Lanham, MD 20706-9954, 1-800-344-9933, (301) 552-0642, fax (301) 552-3762.

RADARSAT

Program Objectives

RADARSAT is the first operational SAR for gathering global data on ice conditions, crops, forests, oceans, and geology. All of Canada can be covered every 72 hours, and the Arctic every day. The system is being designed with no backlog. The instrument is designed to (1) provide detailed information on sea ice and terrestrial ice sheets for climate research; (2) provide radar imagery for surface-based applications; and (3) provide real-time data and products for arctic ocean navigation, including iceberg surveillance.

RADARSAT image.

System Characteristics

Launch

- Date: 1995
- Vehicle: Delta-II
- Site: Vandenberg

Orbit

- ❑ Altitude: 793km - 821km (TBD)
- ❑ Type: LEO, near circular, sun synchronous (14 orbits/day)
- ❑ Inclination: 98.6°
- ❑ Revisit: 24 days (343 orbits); 5 days at equator at maximum tilt revisit cycle (in days)
- ❑ Coverage: Arctic regions daily; typical 500km coverage
- ❑ Equatorial crossing: 1800 hours ascending node
- ❑ Period: 101 minutes
- ❑ Stabilization: 3 axis (0.1°) by reaction wheels

Dimensions

- ❑ Weight: 2750kg
- ❑ Size: 4.2m x 2.8m; solar paddles, 1.5m x 15m

Design Lifetime

- ❑ 5 years

Instruments

- ❑ C-band SAR

Prime Contractors

Spar Aerospace with Ball Aerospace, Canadian Astronautics, Telesat, SED, MDA, Odetics, Dornier, Astro Aerospace

International Participation

Canada (Canadian Space Agency/CCRS); USA (NASA); Spar Aerospace; ASC Inc.; COM DEV Ltd.; First Mark Technologies; CAL Corp.; Calian Communications; MacDonald Dettwiler & Assoc.; Ball Aerospace; MPB Technologies; Prior Data Sciences; SED Ltd.; Dornier; Fleet Industries; SAFT; FRECompOsites; British Aerospace

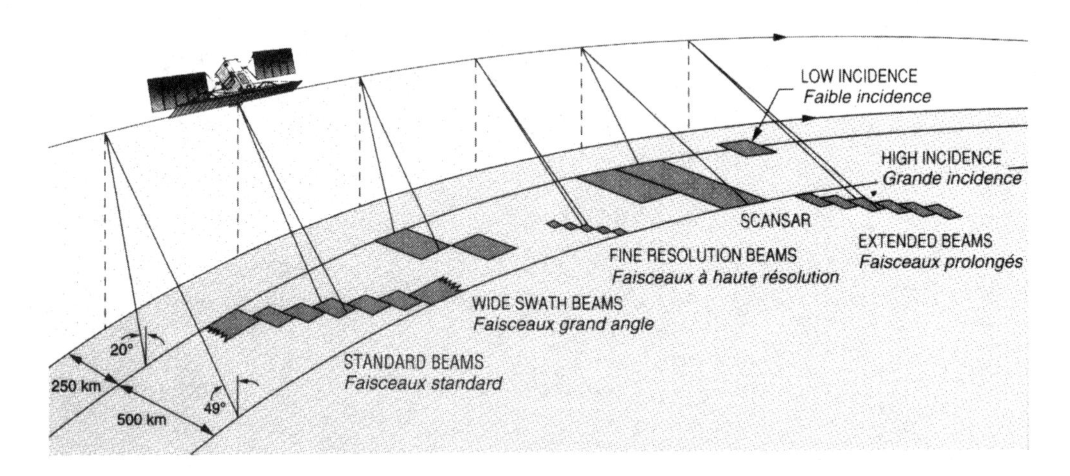

RADARSAT system characteristics.

Data Availability

NASA will receive RADARSAT data through the Alaska SAR Facility (ASF) in Fairbanks. RADARSAT International (RSI) is a private corporation established to process and market Canadian RADARSAT data.

RADARSAT Sensor

Dates of Operation

- 1995-present

Primary Mission

- Map and monitor renewable resources

Sensor Description

- RADARSAT is an imaging radar with a variety of imaging modes. The design includes a choice of three transmit pulses and a selection of beams for a wide range of swath widths, incidence angles, and image resolutions. Special design features include calibration, rapid data processing, the phased array antenna, and the satellite implementation of scan SAR. This latter feature permits an extension of swath width on command by using a set of contiguous beams.

Antenna Description

- 15m x 1.5m phased array

Swath (Scan Angle)

- 45km - 500km selectable (incidence angle 20° - 60° selectable)

Spatial Resolution

- See RADARSAT resolutions, bandwidths, and incidence angles.

Data Availability

Mission control, data processing, and distribution will be from St. Hubert near Ottawa. RADARSAT International and other private firms have distribution rights. Contact: Canadian Space Agency, RADARSAT Program, 6767 route de l'AÇroport, Saint-Hubert, Quebec J3Y 8Y9, (514) 926-4406.

Standard Satellite Digital Data Products

Listed in the following sections are several categories of SPOT Image, EOSAT, ERS-1, IRS-IC, and RADARSAT digital data products. Through their networks, these companies provide global data not only from their satellite systems, but also from others with whom they have data distribution agreements. Only digital products are listed here. Paper prints (color and black-and-white) and maps are also available for some product lines, but these are not directly employable in vector GIS applications.

SPOT IMAGE

http://www.spotimage.fr (Toulouse, France headquarters)

http://www.spot.com (Reston, VA headquarters)

SPOT Scene

- ❑ Full scene: At Level 1A or 1B in panchromatic (Pan) or multispectral (XS). 60km x 60km on a selection of media (CD-ROM, 8mm tape, 1/4" inch cartridge, 4mm tape, DAT tape, or 9-track (CCT)). Level 1 processing includes radiometric and gross geometric corrections providing data that are compatible with image processing software packages.

SPOTView ®

Orthorectified imagery products specifically for easy ingestion into GIS, desktop mapping, image processing, and other systems. They are produced to any map projection to facilitate registration with other data sets and are available in standard map frames. Products are available either for Pan or XS on CD-ROM or any other media. BASIC and PLUS (highly enhanced) versions are available.

SPOT MetroView

These highly enhanced "off the shelf" products are designed for urban GIS and desktop mapping applications. They are custom designed for users as 15 x 15 minute cells for many United States metropolitan areas, and are available in many major projections.

SPOTView BD Carto

These digital image maps of France coincide with the French l'Institut Géographique National (IGN) 1:50,000 map series. Imagery is orthorectified using the IGN elevation database. Products are available in standard IGN map frames (.2gr x .4gr) for regional and environmental planning, mapping, and multitemporal studies.

SPOT LandClass™

SPOT images are used to create digital land use/land cover maps containing 4 to 18 classes. Available for any place in the United States, these maps are geographically registered and formatted for easy ingestion into telecommunications modeling, desktop mapping, and GIS applications.

SPOT Stereo Imagery and Products

- StereoSPOT: A pair of Pan or XS images acquired at different viewing angles. They provide stereo coverage for developing digital elevation models (DEMs).
- SPOT DEM (Digital Elevation Model): A digital elevation data file of any location in the world derived from SPOT stereoscopic imagery, with 20m grid cells (X/Y accuracy = 20m, Z accuracy = 11 to 17m) or with 40m grid cells (X/Y accuracy = 20m, Z accuracy = 12 to 18m).
- 3D-SPOTView: A combination of the SPOT DEM and an orthocorrected SPOTView created to provide realistic land cover representations which can be viewed in three dimensions. Created for geologic mapping, terrain analysis, telecommunications signal propagation modeling, and hazards mapping.

EOSAT Corporation

http://www.eosat.com/

EOSAT Corporation provides data from Landsat MSS and TM. Recently, it has added data sets from the Indian Resource Satellite (IRS) for areas within the North American and Indian region receiving stations. It also distributes Russian satellite photography from the KVR 1000 and TK-350 systems, and the Japanese Earth Resource Satellite (JERS-1). Only the TM and MSS digital data products are listed here.

Path-oriented Products

- Full Scene: TM or MSS data sets (185km x 170km) system corrected on 6250 or 1600 bpi CCTs. TM data are also available on 8mm tape. System corrections include geometric correction for spacecraft orientation and predicted position. There are also several ellipsoid and map projection options available for TM data sets. These include Ellipsoid—International 1909 model; Map Projections—Space Oblique Mercator (default); UTM for areas between 60° north and south; Polar Stereographic for areas between 60° and 82° north and south latitude.
- Subscene: 100km x 100km TM data only and are available as system corrected 6250 or 1600 bpi CCT or 8mm tape. The same ellipsoid and map projection options are available as for full scene.

Map-oriented Products

The following products are available only for TM data sets.

- Full Scene: 185km x 170km scene with system, precision, and terrain corrections, available on 6250 or 1600 bpi CCT or 8mm tape. Precision corrections include data correction using ground control points to correlate predicted spacecraft position with actual geodetic position. Terrain corrections include use of DEMs to correct parallax distortions. Several ellipsoid and map projection options are offered:

Ellipsoid—Clarke 1866 (default); Clarke 1880; International 1909; International 1967; WGS 66; WGS 72; GRS 1980; Airy; Modified Airy; Everest; Modified Everest; Mercury 1960; Southeast Asia; Australian National; Krassovsky; Hough; and Sphere.

Map Projections—UTM (default); Azimuthal Equidistant; State Plane; Gnomonic; Albers Conical Equal Area; Orthographic; Lambert Conformal Conical; General Vertical Near-Side Perspective; Mercator; Sinusoidal; Polar Stereographic; Equirectangular; Polyconic; Miller Conical; Equidistant Conical (Type A & B); Van Der Grinten I; Transverse Mercator; Oblique Mercator (Type A & B); Stereographic; Space Oblique Mercator; and, Lambert Azimuthal Equal Area.

- Subscene: 100km x 100km with all corrections and options noted above.
- Map Sheet: 1/2° latitude x 1° longitude system, precision, and terrain corrected data set available on 6250 or 1600 bpi CCT or 8mm tape. All above options are available.
- Three-band: Corrected as above on 6250 or 1600 bpi or 8mm cassette.

Indian Remote Sensing Satellite

IRS-1C Panchromatic and LISS-3 Multispectral digital products are available from EOSAT Corporation as Path- and

Map-oriented system corrected images on CD-ROM, 8mm tape, or CCT. Pan scenes are 70km x 70km; Pan Junior scenes are 23km x 23km; and Liss SuperScenes are 114km x 114km.

European Resource Satellite (ERS-1)

SAR Annotated Raw Data

Provides ≈ 300 Mbytes of raw data for a 100km x 110km area (5616 samples in range by 27000 samples in azimuth) with nominal accuracy of 0.9km in range and 1km in azimuth. Data are presented in slant range projection on 6250 bpi CCT or as an Exabyte cassette. Intended for use by those who possess SAR echo data processing capability.

SAR Fast Delivery Image

Provides ≈ 63 Mbytes of data for a 100km x 100km area (5000 pixels in ground range by 6300 lines in azimuth) with nominal accuracy of 0.9km in range and 1.0km in azimuth. Pixel size is 20m in range by 15.8m in azimuth with radiometric resolution of 3dB. Media are as above. These are intended as "quick-look" products.

SAR Precision Image

Provides ≈ 131.3 Mbytes of data for a 100km x 102.5km area (8000 pixels in ground range by at least 8200 pixels in azimuth). Pixel size is 12.5m ground range by 12.5m in azimuth. Media are 6250 bpi CCT, Exabyte Cassette, or photographic print. These products are intended for applications oriented analyses, especially those based on multitemporal data sets, or for those deriving radar cross sections.

SAR Ellipsoid Geocoded Image

Provides between 165.8 and 288.2 Mbytes of data for ≈100km^2 rotated according to a map grid. The data set contains 9100 to 12000 pixels in grid easting by 9100 to 12000 in grid northing with a pixel size of 12.5m in both easting and northing. Locational accuracy ± 150m in grid easting and northing. Projection GRS 84 transformed to UTM for latitudes between -80° and +84°; UPS for other latitudes. Intended for use in GIS applications calling for geocoded data sets.

SAR Terrain Geocoded Image

Provides much the same product as the ellipsoid geocoded data set with additional quality control.

RADARSAT International

SAR Georeferenced, Coarse

Available in full swath or selectable on a ground range projection oriented to the satellite track. It is intended as a "quick-look" data set provided in a few hours of acquisition on a CCT, ITN, Exabyte cassette, or in film format. Ground resolution is 100m x 100m with pixel size of 50m x 50m.

SAR Georeferenced, Full Resolution

Provided as full swath or selectable swaths on a projection oriented to the satellite track. Data are presented as CCT, film, ITN, or Exabyte cassette. Ground range resolution ranges from 25m (range) by 28m (azimuth) to 11m (range) by 9m (azimuth) with pixel sizes from 12.5m x 12.5m to 6.25m x 6.25m.

SAR System Geocoded

Area of coverage is selectable with a four-NTS map area or smaller. Data are projected onto ground range UTM, Lambert, or Polyconic projections, or are oriented to map projection using predicted orbit data. Media include CCT, ITN, film, and Exabyte cassette. Ground range resolution and pixel sizes are as given for SAR Georeferenced, Full Resolution.

SAR Precision Geocoded

Similar to system geocoded products.

SCANSAR (Narrow and Wide) Georeferenced

Provide full swath or selectable swath projected to ground range display oriented to the satellite track. Wide provides 100m x 100m ground range resolution with 50m x 50m pixel size, while Narrow provides 50m x 50m ground range resolution with 25m x 25m pixels.

Raster Storage Formats and Media

While image data from satellite sensors are usually provided in the formats indicated above, commercial image processing software has specialized formats which specify internal file formats. Most software vendors provide processing conversion subroutines for data input. Future data may also be available in a compressed data format, and this will require decompression before using.

MSI data is commonly stored on tape in one of three formats (Plate 16):

- Band interleaved by line (BIL) format stores image data for all bands consecutively, line-by-line within the file (i.e., Band 1 Line 1, Band 2 Line 1, Band 3 Line 1, and so on).
- Band interleaved by pixel (BIP) format stores image data for all bands consecutively, pixel-by-pixel, one after the other within the file (i.e., Pixel 1 Band 1, Pixel 1 Band 2, Pixel 1 Band 3, and so on).
- Band sequential (BSQ) format stores each band of data as a single file of row and column data.

Once formatted, data are available from vendors in a variety of media stored as bit-structured data. There are five common media in use today as tape or CD-ROM. SPOT image data are available on CD-ROM, 1/4" tape, 8mm tape, 4mm digital audio tape (DAT), and 9-track tape. Landsat image data are available on CD-ROM, 8mm tape, 1/4" tape, or for some products, on floppy disks for use on PC-based systems. Some data for SPOT and Landsat are also available via electronic network.

SPOT image data are stored in BIL format, Landsat in BSQ format, and IRS image data in BSQ format. For image data stored on CD-ROM, vendor documentation is required to determine storage parameters; however, the data are normally in the same format as tape.

Modeling Techniques

Business applications represent a broad and relatively new community of GIS developers and applications. This chapter presents case studies that characterize this breadth of opportunity while at the same time illustrating the benefits of combining raster and vector data sets. Optimal business siting, pipeline route selection, highway and transportation viewsheds, cellular telephone transceiver site selections, and change detection for tax assessments are typical of the many ways GIS is penetrating day-to-day operations. Supporting these and many other applications in subsequent chapters is a series of GIS modeling scenarios designed to enhance product generation. Among the case studies are techniques for combining raster and vector data in cartography, performing automated change detection, visualizing terrain without contours, and building attribute tables for raster files. The message throughout is that raster data enhance, augment, and support the complete range of GIS applications.

Economic Applications

Evaluating Franchise Locations

Bradley Cullen and Richard Curtiss, Department of Geography, University of New Mexico
Paul Neville and Teri Bennett, Earth Data Analysis Center, University of New Mexico

Challenge

Franchising in the United States is flourishing. The *Statistical Abstract of the United States: 1994* indicates that the number of franchised establishments increased from 442,400 to 542,500 between 1980 and 1991 and accounted for over one-third of all retail sales. The success of this marketing concept is largely attributable to its low closure rate. According to Janean Huber of *Entrepreneur*, less than five percent of franchised businesses per year have been discontinued since 1974. This rate could be reduced if franchisees would conduct an exploratory spatial analysis before signing a contract with the franchiser. One-fourth of the sites selected by surveyed franchisees in Albuquerque, New Mexico, failed to evaluate site market potential prior to purchase. This study shows how a GIS-generated distribution of retail establishments superimposed on a SPOT image could aid franchisees in evaluating potential site locations.

Franchising eliminates many of the risks associated with opening a new business. Franchisees can take advantage of the experience of the franchiser, economies of scale associated with multiple established organizations, training programs, name recognition and an established reputation, and business synergy. But simply buying a franchise is not a guarantee of success. As with most retail establishments, the profitability of a franchise is influenced by location. Factors such as site selection apply generally to retail establishments; other factors, such as territorial considerations and agreements, are fairly unique to franchising. Many of the location factors that determine a franchisee's success can be remotely evaluated with GIS. Remote evaluation eliminates the time and cost involved in conducting a field survey of each site. The potential of such a system is demonstrated for a new copy and duplicating franchise of Company X in northeastern Albuquerque.

The first question a prospective franchisee needs to have answered is: Who selects the site? In a survey of franchisers in Albuquerque, only one-fourth of the sites were selected by the franchiser; 75 percent of the franchisees were themselves responsible for finding the site for their franchise. But even if the franchisee is not directly responsible for site selection, s/he should carefully evaluate its potential. An example of a problem faced by a franchisee who failed to evaluate the site was reported in the *Wall Street Journal*. An AlphaGraphics franchisee was told by the franchiser that only 11 competitors were located within four blocks of a proposed Miami site. But only those existing establishments visible from the street were counted. After agreeing on the site, the franchisee found that he actually had 22 competitors in his trade area. As a result, the profitability of the site did not meet expectations. The shock of discovering that the franchiser had not done his/her homework could have been avoided had the franchisee accurately plotted existing establishments on a map.

Data and Methodology

Plate 24 shows the 1996 distribution of all copy and duplicating shops in northeastern Albuquerque. The addresses were obtained from the US West *Yellow Pages*. ARC/INFO was used to match the addresses of copy and duplicating shops with street addresses on the city's TIGER file. The distribution was subsequently superimposed on a SPOT image of northeastern Albuquerque.

Two SPOT images were used in this study. All image processing was accomplished using ERDAS IMAGINE software. Both the 20-m SPOT high resolution visible (HRV) and the 10-m SPOT panchromatic (PAN) images were map rectified. The HRV scene was converted to a pseudo-natural color scene using the following formula:

$$Red = HRV2$$

$$Green = \frac{(HRV1 \times 3) + HRV3}{4}$$

$$Blue = HRV1$$

The pseudo-natural color HRV scene was merged with the PAN scene using the principal component method for merging images of different resolutions. The near natural appearance of the urban image does not require the user to be knowledgeable in spectral response patterns in order to interpret it. In addition, at a 1:15,000 scale display, the user is easily able to tell the difference between various types of commercial and residential developments.

Results and Discussion

As seen in Plate 24, a string of copy and duplicating shops is already located along Montgomery Boulevard. An ideal location for a new Company X franchise can be identified near the western end of the street (see the yellow star on Plate 24). The location is far enough from the existing Company X franchise that competition for customers can be avoided.

The franchisee generally purchases from the franchiser an area or region in which the facility may, or must, be operated. But not all franchisers delimit territories and define territorial rights in the same way. Thirty-one percent of the franchisees responding to a recent survey indicated that their franchiser based territory on population size. In many cases, franchisers using this method divide the total population of a region by the threshold population, which is the minimum number of people needed to support that activity. The quotient is the number of facilities a region can conceivably support.

When population is used to delimit territory, it is often up to the franchisee to select the specific site. If the approximate distance an individual is willing to travel to obtain the good or service in question is known (range of the good), then ARC/INFO's BUFFER command can be used to construct circles with a radius equal to the range of the good around the potential site and the locations of the competition. In Plate 24, it was assumed for demonstration purposes that the the range for a copy and duplication shop was one mile. The SPOT image allows the franchisee to inventory the contents of the circle. S/he can remotely identify barriers that might impede access to the site (land use, street patterns) or otherwise influence the site's profitability. Without the SPOT image, a costly and time-consuming field survey would be required to evaluate the site's potential.

When there is another franchise of the same company or a competitor that sells a similar product within the range of the potential site, then the territory must be apportioned between the competitors. If we assume that most customers will frequent the nearest establishment that offers the particular good or service, then the intervening distance can be divided between the competitors. If there is more than one competitor, the intersection of the dividing lines will define the theoretical trade territory for the potential site, which can be inventoried to assess its potential. Similar procedures can be followed by franchisees whose franchiser does not sell territorial rights (11 percent of the franchisees surveyed).

Forty-four percent of survey respondents indicated that their franchiser defined the trade area geographically. A franchisee is, therefore, sold the right to service an area covering a specified number of blocks or square miles. To visually evaluate the territory, the franchisee can delimit the proposed territory with ARC/INFO's draw tool and make an inventory of the trade area. For the 14 percent of the franchisees whose franchiser used specific company criteria to define territory, a GIS that focuses on those criteria would have to be created.

Benefits

Three important factors determine a retailer's success: location and location and location. In retail geographic terms, the region, trade area, and site selected for a retail establishment will determine its success. GISs are invaluable time and cost saving tools for remotely evaluating the potential of each. This example demonstrates how a simple GIS can be used to evaluate potential franchise locations. Others have shown how more sophisticated GISs can be used for site modeling, while incorporating analog, regression, spatial interaction, and location allocation models. GIS applications in market research are only beginning to be recognized. Their potential is enormous!

Cellular Phone Transceiver Site Selection

John R. Jensen, Xueqiao Huang, and Derek Graves, Department of Geography, University of South Carolina

Richard Hanning, Fluor Daniel Siting and Consulting

Challenge

Intervisibility analysis for urban areas is a tool in many civil engineering applications. In the past, these analyses were prepared from contour lines and/or profiles. The advent of the GIS has enabled them to be performed automatically using digital elevation models (DEMs). However, even though such analyses are now more easily performed, obtaining an accurate and precise DEM can be difficult. Advances in softcopy photogrammetric technology have made it possible to create accurate DEMs from digitized aerial photography using desktop softcopy photogrammetric workstations. GIS analysis can frequently be conducted on the same workstation.

A joint project between the University of South Carolina and Fluor Daniel Siting and Consulting was completed through the NASA Stennis Space Center Visiting Investigator Program. This project examined a DEM generated from softcopy photogrammetry to support the identification of optimum locations for cellular phone transceivers. This project also introduced the general procedure for urban DEM creation using softcopy photogrammetric technology.

Data and Methodology

Softcopy photogrammetric techniques use the same mathematical methods to obtain point elevations in a DEM as the traditional analytical photogrammetric approach, but are completed using a computer workstation. To create accurate DEMs using this method, three principal components are required in addition to the relevant software and hardware: data in a digital format, aerial camera calibrations, and suitable ground control points (GCPs). Numerous software packages currently available have the ability to complete softcopy photogrammetry. The software used for this project was Vision International and ERDAS OrthoMAX.

The study focused on downtown Columbia, South Carolina. Vertical black and white panchromatic metric aerial photography was acquired on March 30, 1993 at a nominal altitude of 900m (3,000 ft) using a Wild RC 10 camera. Two original 22.5 sq cm (9 sq in) stereopair negatives were scanned at 500 dpi using a linear array charge coupled device (CCD) Vexcel VX3000 film scanner to produce digital data. This resulted in digital images with a pixel size of approximately 30 sq cm.

The aerial camera's elements of exterior orientation (i.e., the camera's position in Cartesian space and its attitude, as in roll, pitch, and yaw) are determined in the process of aerial triangulation. The aerial camera's elements of interior orientation, such as calibrated focal length, radial lens distortion, coordinates of principal points, and fiducial marks, as well as the GCPs and image tie points, are also input during the process of aerial triangulation. Because there were no geodetic vertical/horizontal benchmarks in the study area, the 12 horizontal/vertical GCPs used during the project were collected and processed using global positioning system (GPS) receivers. The GPS data were differentially corrected using data from a local base station and were located in the digitized photography to within +1 pixel.

1:6,000 black and white panchromatic stereoscopic aerial photography obtained on March 30, 1993, at 3,000 ft above ground level.

Results of least squares adjustment aerial triangulation are necessary to build a stereomodel. From the stereomodel, elevations of DEM points are measured or interpolated by the computer program. Both the stereomodel and DEM generation in softcopy photogrammetric applications are based on computer stereo image matching techniques. The stereomodel was viewed three-dimensionally using a stereo-ready monitor, an infrared (IR) signal emitter, and a pair of special stereo glasses with an IR receiver.

Due to the non-uniform nature of the urban environment, there are some distinct problems with softcopy photogrammetric techniques. One such problem is the difficulty in image matching. Due to the occlusions (shadows) produced by buildings and to other factors affecting image quality, such as low contrast, the stereo matching (correlation) may not be good for some points. Thus, the elevations of those points cannot be directly computed. Instead, they must be determined by interpolation of neighboring computed elevation points. In urban areas, interpolation tends to turn the vertical walls of a building into more gentle slopes. This introduces distortion in the resultant DEM.

Another dilemma in urban DEM generation is the effect buildings and trees have on the computation of differential parallax and height measurements. The DEM algorithms assume these features are terrain, but this assumption is not acceptable for slope computation. A DEM of nominal terrain without buildings and trees is often desirable. On the other hand, applications like intervisibility analysis require a DEM with buildings and trees. The creation of both types of DEM requires manual editing. The OrthoMAX Stereo Editing function provides the manual editing ability necessary to correct DEM errors. With careful operation, the DEM error can be eliminated or greatly reduced. However, because it is not an automatic process, DEM editing can be very time-consuming and may offset some of the advantages of softcopy photogrammetry. Automation of this process should be an important issue for future softcopy photogrammetry research.

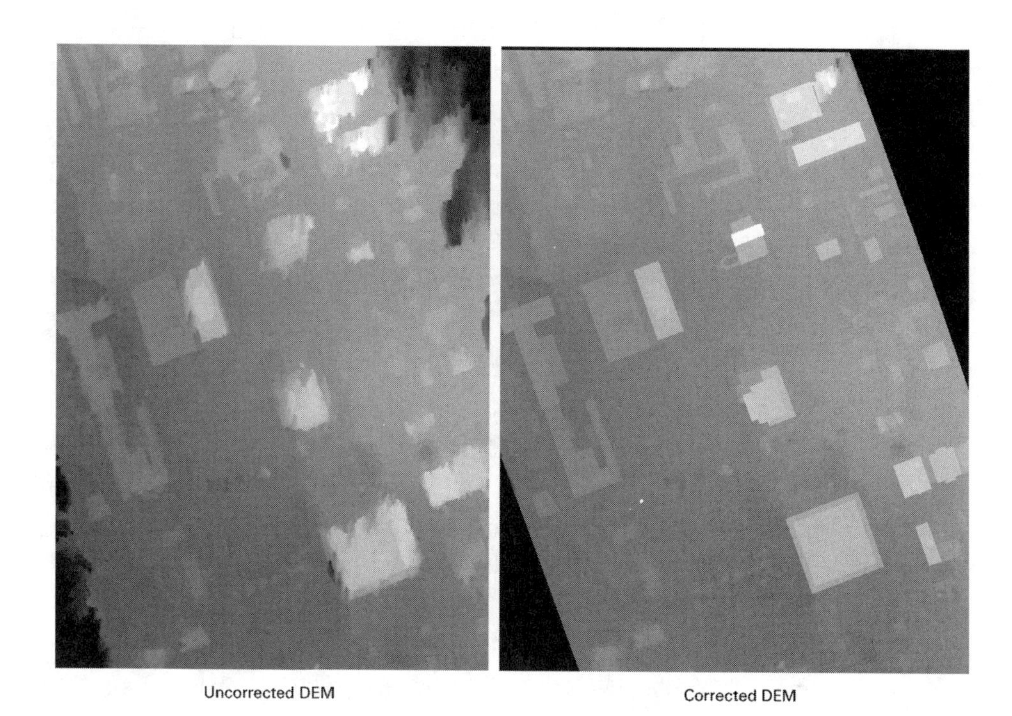

Digital elevation model (DEM) extracted from stereoscopic photography using softcopy photogrammetric techniques.

Results and Discussion

By using corrected rather than uncorrected DEMs, it was possible to produce orthophotographs that depicted the buildings in the study area with sharp and distinct edges. Orthophotography, a highly beneficial by-product of the softcopy photogrammetry, was draped over the DEM to create perspective images. This provided the three-dimensional visual representations of the study area shown in the following illustrations.

Orthophotographs generated from uncorrected and corrected digital elevation models.

from uncorrected DEM

from corrected DEM

Intervisibility analysis allows identification of all areas in a DEM that are visible from one or more locations. In this project, it was necessary to analyze numerous buildings to determine which could provide the largest ground coverage for one or more cellular communication transceiver towers. The determination of intervisibility is based on the height of the observer position, as well as the surrounding environmental feature heights, and can be conducted in many GIS systems. Generally, the more accurate the DEM, the more accurate the intervisibility determination results. The high resolution DEM created using softcopy photogrammetric techniques was easily accessed by a GIS.

Digital photography draped over a photogrammetrically derived digital elevation model of Columbia, South Carolina.

A procedure to conduct intervisibility analysis using a grid DEM and ESRI's ARC/INFO and ArcView was developed in this project. ARC/INFO provides a function, VISIBILITY, from which it is possible to complete intervisibility analysis. However, certain limitations complicate intervisibility analysis and make it difficult to conduct interactive, on-the-fly selection of observer points, visibility analysis, and real-time display of results. To solve this problem, several ArcView Avenue and ARC/INFO AML programs were written to combine ArcView, which functions as the user interface for observer selection and display, with ARC/INFO, which runs the VISIBILITY command in the background. This provides a user-friendly graphical interface for the visibility analysis and integrates observer point selection, visibility analysis, and display into one seamless process. It also provides the ability to compare the results from different site selections.

The VISIBILITY command in ARC/INFO allows the user to define more than one observer point per iteration, thereby providing a capability for studying a network of cellular communications transceivers simultaneously or any one individual transceiver independently. Because of the high quality of the DEM that is created through softcopy photogrammetry, it is possible to obtain very accurate intervisibility analysis results. Comparing the visible areas from different potential locations of cellular phone transceivers, an engineer can make better decisions about the most cost-effective site location. The results of an intervisibility analysis can be seen in Plate 25.

Benefits

Use of stereoscopic aerial photographs and softcopy photogrammetry provides an economic approach for developing large scale DEMs in a format easily accessible by a GIS. The cost of such an approach is significantly lower than traditional analytical photogrammetry in part due to the low cost, ranging from $25,000 to $50,000, of many softcopy photogrammetric workstations. However, as these procedures confirm, serious errors can develop in creating a DEM that describes an urban area. Manual editing with stereo display mode is recommended as one possible solution but is, in itself, a very time-consuming process. A method for automating the DEM editing process is needed. Despite the drawbacks, softcopy photogrammetric generation of DEM and GIS layers provides a very effective approach to intervisibility analysis for a cellular transceiver site location study.

Finding a Least-cost Path for Pipeline Siting

Sandra C. Feldman, Bechtel Corporation

Ramona E. Pelletier, NASA Stennis Space Center

Ed Walser, James C. Smoot, and Douglas Ahl, Lockheed Martin Stennis Operations

Challenge

The cost of building an oil pipeline varies with the type of terrain and land cover the pipeline must cross. Construction engineering can include months of costly site surveys to identify the most cost-effective corridor. Working with the National Aeronautics and Space Administration's (NASA) Commercial Remote Sensing Office at Stennis Space Center, Bechtel Corporation has developed a procedure that uses satellite imagery and GIS analysis to find a least-cost pipeline route.

From late 1992 through 1994, Bechtel Corporation of San Francisco, California, worked with a consortium to design and build a pipeline to carry oil from the giant Tengiz Oil Field in Kazakhstan on the Caspian Sea to the Russian port of Novorossiysk on the Black Sea. The proposed 700km (438 mi) pipeline would link with an existing pipeline and carry up to 1.5 million barrels of oil per day to international markets.

After examining eight options, the consortium chose a general route for the new pipeline. This route, which passes through terrain with predominantly low relief, was deemed the most politically and economically attractive.

For the pilot project, a 50km-long section within the general route was selected to conduct the least-cost path analysis. An area at the northern end of the Greater Caucasus Mountains where both terrain and geology vary was picked for the pilot project. This area was chosen so that numerous geography-related cost factors could be included in the analysis.

Caspian pipeline corridor, Novorossiysk, Russia. Shown on Band 1 of Landsat TM and SPOT merged image.

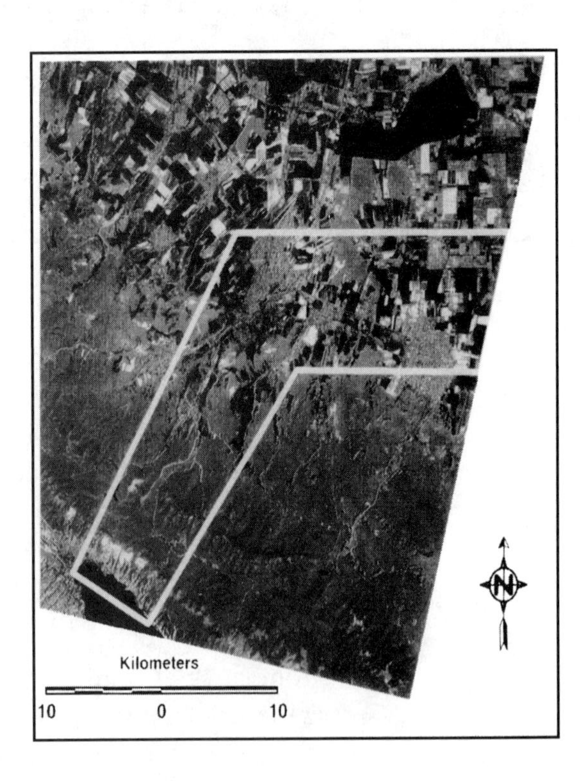

The pilot project was performed under a Domestic Nonreimbursement Space Act Agreement between NASA, John C. Stennis Space Center, and Bechtel Corporation.

Data and Methodology

In theory, the cheapest route for building a pipeline is a straight line between two points. In reality, of course, many factors affect the cost, and therefore, the final route of the pipeline. Aside from the personal experience of engineers involved in such projects, accurate methods of balancing engineering concerns and construction costs against environmental costs and other pertinent factors are not well defined.

The usual cost factors considered are distance, topography, near-surface geology, river and wetland crossings, road and railroad crossings, and the proximity to large population centers.

The impact of distance on cost is obvious. Topography is considered because high relief terrain increases the cost of pipeline engineering and construction. In terms of surface geology, unconsolidated fine-grained material is preferred because it does not require blasting. Rivers, roads, and railroads are obstacles that require the building of bridges or supports to cross. Wetlands, agricultural lands, and other environmentally sensitive areas also must be considered. Crossing them may mean the completion of lengthy environmental impact studies.

Cost factors used in the least-cost path analysis were calculated from actual construction expenses on other Bechtel pipeline projects. These were normalized to a baseline cost for comparison in the pilot. The baseline cost was the expense of pipeline construction per unit length excluding special modifications for terrain, geology, land use, and other factors.

Bechtel calculated incremental costs (over the baseline costs) for construction activities related to these modifications for geographic factors such as trenching in consolidated rock, clearing brush and trees, crossing rivers, railroads, and wetlands, or passing through agricultural land.

Terrain slopes were divided into seven categories commonly used by pipeline engineers: flat, slightly rolling, rolling, sharp, choppy, rough, and mountainous. Incremental construction costs were calculated for each of these terrain categories.

Urban and industrial areas were highly undesirable for pipeline siting in this case, so these areas were assigned costs ten times the baseline cost.

A critical step in the study was the collection of terrain, land use, and land cover information upon which associated construction costs could be calculated. SPOT panchromatic (10m spatial resolution) and Landsat TM (30m resampled to 25m) imagery were used in combination with 1:500,000 scale topographic maps and 1:500,000 scale geologic maps to gather the necessary information.

The value of multiple data sources was underscored by working in the former Soviet Union. Although 1:200,000 scale contour maps were available, some information on them had been erased to mask strategic facilities. The acquisition of remotely sensed data is a partial solution in the absence of quality maps.

With all of the resource materials collected, project analysts employed a variety of raster processing tools to extract information from the image and map data and incorporate the results into the GIS environment for analysis. (Trade names are mentioned here for the benefit of the reader and do not indicate an endorsement by Bechtel Corporation or NASA.)

ERDAS IMAGINE Spatial Modeler was used to merge the TM data with the SPOT imagery. The images were coregistered and an intensity hue saturation transformation was performed so that the land cover information found in the spectral data would be clearly defined by the spatial detail of the SPOT imagery. The resulting merge map would be used as the base map upon which analytical results would be overlaid for display purposes and visual inspection.

The ELAS image processing program (Earth Resources Laboratory Software, developed at Stennis Space Center) was used to extract land use information. A color composite was made of TM bands 3, 5, and 4 as blue, green, and red. These bands distinguish key land use classes and were used as input for a supervised classification (see Plate 26). Training areas for specific land use categories, such as wetlands, agriculture, native vegetation, water, urban, and industrial, were selected, and the classification was run in ELAS. It produced a map that identified and distinguished farm land, urban centers, and industrial areas. Results from the classification were checked against other maps and field surveys.

ARC/INFO was then employed to digitize other key features from the paper maps, such as roads and drainage features. Topographic contours within the proposed corridor were also digitized and used to produce a digital elevation model. This information was used to create the terrain slope map for the cost analysis.

Boundaries between geologic units were extracted in digital form from the 1:500,000 scale geologic maps and, along with all other layers, were converted from vector to raster format within the GIS database. Based on lithologic descriptions in the map legends, each rock unit was classified as "consolidated hard rock" or "unconsolidated." This information was added as attributes to the GIS files.

The land use information, slope map, and other geographic features which had been digitized and converted to raster format were loaded into the GRASS (Geographic Resource Analysis Support System, developed by the U.S. Army Corps of Engineers) GIS software package. Each feature type comprised a separate layer in the GIS.

Project analysts utilized a variety of common GIS analysis functions in GRASS to carry out the least-cost analysis procedure. This analysis is a modification of algorithms commonly used in GIS for drainage basin analysis, with the path constrained to flow through specific nodes to specified endpoints.

In the pilot area, pipeline engineers picked four points through which the Caspian pipeline had to pass. The objective of the analysis was to compare the cost of a straight line route with a least-cost pathway between the preselected points.

The procedure involved resampling each of the raster layers into 25 meter cells. In the course of the analysis, the cells in each feature layer were assigned cost values based on predetermined construction costs associated with crossing each feature (i.e., stream, road, hardrock, or slope). For example, any cell in the land use layer identified as urban or industrial land was assigned a cost ten times the baseline construction value.

In this analysis, the least-cost algorithm added the cost values of each cell to arrive at a cumulative incremental cost calculation. For instance, if one cell contained hard rock and a steep slope, the costs associated with each factor were added together.

The GIS added the cost values for all of the cells in the pilot area, creating a cumulative cost surface map (see Plate 27) that would resemble a topographic map if displayed. Peaks represented the most costly areas and valleys the least costly areas to build upon. Any cells outside the study area were masked.

A straight line was drawn between each of the points in the pilot area to determine the total cost for the shortest route. Total cost was calculated by adding the values of each cell in the route. An additional routine in the GIS then searched for the least expensive path between the same points. This path had the lowest cumulative cell value, and therefore was the least costly.

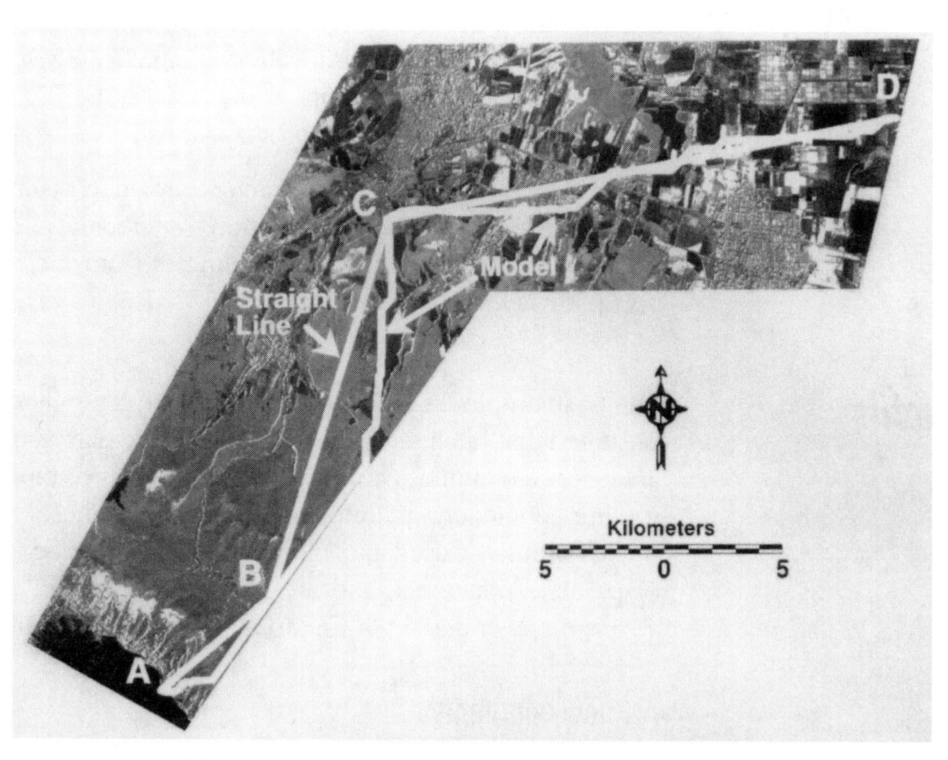

Pipeline route derived from least-cost pathway analysis (model) compared with straight line route.

Results and Discussion

This project demonstrates that the shortest route between two points is not always the most cost effective. At 42 km, the straight line was shorter than the 51 km least-cost route, but its overall expense was calculated at 14 percent higher as a result of crossing a greater number of high-cost features.

The GIS compared the relative costs of the straight line route and the least-cost route by counting the total number of cells in each that exceeded the baseline cost. For the straight line route, 1,667 cells were above baseline, while only 1,478 cells were above baseline in the least-cost pathway.

A comparison of the frequency with which each route crossed various cost-related features revealed that the straight line path traversed more roads and urban and industrial areas. These features added significantly to the relative cost of the straight line route.

In future projects, Bechtel plans to refine the model on a per project basis to account for actual project costs in the geographic area under consideration. Methods also need to be developed to remove sharp angles from the least-cost pathway.

Benefits

The Caspian pipeline least-cost analysis successfully demonstrated that satellite based remotely sensed data and GIS analytical techniques can facilitate the process of pipeline routing, engineering, and cost estimating. These techniques should be used in conjunction with field experience of pipeline personnel, and are especially useful in areas where maps or aerial photography resources are limited.

Acknowledgments

This work was funded by Bechtel Corporation and NASA. Parts of this work were supported by the NASA Office of Space Access and Technology, Commercial Remote Sensing Program Office, under contract number NAS 13-650 at the John C. Stennis Space Center, Mississippi.

Visualizing the Albury Bypass

Michael Byrne, Roads and Traffic Authority of New South Wales, Rosebery NSW, Australia

Challenge

One of the goals of modern government is to be more responsive and relevant to the community. The uncompromising bureaucracies of previous decades have given way to authorities that actively encourage public involvement in deciding how communities evolve.

One area with a history of major impact on the community is public transport, especially roads. Roads continue to be essential, but issues concerning size, purpose, and location also continue to be sensitive. Conflicting views arise which require mediation and compromise. The net result is a protracted and sometimes expensive development. Another goal of present government is to reduce inefficiencies in public spending. Consequently, gaining consensus on a proposed development while simultaneously reducing or maintaining public spending levels becomes a difficult task.

The Roads and Traffic Authority (RTA) of New South Wales has been rethinking its approach to constructing roads and freeways. Because freeways often have the greatest impact on communities and the environment, the challenge is to identify common ground where all stakeholders can agree.

The enterprise-wide introduction of visualization tools, such as GIS and image processing software, into the RTA has offered new ways of getting the public involved in planning a road development at a much earlier stage than was previously possible. This study describes early attempts by the RTA to change the way a road development is presented to the public. The development is the proposed freeway at Albury, a vital missing link in the Hume Highway between Sydney and Melbourne. This project has a difficult history in terms of attaining agreement for freeway location. The aim of RTA's new attempt at seeking agreement on location is to provide understanding of proposals by simulating a "fly-through" of the project for public exhibition.

The absence of an effective corridor selection process that takes account of environmental and social issues is the most significant weakness in current planning procedures. The present approach is essentially an extension of techniques used to define a route within the corridor. Route selection involves strict engineering design. The selection of suitable corridors is based on the engineering feasibility of trial routes, and requires a significant amount of preparatory work by the engineers and planners.

To assist in the design process, the RTA and other public agencies are developing an expert system software package for route selection called ALIGN3D. The software automatically generates a least-cost road alignment by optimization against input data such as terrain, road design standards, and restricted areas. The use of the ALIGN3D software in corridor selection provides a process in which engineering design is tested against environmental and social constraints. Most non-engineers have difficulty understanding the design process. The software is an attempt to make the design process more relevant to the affected community by incorporating environmental and other concerns presented by the public.

Data and Methodology

The site of the proposed national highway link is on the Hume Highway at Albury Wodonga on the border between New South Wales and Victoria. This link will join an existing freeway north of Albury to south of Wodonga in Victoria. It is a vital segment for travel and transport between the state capitals. The timber industry in the area requires road transport as well. All traffic must pass through the city center.

The Road Technology Branch of the RTA initiated a project in 1994 to test the applicability of the ALIGN3D software. The software requires terrain models and satellite imagery.

The goal of this visualization was to create a simulated fly-through of preferred routes for public exhibition and comment. These routes were generated from the ALIGN3D system on previously determined corridor options. It was believed that if the visualization were acceptable, then the process could be used earlier in the concept stage to determine a preferred corridor. The visualization requires four data components: terrain model, satellite site image, route paths, and sections of route photography.

❒ The *terrain model* was provided by a digital elevation model (DEM) derived from two pairs of stereo SPOT panchromatic images covering an area of 8,000 sq km. These images were captured in March 1991, and the flight paths were six days apart. For georeferencing, 22 ground control points were used from seven topographic maps. The resolution of the DEM was 20m with a stated planimetric accuracy of 50m and a vertical accuracy of 10m.

  A constant was added by SPOT to each pixel value of the DEM. The constant was a function of an offset value and a scale value, and resulted in a false height of about 10,000m. Fortunately, the original value of specific pixels was provided for checking purposes.

❒ The *satellite image* was a combined SPOT panchromatic (PAN) and multispectral (XS) scene of the same site. The PAN scenes were the same used in creating the DEM. The XS scene was captured in October 1992.

❒ The r*oute paths* were created using the ALIGN3D system, and consisted of vector alignments of the surface formation. The lines were exported in a DXF format.

❒ *Aerial photography* was available for parts of the routes, and provided scope for better resolution in specific sections as well as a natural color look.

ERDAS Imagine Version 8.2 was used for the bulk of the processing and composition of the perspective scenes. ARC/INFO Version 7 was used to import and process the DXF files generated by the route selection software, ALIGN3D.

Most of the processing was carried out by the Sydney office of ESRI-Australia and required about ten days to complete. The creation of the fly-throughs can be separated into two major processes: composition of the combined image model, and generation of the perspective views and subsequent image sequencing.

Composition of the Image Model

This phase involved importing, cleaning, and aligning data. Both the DEM and the image scene were imported directly into ERDAS using the default SPOT format. Both data sets were in the same AMG coordinate system and needed no transformation. However, a constant was added to the DEM which resulted in a false elevation.

The Spatial Modeler in IMAGINE was used to remove the constant according to a function noted in the data documentation. This process was checked against the seed pixel value which was also provided in the documentation.

Route data were initially loaded directly into IMAGINE using the default DXF format. The native Vector Module was used to overlay data onto the image in cases where the data contained the complete route formation, including both longitudinal and cross sectional carriageway formations. This was considered overly complex for a concept visualization process. Only the basic formation was needed, and ARC/INFO was considered a better environment to amend the line work. In addition, ARC/INFO was used to transform the route information into AMG from the Integrated Survey Grid (ISG), the predominant coordinate grid system used in New South Wales.

The generalized route was loaded back to IMAGINE directly from the binary ARC/INFO coverages. When the route alignment was loaded with the image data, an apparent shift was detected at the route terminals. Because the shift was small, and the objective was a generalized visualization, the shift was removed via a simple linear shift using the drop point tool to move the image to align with the arc vectors.

The extent of the satellite data was well in excess of the bypass site, and the size of the combined data was in excess of 150 megabytes. To make the processing more manageable, the project area was excised from each data set via an area of interest (AOI) defined with the use of a rectangle.

Aerial photographs for sections of the routes were scanned at a density of 300 dpi and imported into ARC/INFO for georeferencing. These TIFF images were sent to the ESRI office and imported into IMAGINE.

All images were clipped of overlap areas and mosaicked together. The area to be clipped was small because only alternate photos were used. As the clipped areas corresponded to pixel rows, the image-to-image matching was simply a matter of overlapping several columns at the edges.

Finally, an annotation layer was created to place the names of towns and other landmarks on the scene. This annotation layer was added to the image stack. In addition, the line styles of the alternate bypass routes were colored and sized to give the appearance of a broad planning approach, rather than a highly engineered proposal.

Merging or "burn in" of all data into a single image is ideally a matter of loading all the data into the same IMAGINE Viewer and applying the View to Img command from the

View menu. This procedure merges all the displayed data together. The output scene adopts the largest resolution unit of the merged data, which in this case was 20m.

This is an important step because the resultant resolution greatly affects the visual appearance of the scene, especially in the foreground of a perspective view. An alternate approach was explored to improve final image resolution. Map Composer was used as the composition area to build the combined image and print the map to an image file. There was little observable difference between the two methods. The determining factor in the final image resolution was the input data.

Perspective Views and Sequencing

The Perspective View Module was used to create the perspective views. The views were created by a pair of linked viewers: a plan viewer, which allows the placement of an eye and target, and a perspective viewer. The most efficient approach was to use only the DEM to get the terrain perspective right, and then drape the merged image onto the DEM. A trial and error approach was used to determine the best combination of viewing parameters such as the height of the eye point, vertical exaggeration, and width and depth of the scene.

The purpose of the visualization was to show a series of static views from topographic high points along the routes as well as the fly-throughs along the routes. Selecting the static views to be used was based on appeal, and the views were generated to image files.

The selection of the flight path is perhaps the most important part of the visualization process because it determines how realistic the fly-through will look. The limitation imposed by the coarse pixel resolution of the merged scene made a ground level "drive-through" appear too blocky. The height selected was 600m above ground level. Along with the slight exaggeration of the vertical (1-1/2 normal), the selected height was believed to provide the best continuous view over the length of the routes.

To avoid jerkiness in the flight path, a generalized path was selected in an attempt to average out the bends similar to how an aircraft would set headings along its flight path. The first attempts at generating the individual scenes (or frames) used a standard distance between scenes, which was based on a subdivision of the total length of the route by a set number of scenes. The individual scenes were created manually from the draped image as described above.

Sequencing was then managed by the IMAGINE Movie option from the Utilities menu. The individual images can be selected and added to the sequence and then replayed at desired frame speeds.

Later work was aimed at using a batched macro to generate the views at a smaller separation. A greater number of frames allows a smoother appearance to the final fly-throughs.

Results and Conclusions

The fly-throughs produced thus far have been well-received, but have not yet been placed into a viewer for public display. Public display will require using a PC style viewer that can allow a simple set of user selection buttons to activate a fly-through or a static view. This will be much the same as a touch-screen style information system.

Perhaps the most important aspect of the response to the initial fly-throughs was the difficulty most people had in understanding the color composition of the scenes. The use of aerial photography was able to provide the natural color look in merged sections, but generally some explanation was required. With the use of natural color composite satellite imagery, the contextual information contained in those scenes will be more readily understood.

Generally, acceptance of the results indicates that this approach has real potential to provide a low-cost method for concept design that can be widely understood. The key to this approach is in creating a common view of an idea or need. Visual demonstration promotes understanding more than any other method. In other words, "I see, I understand."

The particular strengths of the project are that a large area can be viewed continuously, which is suitable to the concept planning stage. The discussion generated by viewers of the fly-throughs is also considered a strength because project objectives are to create a common understanding and stimulate discussion among the stakeholders.

Another strength is the compatibility between the ERDAS and ARC/INFO software. The RTA has highly developed GIS skills and has accumulated well over ten gigabytes of spatial data. Until recently, the RTA had little interest in using satellite imagery. The width of most road formations is less than 20m (66 ft), and satellite image resolution has not been very useful.

This project, however, established the usefulness of satellite imagery in its own right. But the real benefit will derive from using the existing GIS data with the imagery. Future uses would include draping GIS data into the merged scenes to provide greater definition of concept projects as well as development projects.

Current developments of the ERDAS software toward real-time manipulation of raster and vector data, such as the VirtualGIS module due in the next release, may have enormous potential for future fly-through simulations. Real-time manipulation could allow interactive sessions where all stakeholders in a proposal may contribute location information, such as areas of environmental or cultural importance, and permit everyone to see the relationships between all influences. The manipulation or placement of a corridor may be generated during this type of forum and effectively facilitate public involvement in the process.

Acknowledgments

The author wishes to thank Peter Rufford (Manager Locations, Road Technology Branch, RTA Technology) and Mark Quill-Williams (ESRI-Australia) for their help in this project.

Choosing Efficient, Cost-effective Transportation Routes

Dana Nuñez Brown, Parsons Brinckerhoff

Challenge

Gridlock is the best term to describe travel in and around the Washington, DC, area. The Virginia Department of Transportation (VDOT) has proposed a western DC bypass in Northern Virginia that would ease congestion on the Capital Beltway.

VDOT has funded a major investment study to determine exactly what the bypass project will include. The agency is open to considering several modes of transportation, such as high-occupancy vehicle lanes, rail transit, exclusive bus lanes, or a combination. The bottom line, however, is that VDOT wants the transportation mode and route (within a defined corridor) that will most efficiently and cost-effectively move the greatest number of people.

Determining the best combination of route and transportation mode requires analysis of an incredible number of engineering, environmental, demographic, and cost factors, as well as their interrelationships. Satellite imagery and GIS are the best tools for collecting and analyzing this diverse information.

VDOT contracted Parsons Brinckerhoff, an international planning and engineering company, to perform the major investment study for the bypass project. This study will examine all pertinent transportation options and present the pros, cons, and costs associated with each for consideration by VDOT, federal agencies, local governments, and the public.

The study focuses on a preselected corridor located west of Washington. Stretching through five Virginia counties, the corridor would contain one or more transportation routes for use by local commuters in the suburbs and allow travelers on Interstate 95 to avoid mixing with Beltway traffic.

Parsons Brinckerhoff has developed a systematic procedure for performing investment studies for transportation projects. Key to these studies is consideration of the financial, technical, and legal impact of every factor related to each transportation alternative. The factors typically considered are related to natural, social, and cultural resources, as well as land use and engineering concerns.

The impacts on resources can be numerous. For instance, construction through a wetlands region might require costly replacement of lost acreage at a three-to-one ratio. From an environmental justice perspective, displacing low-income populations should not occur at a rate disproportionate to population demographics. In addition, basic engineering limitations on railroad grade or highway slope must be compared against the existing topography.

Impacts that might have positive effects on transportation patterns are equally important. The volume of people served by a bus station, the number of cars removed from the road if a rail link is established, and the average number of minutes saved in travel between two dense suburban areas are all beneficial impacts that must be modeled and quantified in a transportation study.

In the early phases of an investment study, broad-scale analysis of the study corridor is conducted to eliminate specific areas that would be too prohibitive to traverse. Several candidate routes are then chosen for closer scrutiny. Hundreds of factors and potential impacts are identified, and costs quantified.

Data and Methodology

Collection of land use, natural resources, demographic, and cultural information for a major transportation study requires diverse data sources. In many cases, paper and vector maps already exist. In others, geographic information must be extracted from satellite imagery.

In the Washington project, SPOT panchromatic imagery with 10m resolution served a variety of functions. Project analysts extracted land use classes and calculated housing development densities using the imagery. For perspective analysis, the raster SPOT imagery also served as the GIS base map.

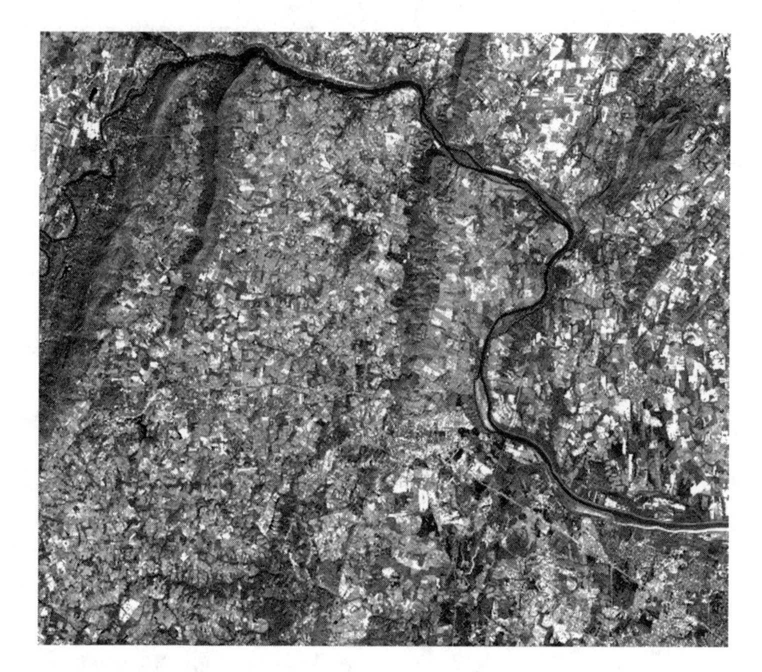

Western Washington, DC portion of bypass study area in Virginia. SPOT satellite imagery was instrumental in mapping land use and housing densities for the proposed bypass project.

Investment studies also typically involve use of National Wetlands Inventory mapping data, TIGER demographic and street centerline files, state soil maps, local planning maps, hazardous materials sites, and cultural feature locations. For many areas, these maps can be obtained already digitized in ARC formats.

Transportation siting projects can require a variety of data analysis techniques. The following is a composite description of several methods employed by Parsons Brinckerhoff.

ERDAS IMAGINE software is used to enhance, analyze, and prepare information from the raster satellite imagery for further GIS analysis. Although procedures may vary for a given project, the typical first step performed using GIS imaging software is to rectify the imagery to the map coordinate system appropriate for a given project. In the bypass project, the Lambert Conformal Conic coordinate system was used.

Unsupervised classification routines are usually run on the satellite imagery to obtain a broad overview of land cover. In the Washington bypass project, however, numerous correlative sources of up-to-date land use information made supervised classification the preferred alternative.

Because an understanding of population density plays such a key role in choosing transportation routes for the Washington project, Parsons Brinckerhoff planners performed classifications on the SPOT imagery to obtain this information. The 10m spatial resolution allowed the planners to count individual dwellings to determine housing densities per acre.

Housing density has been found to create a pattern of identifiable signatures in satellite imagery. Therefore, planners have to manually count only a few neighborhoods before they have the training site data necessary to run supervised classifications throughout a corridor to categorize neighborhoods as low, medium, or high population density.

Choosing training sites for supervised classifications is an easier and more precise process when the ERDAS IMAGINE Vector Module is implemented. This module enables ERDAS users to import vector data directly for integration, updating, and manipulation within the image processing environment. Although the user works on the vectors and their attributes in the raster environment, the vectors remain in ARC format.

In many transportation projects, the National Wetlands Inventory mapping data in ARC vector format are overlaid on the satellite images. Planners locate wetlands on the vector map and then delineate them on the satellite image as training sites for the classification. In similar fashion, the ARC/INFO TIGER files are overlaid on the images to assist in the population density classification.

Attribute query functions in the aforementioned software packages are critical for quantifying the costs of various transportation route options. Information that will later help in assigning real cost figures to each feature is added as attribute data to the vectors. For instance, neighborhoods must be categorized as having low, medium, or high real estate values. The level and type of potential contamination at hazardous waste sites must be specified. Neighborhoods that will have to be protected by sound walls near a new highway must also be noted. All of these factors can be translated into dollar values.

In the later stages of the study, after major problem areas have been eliminated and actual route options are being evaluated, the engineers choose several potential route segments between various points. These segments are drawn as polygons in ARC/INFO. The vectors are usually brought into the IMAGINE Vector Module and overlaid on the classified satellite image for analysis.

GIS routines add up the acreage of wetlands, farmland, and hazardous waste sites; count the number of houses that would be displaced; and measure the length of noise barriers that would have to be erected for each route scenario. These quantified factors are then used to compute the dollar amounts that are multiplied by acreage, mile, house, or other relevant variable. The results are the potential impacts and dollar costs of each route option. Each proposed route then has its own attribute file quantifying the environmental impacts and cost of each geographic feature that must be crossed, disturbed or replaced if the route were chosen for development.

Financial costs and environmental impacts are not the only factors considered in choosing a transportation mode or route. Effectiveness is even more important. As mentioned earlier, the cost of each option must be weighed against its potential benefit, which sometimes can be difficult to quantify economically. For instance, although it may be more expensive to build a commuter rail station in the middle of a dense suburb, that location is the better option because it will serve more residents than a less expensive station built farther from the population center.

Parsons Brinckerhoff uses the Spatial Modeler in ERDAS IMAGINE to compare these options. The Spatial Modeler contains GIS modeling algorithms that allow the user to draw relationships among thematic data layers or maps. Factors in each data layer are weighted on the basis of considerations that may or may not be related to a quantifiable variable such as cost or impact.

For example, in an area such as Northern Virginia, it may actually be less expensive to displace low-income homes to build a new highway than it would be to purchase prime farmland. However, due to the environmental justice issues involved in disproportionately affecting low-income areas, the farmland route may actually be the best alternative. In the model, the low-income homes would be assigned higher negative weight.

As a way of combining quantifiable and nonquantifiable factors in the study, the planners add weight to each option as appropriate and then run the Spatial Modeler to produce an image that displays the overall impact of potential construction in each route segment as positive, neutral, or negative. A significant advantage of using ERDAS IMAGINE software instead of a package only capable of image processing is that the GIS software has tools available for application in the raster environment. This adds tremendous convenience to an investment study in which so many sources of data in both raster and vector formats are employed.

In GIS analyses, Parsons Brinckerhoff utilizes both ERDAS IMAGINE and ARC/INFO. The systems complement each other and allow data to be transferred transparently.

Results and Discussion

Although the Washington, DC project is far from complete, Parsos Brinckerhoff has successfully used the described procedure in numerous transportation corridor siting studies, including the California High Speed Rail Corridor Project for the California Intercity High Speed Rail Commission, the first such project in the United States.

Other major transportation projects for which Parsons Brinckerhoff used ERDAS IMAGINE, ARC/INFO, and satellite imagery include the High Speed Rail Feasibility Study GIS Pilot Study, Los Angeles to San Diego (see Plate 28); the Northeastern Corridor HSR Feasibility Study for the Volpe National Transportation System Center; and the State Route 68 Tier I Environmental Impact Statement in Monterey, California, for the California Deptartment of Transportation (see Plate 29).

Based on years of experience with satellite imagery and GIS, Parsons Brinckerhoff has concluded that without these technologies it would be cost prohibitive to consider all of the viable routes, transportation modes, environmental impacts, and costs for a regional transportation project. With these technologies, the results are more detailed and arrived at more quickly.

Assessing Tourism Potential: From Words to Numbers

João Ribeiro da Costa, Universidade Nova de Lisboa, Portugal

Challenge

UNINOVA, a Portuguese research institute associated with the New University of Lisbon, coordinated and directed a Research and Development project partially funded by the European Union MEDSPA Programme (MEDSPA-COVEPLAM project). The aim of the project is to develop tools for environmental planning in the Mediterranean areas of Europe. These areas are usually economically depressed and have considerable social problems, including aging populations, high illiteracy levels, and chronic unemployment, resulting from lack of economic activity and low levels of technical skills among the population.

Questionnaires to local municipalities showed that tourism was the common positive expectation regarding economic activity throughout the area. The question was simple: is this expectation based on realistic factors? The challenge was to quantify in an objective manner the potential of tourism over a large area in the Mediterranean region of Europe.

Data and Methodology

To conceive an objective measure of touristic potential, the study was structured as a marketing problem. On the one hand there are services to be sold, and, on the other, there are clients buying those services. Services in this particular case are the area's diverse tourist attractions, such as excellent cuisine and wine; picturesque villages and towns, some of which date to the Roman and Arab occupations and maintain their original style; rivers and dams with good potential for water sports and fishing; natural parks with rare species and beautiful landscapes; and archaeological/historical sites. In terms of potential market, there are two major sources of tourists: the metropolitan area of Lisbon (including all the tourists entering Portugal through Lisbon airport), and the Spanish border in Badajoz.

An attempt to combine cuisine-based attractions with historical sites and parks led to the selection of travel time as the common attribute. A place in a beautiful landscape near villages where the tourist can find good food with a nearby park and archaeological sites worth seeing certainly has high touristic potential. Travel times were computed for each point in the area from the nearest village with good food and wine, from the nearest park, and so on. Tourist potential was then defined as a function of travel times, assuming that the willingness to visit each of the categories based on time would not be equal for all. Transfer functions were developed for each category, transforming time into a tourist potential index, ranging from 0 to 100. Five indices were combined into a single final value.

The selected study area was the Guadiana river basin in Portugal, an 11,000 sq km area in the southern part of Portugal. The Guadiana river is the third largest river in the Iberian Peninsula, draining a total area of 81,000 sq km (31,274 sq mi). The climate in the area is Mediterranean, with average rainfall ranging from 400 to 500mm per annum, average temperatures of 10° to 12° C in winter and 24° to 28° C in summer. Most of the area is comprised of agriculture and forest areas. Population densities at an average 20 inhabitants/sq km are among the lowest in Europe. The lowest population density occurs in Monforte at 9 inhabitants/sq km. (The mean density for Portugal is 107 inhabitants/sq km, and for the European Union, 146 inhabitants/sq km.)

The Guadiana Information System (G4) is one of the end products of the MEDSPA-COVEPLAM project. This system, developed using ARC/INFO, ArcView 2, and ORACLE, stores all information compiled during the two years of project development.

G4 includes various types of geographic and point measurement data. Base geographic themes, such as river network, topography, towns and villages, administrative boundaries at various levels, and roads were digitized from existing base maps. A scale of 1:100,000 was selected. Using the base map and GIS capabilities, new maps were produced for slope and aspect, rainfall over the entire area, and other themes. Vegetation and land use were obtained from LANDSAT TM image processing. Socio-economic data, mostly counts, are related to administrative entities at various levels. ArcView 2 was used to combine the data stored in ORACLE with the graphic representation of the entities (mainly polygons).

Results and Discussion

Having defined the goals and the basic methodology, it was necessary to select either a vector or raster approach. Limiting the computation of travel times to the road network using network functions was considered highly restrictive in an area with a low road density network. Given this feature of the area and the objective of computing travel time from every point in the area to the nearest site of interest, the vector approach was not feasible.

Using GRID's Map Algebra capabilities, to compute friction and cost surfaces appeared to be a better solution. These surfaces allow each pixel in the grid to be defined in terms of the time it takes to traverse it. Once the surface friction is computed, estimating the time of travel to any set of pixels (the cost function) is a simple raster operation.

The surface friction was computed based on five different themes: road network, land use, slope, river network, and nature conservation areas. The vector maps were rasterized using a 100m cell size. Themes were then divided into two classes: those which are translated directly into time of travel through the cell and those which increase the time of travel. The first class includes road network and land use. The road network was reclassified by converting average velocities along the road to the time it would take to traverse a 100m cell at that velocity. Land use was reclassified into categories (forest, agriculture, pasture, and urban), and time of travel through each land use type was estimated based on field experience. Roads took precedence over all land uses except urban.

Slope was used as an aggravating factor. The slope map was reclassified into four classes and combined with the map resulting from merging the previous two themes. First order rivers were considered as barriers, forcing the analysis to use the nearest bridges to pass from one bank to the other. Whenever the area crossed a nature conservation area, speed and access limitations were applied that further increase the travel time.

The final result of the process was a friction map showing the time it would take to cross each 100x100 sq m cell in the 11,000 sq km area. This map is the basis for computing times from any cell to objective cells. The first attempts to compute the cost maps with the 100m resolution showed that it would take days of computing per map. Consequently, the map was generalized to 1,000m raster resolution.

The next step was to compute the cost surface for each theme, or the time it would take to travel to the nearest feature of interest, including cuisine and village, body of water, archaeological or historical site, and nature conservation area. The villages with good locations for cuisine and wine were rasterized, and the resulting map used as the target map to compute time from each cell to the nearest cell with these characteristics. Plate 30 presents the results for the villages and cuisine feature. This operation was repeated for each of the other themes; bodies of water with potential for fishing and water sports were rasterized and used as targets, as were the nature conservation areas and archaeological and historical sites.

At this point there were five cost maps with travel times. Each of the maps was converted to a tourist potential index using special step functions. An example appears in the following illustration. These functions aim at translating the tendency of a tourist to visit a given place depending on the time it takes to get there. Longer drives are more likely to be acceptable if the destination is a nature conservation area rather than a place for dinner or having a glass of wine. Using GRID's Map Algebra, the five maps were transformed into maps of potential tourist value, a dimensionless variable varying between 0 and 100 percent.

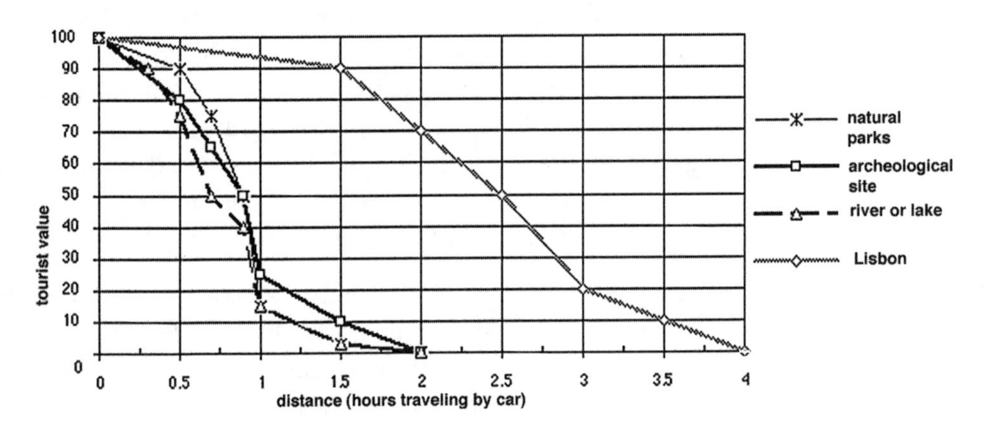

Tourist potential index using special step functions.

The final tourist potential was computed as the weighted average of the five maps. This means that different people may value each of the themes differently and obtain slightly different final results. Plate 31 presents the areas with tourist potential of over 50 percent. It is evident that local authorities are correct, and that the area has considerable potential for various forms of tourism.

Depending on the type of tourism and region, the selected themes and functions to compute tourist value indices can be easily changed. The approach, however, is general and allows the analysis to move from words to numbers.

Georgia Tax Pilot Study

D. Kasouf, Pacific Meridian Resources

Challenge

When the Georgia state legislature changed the method of taxing timber landowners in 1992, they solved one problem. But state and county officials, along with organizations such as the Georgia Association of Assessing Officials (GAAO) and the Georgia Forestry Association, soon suspected that they were facing another problem. This time, however, instead of making another change in the tax laws, they chose to explore a high technology solution.

When Georgia state officials changed the method of taxing timber landowners, they replaced an ad valorum tax with a yield tax. Under the ad valorum tax method, as trees grew in height, they grew in value. And as the trees grew in value, the owner's tax bill increased accordingly. This encouraged some timber landowners to harvest trees prematurely, thereby reducing their tax obligation. The new timber tax law addressed this problem. Under the new legislation, a timber landowner's trees are taxed when they are harvested. The landowners are responsible for reporting harvest yields to county tax officials. Verification of voluntary reporting of timber harvests is the responsibility of each county's tax assessor.

In many Georgia counties, particularly in the southern region of the state, the number of timber harvests, both large and small, can easily exceed a dozen a week. If even a small portion of harvest activity goes unreported, the lost revenue to the counties could be substantial. The problem many counties face is a shortage of resources to verify the accuracy of the harvest reports and to uncover unreported harvests.

Georgia state officials began exploring the feasibility of using high tech methods to monitor timber harvests on an annual basis. In Spring 1995, Pacific Meridian Resources was requested to set up a satellite imagery pilot project in the southern region of the state.

Data and Methodology

Pacific Meridian Resources designed a pilot project to explore the use of vegetative change maps based on satellite imagery to monitor timber harvests. The change detection technique was initially used by the military and has proven to be extremely useful in the private sector.

EOSAT and the government of India agreed to provide IRS 1-B satellite data for the pilot project, while the National Aeronautics and Space Administration's Commercial Application Program (EOCAP) provided matching funds.

Pacific Meridian Resources obtained two IRS images of part of southern Georgia where a large amount of timber is produced. These raster data were captured on two days roughly one year apart in May 1994 and April 1995. In order to focus the change detection effort on timber harvest, Pacific Meridian created a vegetative index for each image. By measuring the ratio of near infrared reflectance to red reflectance, a vegetative index provides a visual indication of different types of ground cover.

The two images were registered to produce an image that would show areas where timber harvest had occurred. The potential clearcut areas were indicated by changes in reflective value and were highlighted in magenta on the change detection image. Large-scale plots of the change image were produced for the two counties chosen for field investigation. These plots showed large areas of clearcut timber land in southern Wayne County and northern Brantley County.

To assist county tax assessors in locating potential timber harvest areas, Pacific Meridian Resources overlaid roads and hydrography from USGS 1:100,000 scale Digital Line Graphs (DLG) onto the change image. The resulting plots were provided to each county's tax assessor for review.

Assessors from Brantley and Wayne counties compared the information with county tax records of reported timber harvests for the 1994-1995 time period. A pilot project field team comprised of state and county officials, representatives from the GAAO and the Georgia Forestry Association, and from Pacific Meridian Resources, spent two days in the field verifiying the change detection image.

Results and Discussion

The results for Wayne County were impressive. Hundreds of sites were identified as potential timber harvest areas. The pilot project field team spot checked 11 of the sites, confirming that all 11 were, indeed, sites where timber had been harvested. Of the 11 harvest sites investigated, three were confirmed as unreported clearcut areas. The three unreported parcels—the smallest of which measured 15x70 yards—represented over $2,000 in uncollected tax revenue for Wayne County. The Wayne County tax assessor stated that the additional taxes collected as a result of finding those three sites alone would have funded his county's participation in an annual statewide timber harvest verification project.

The pilot participants were pleasantly surprised by the high quality of the results from the IRS 1-B satellite data. Normally, they would use higher resolution images, but they found that the image data results were adequate even with the use of lower resolution data from the IRS satellite.

Benefits

Establishing an annual statewide timber harvest verification project has now become a priority for the Georgia Tax Pilot project participants. Based on the success of the pilot project, the current chairperson of the Georgia state House of Representatives Ways and Means Committee, along with two additional state representatives, sponsored a bill in the Georgia General Assembly that would give the state tax commissioner the authority to "actively seek out technological advancements and systems that will improve the uniformity, fairness and efficiency of property valuations and assessments."

Local governments have traditionally been slow to recognize the advantages of spatial data technology because the costs are perceived as too high. However, the rapidly declining costs of such technology, combined with increasing data accessibility, have put it within reach of many local governments. The Georgia Tax Pilot project clearly demonstrated to state and local officials that satellite digital data and GIS applications are economically feasible. Pilot participants agreed that spatial data technology was economical, and served as a means of increasing tax revenue for the state without raising taxes. Through this approach, local governments can detect and collect previously undetected and uncollected tax revenue while ensuring that all timber landowners are paying their fair share.

Acknowledgments

The author wishes to thank The Georgia Association of Assessing Officials, The Georgia Forestry Association, Earth Observation Satellite Company (EOSAT), and the National Aeronautics and Space Administration (NASA).

DIVERSE MODELING SCENARIOS

Editing and Updating Vector GIS Layers Using Remotely Sensed Data

Douglas Stow and Dong Mei Chen, Department of Geography, San Diego State University
Robert Parrott and Sue Carnevale, San Diego Association of Governments

Challenge

This case study details procedures for using digital remotely sensed data to edit and update land use and land cover and other polygonal thematic map layers in vector GISs. Digital imagery are useful for providing spatial analysis with sufficient information to delineate and identify thematic map polygons. Integrated geographic information systems enable processing, displaying, and analyzing digital image data, as well as geographic data that are spatially encoded in both raster and vector formats. Once image data are referenced to geographic coordinates they may be used to revise existing layers.

Data and Methodology

Vector Editing Using Raster Imagery

The most common modern approach to revising vector GIS layers is to interpret digital imagery on a display monitor and encode revised vectors through a process called "heads-up" (on-screen) digitizing. On-screen digitizing may be achieved with ARC/INFO using the Image Integrator or with ERDAS using the Vector Module. The key is that digital image data are georeferenced and usually more current, such that the location (or polygon boundaries) on the image provides the basis for digitizing and geographically encoding the GIS layer to represent more accurate and up-to-date conditions. In addition to digital images, existing GIS data layers may be exploited to facilitate the interpretation and category labeling processes. More automated solutions may be achieved via classification and vectorization routines of raster maps. Automated approaches are most appropriate for revising data layers of limited complexity, which cover very large geographic areas.

Some locational error will occur in editing digital polygon boundaries of thematic GIS layers. The amount of error depends on many conditions such as accuracy of the source data, interpretations of the data, and automation procedures used to generate the vector layer. Absolute locational errors in boundaries are generally not problematic for most GIS applications, unless errors are extreme or precise boundaries must be known for legal purposes. More significant influences on the absolute error are the relative boundary discrepancies between multiple GIS layers, such as misregistration, which may pose a problem for GIS models based on the multilayer approach.

By providing rectified and georeferenced digital images, map-like visual depictions representing locations and relationships of surface features enable boundaries of many polygonal thematic layers to be corrected. They certainly allow multiple vector layers to be registered to a common source. Appropriate image sources include digital orthophotographs (from scanned air photos or digital camera images) and georeferenced and terrain corrected images from airborne or satellite sensors like SPOT HRV.

Prior to attempting to improve the accuracy of polygon boundaries, evaluate the necessary locational accuracy of the image required to achieve accurate co-registration of data layers. The assumption made when using digital images for boundary editing is that the locational accuracy of the image should be more accurate in most cases than the locational accuracy of the existing GIS layers. By assessing the accuracy of georeferenced and orthorectified SPOT Panchromatic (Pan) data, locational errors were characteristically found to be 10m or less. Polygon layers in the ARC/INFO GIS database of the San Diego Association of Governments (SANDAG) contained errors in boundary locations ranging from 25 to greater than 100m. These results led to the conclusion that the geometrically processed SPOT Pan data could serve as an excellent base for (1) correcting the locational accuracy of SANDAG's existing vector coded land use layer, and (2) registering multiple layers in a regional scale GIS.

With the use of georeferenced image data, a number of approaches can be applied to correct the location of polygon boundaries and attributes in a vector GIS layer. These procedures may be applied to correct point and curvilinear data such as well locations and transportation features. One approach involves rubbersheeting the polygon layer using standard vector GIS functions like the ARC/INFO ADJUST command in combination with ground control points (GCPs). For example, a SPOT Pan image subset is displayed as an image on a monitor with land use corresponding to the same area displayed as an overlay. GCPs are selected based on features that can be identified in the image in the corresponding GIS layer, and whose location can be accurately defined. The location of GCPs provides the basis for a polynomial transformation registering the GIS layer to raster image. Several iterations are usually required to achieve a reasonable fit. Some attribute label points of the uncorrected layer will fall outside of the adjusted polygon boundary, and new points should be added.

The other major approach is to adjust specific mislocated arcs of a polygon relative to the image within a distance greater than a prespecified error tolerance. As with the first approach, a subset image was displayed and the land use layer was overlaid and edited using the image as reference. However, with the second approach, individual features of the land use layer are adjusted rather than the entire layer. Adjustments can be made in two ways. Vector boundary features can be dragged from their current location to align with the image base. The other choice is to redraw an arc or an entire polygon, and then dissolve the old feature. The most efficient and accurate approach is generally one that involves interactively pointing to the vertices of misaligned arcs and dragging them to the location of the corresponding imaged feature.

Vector Updating Using Raster Imagery

Several image processing and interpretation strategies were developed for updating thematic polygon layers in vector format: (1) direct editing of vectors overlaid on enhanced geocoded images; (2) automated detection and editing of land use change boundaries with visual identification of land use categories; and (3) automated editing of boundaries and categorical attributes for land use change parcels. The first approach, which requires more visual interpretation and interactive digitizing, is generally required for updating most polygonal thematic layers due to their inherent spatial and categorical complexity. An example of this approach is described below and illustrated in Plates 32 and 33.

A polygonal land use layer in ARC/INFO GIS format can be updated to reflect more current conditions using a two-phase procedure that exploits recently acquired, georeferenced satellite image data and the graphic overlay capabilities of the ERDAS and ARC/INFO image integrator module. In the first phase, change detection analysis is performed based on multidate SPOT Pan image data and the boundaries of new land use polygons are digitized. The categories of the new land use polygons are identified from merged SPOT Pan/Landsat TM subsets and the categories are coded as INFO database attributes in the second phase.

Phase one of the updating methodology starts by displaying screen-size subsets of the dated (red planes) and recent Pan (blue and green planes) data into the image memory, and corresponding portions of the dated land use file in the graphics plane of the display. Land use changes are enhanced in ERDAS by (1) displaying both Pan images as a multidate color overlay image, and (2) flickering back and forth between the two Pan images. More automated change detection procedures are useful as enhancements, but may distort the boundaries of land use change features. Once an area of apparent change is confirmed to be a land use change, the boundary of the new land use parcel is interactively digitized in ARC/INFO using the recent Pan data as the image base. The interpreter proceeds systematically through the subset by visually traversing from side to side in search of land use changes. Polygons of urban-type land uses in the dated layer are shaded using ARC/INFO commands to assist the interpreter in focusing on other areas that are more likely to have incurred a land use change. Label points are generated for all new polygons. An ARC/INFO Macro Language (AML) routine may be created to streamline the polygon digitizing and label generation procedures. In this phase, the land use attribute is only coded when a clear identification could be made from the recent satellite image.

Phase two, the land use identification phase, is started by simultaneously displaying merged SPOT Pan XS data as false-color infrared composite subimages and the newly delineated land use polygons in the graphics plane. The merging of high spatial resolution Pan data and XS data with their multispectral discrimination attributes combines detail and color enhancement in a single image. The interpreter again searches in a systematic fashion across the color display monitor to identify new land use polygons. In doing so, new land use polygons may be detected that were missed in the first pass. For each new polygon, a land use category is identified by interpreting the merged Pan XS image. Air photos, street and thematic maps, and other GIS layers, such as ownership and general plans, may also be used to assist the classification of land use types. The interpreted land use category is then added as an INFO attribute associated with a particular label point that was added in phase one. An AML program may be written to facilitate quick, interactive coding of new land use codes.

The final step of the land use updating methodology is to perform standard topological building and editing functions in ARC/INFO. Once quality assurance is completed for the revised land use layer, two types of products may be derived. Tabular statistics of land use change amounts can be generated by category and geographic subarea. Land use maps portraying updated land use conditions or areas and types of land use changes may be generated with computer printing devices.

Other data contained in a GIS database can aid in the updating of land use or other dynamic layers. These ancillary, non-image sources are often acquired for purposes and at spatial and temporal scales that differ from those for land use data. To use these ancillary data reliably in the update process, the relationship of the data to land use must be determined. Also needed are procedures to incorporate the ancillary data efficiently and consistently during visual interpretation of the image data.

Modeling Vegetation Distribution in Mountainous Terrain

Diego Fabián Lozano García and René G. González Murguía, Centro de Calidad Ambiental, Instituto Tecnológico y de Estudios Superiores de Monterrey, México

Challenge

The Cumbres de Monterrey National Park (the biggest park in the Mexican system) has been subjected to major pressure by land developers, agricultural activities, and illegal forest clearcuts. In the early 1990s concern over the destruction of natural resources in the park initiated a campaign to achieve two objectives: (1) redefine the limits of the park because almost 80% of the city of Monterrey (over 3 million inhabitants) lies within the limits of the park; and (2) define new management strategies for the park's natural resources.

In 1993, the Monterrey Institute of Technology initiated a GIS to address the above objectives as part of a joint effort by the federal and state governments. Early in the project, it became evident that the 1970s maps of the area published by INEGI, the Mexican national mapping agency, were obsolete. Thus, the first task of the GIS managers was to develop a methodology to map land cover types in mountainous terrain by integrating physical factors that determine vegetation distribution.

Data and Methodology

The study area is part of the Cumbres de Monterrey National Park, located south of Monterrey. Land use classes described by INEGI are listed in the table below.

Abies-Pseudotsuga forest	Desert scrub
Pine forest	Microphyllous desert scrub
P. Cembroides forest	Grasslands
Oak-pine forest	Agriculture
Oak forest	Bare soils
Chaparral	Urban areas
Submontane scrub	Secondary forest

A Landsat Thematic Mapper (TM) image (Plate 34) was rectified and registered to a UTM grid. The study area was extracted using the polygon of the Park. In addition, a DEM and various thematic maps were digitized and incorporated into the GIS. All image processing tasks were performed using a network of IBM RS/6000 workstations running ERDAS IMAGINE and ARC/INFO.

The following table presents a few of many variables that determine vegetation distribution. Because the range of values for these variables was often contradictory, the existing vegetation and land use maps were used to digitally compare with other GIS coverages. Results were reported as area estimates by cover type for each of the following variables: elevation, aspect, soils, geology, temperature, and precipitation. The calculations were transformed into total area percentages and plotted onto individual graphs, first by cover type, separating the forest classes, and then in combination with elevation, aspects, and area.

Ecological Conditions Controlling Vegetation Distribution

Vegetation types	Elevation min-max (m)	Mean annual precipitation (mm)	Mean annual temperature (°C)	Climate (Köppen)	Aspect
Pseudotsuga forest	2500–3500	1000–1500			North
Abies forest	2800–3000	above 1000			North
Pine-oak forest	1500–3000	600–1500	10–16	Cw, Cf	
P.cembroides forest	1700–2600	400–700	13–16	Cw, BSk	West
Oak forest	800–1900	600–1200	10–26	Aw, Am, BS	North
Semi-deciduous	800–1500				
Semi-evergreen	1500–2800				
Chaparral	1700–1900	350–750		BS, Cw	Southeast
Montane scrub	600–1700	450–1900		Aw, Am, BS	Northeast
Desert scrub		200–500		BS	
Microphyllous desert scrub		150–500		BS	
Riverine forest	0–2800			Am, Aw	

The first classification was carried out using image data alone in an approach that led to 184 training fields representing 14 classes. The second classification was performed on the stratified image using a standard maximum likelihood algorithm for individual subscenes that were later merged into a single image. Final evaluation of both classifications was performed using 178 test fields (see following table).

Classification Performance of Single and Stratified Landsat Images

Cover types	# of test pixels	Single image (%)	Stratified image (%)
Abies forest	2064	81.4	89.8
Pine forest	2036	26.2	89.6
P. Cembroides forest	1696	0	88.2
Pine-oak forest	4782	74.7	85.5
Oak forest	5429	99.9	99.9
Chaparral	1059	24.3	86.7
Submontane scrub	1772	99.5	99.5
Grasslands	161	100	99.4
Desert scrub	586	84.1	92.0
Microphyllous desert scrub	740	48.8	60.5
Bare soils	896	100	100
Urban areas	734	15.7	97.3
Agriculture	32	0	84.4
Secondary forest	106	0	100

Results and Conclusions

The following four figures show relationships between the main vegetation types and the two more significant variables that control their elevation and aspect. The distribution of forest classes is controlled by the combined effect of elevation and aspect being dominant above the 1,500m mark on north- to east-facing slopes. Chaparral vegetation is more common in areas between 1,500m and 3,500m on slopes with southern aspects. Desert scrub is restricted to an elevation range between 700 and 2,500m, and is dominant on southern slopes. Submontane scrub occupies areas between 600 and 1,500m, with a slight preference for northern slopes.

The latter two figures illustrate the effect of elevation and slope on three forest classes. Pine forest is distributed between 1,500 and 2,800m, with a slight preference for the north-facing slopes; oak forest appears between 600 and 2,400m on east- and west-facing slopes; and *Abies* forest appears in small areas around 3,000m.

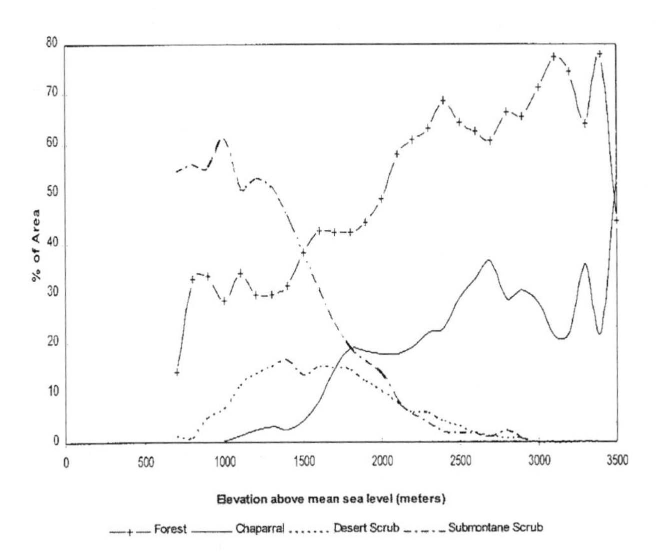

Elevation ranges of main cover types.

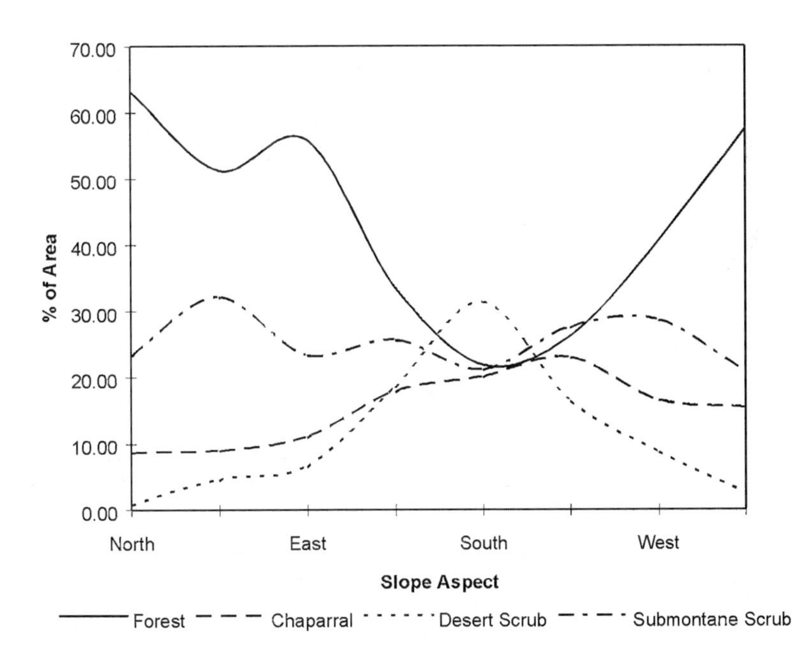

Aspect ranges of main cover types.

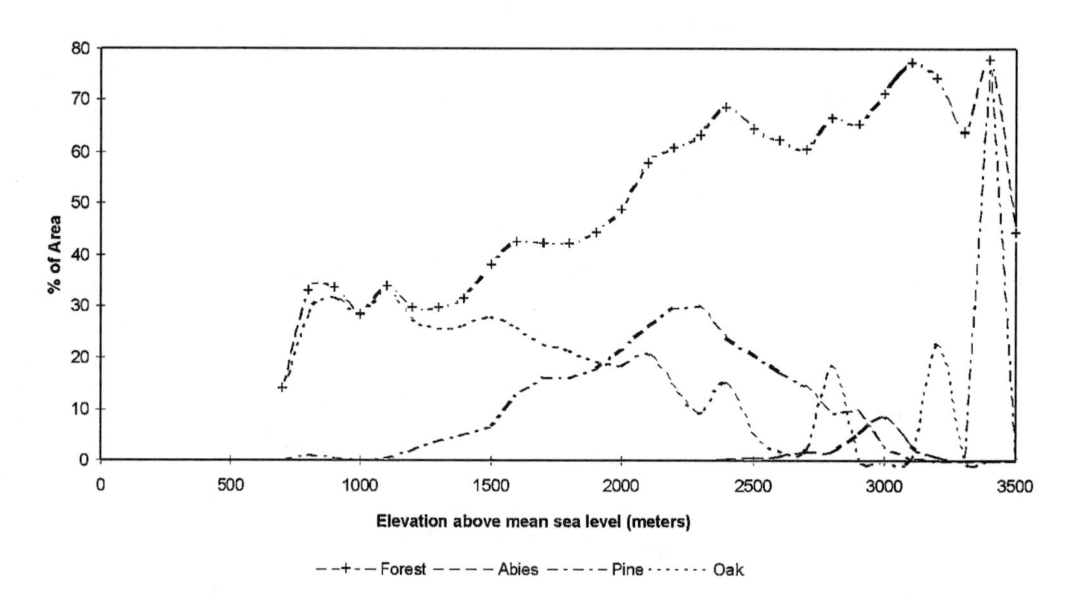

Elevation ranges of forest cover types.

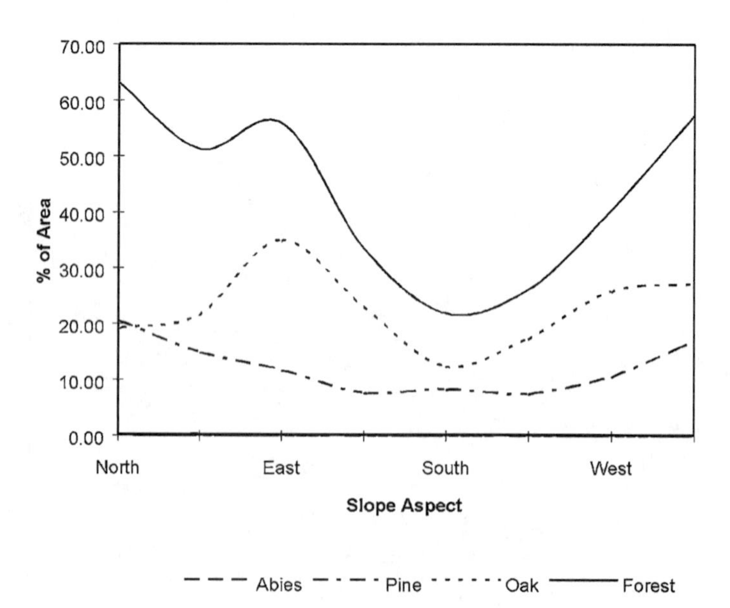

Aspect ranges of forest cover types.

Based on the main environmental factors that affect vegetation distribution, the Landsat TM image was subdivided into three subscenes: areas below 1,300m; areas above 1,300m with northwest-, north-, northeast-, and east-facing slopes; and areas above 1,300m with southeast-, south-, southwest-, and west-facing slopes, and flat areas. Training fields for the three subscenes provided a more consistent set of training statistics and made it possible to identify new cover types: *P. Cembroides* forest (3% of total area), agriculture (4%), and secondary forest (0.1%).

For the classification of individual subscenes (Plate 35), all classes approached 90% accuracy. The overall accuracy of the stratified classification was 91%.

The use of environmentally based criteria to "subdivide" a remotely sensed image in mountainous terrains proved to be an adequate approach. The final stratified classification provided high accuracy results, as well as the differentiation of some informational classes that were not easily separable by classifying the Landsat image alone. The analytical procedure employed to derive the values of the environmental variables (that defined the three subscenes) had the advantage of incorporating existing digital maps at the beginning of the process. With this information, it was possible to carry out extensive analysis of the data and eliminate decisions based on judgment or "expert" opinion.

An Urban Mask Raster Image for Vector Street Files

Jeffrey T. Morisette, Heather Cheshire, and Casson Stallings, Computer Graphic Center, North Carolina State University

Siamak Khorram, Computer Graphic Center, North Carolina State University and International Space University, Strasbourg, France

Challenge

Among the major requirements for classifying land use/ land cover on the North Carolina coastal plain is distinguishing high density and low density developed land. Standard image classification techniques reliably determine large scale developed areas such as industrial sites, shopping malls, airports, and hospitals. Difficulty often arises, however, in determining moderate to low density developed areas. To address this problem a mask has been developed using TIGER road networks and a raster based algorithm in ARC/INFO to define developed areas a priori. The developed areas were cut from the original image using ERDAS Imagine software. This cutting resulted in a separation of the image into areas likely to be developed and areas not likely to be developed. Standard classification techniques were then applied to the two images.

Consider, for example, the difficulty of using low resolution imagery alone to distinguish an undeveloped oak forest from an old neighborhood with large oak trees lining the streets, or separating paved areas from bare soil and sand. It is rather easy for a human interpreter to evaluate the pattern of a line of trees adjacent to squares of grassy areas and classify it as residential. It is also simple for a person to see a strip or square of bright area and know that it is paved. These seemingly simple human interpretations involve complex pattern recognition that goes beyond spectral classification techniques. Common computer-based classifications label most pixels in a neighborhood as vegetated and many paved areas as bare soil or sand. It is true that the individual pixels within a residential or low level density development area are indeed tree- or grass-covered and some of the bright areas may be bare soil. However, if the area is even somewhat developed, most users prefer that it be classified as residential or as a low intensity development; or at least they would like such a classification option.

Data and Methodologies

The goal was to create an objective and reproducible method for separating an image into regions most likely to be developed and regions most likely to be undeveloped. In classifying these two regions, the undeveloped area would be classified using traditional methods and the developed area would use standard methods augmented by each pixel's attribute as lying "within a developed area." For example, an oak pixel in the developed image would be labeled as "oak-within-a-developed-area." This type of labeling retains a maximum amount of information for each pixel allowing the end user freedom to treat developed areas either as residential areas or as vegetated surfaces. The initial separation of the image also helps maintain objective interpretation of confused or mixed pixels. Pixels in the developed image which contain some vegetation and some bare surface are considered to be low density developments. The same type of pixels in the undeveloped areas are considered to be mixed, bare soil/vegetation pixels.

The ambiguity of using the image information alone led to a search for ancillary data that could help predetermine the likelihood that a pixel lies within a developed area. Using city limits alone is unacceptable in eastern North Carolina because there are many undeveloped areas within city limits. Tax maps and zoning maps were not available in digital form for most of the study area, so vector based TIGER data describing census tracts, county boundaries, and street networks were obtained. The challenge was to convert the street files into a binary raster image indicating developed and undeveloped areas.

ARC/INFO's GRID module was used to create the urban mask. The next step involved importing files into ERDAS IMAGINE to perform the image processing. Initially, the roads were imported into ARC/INFO for each county. The road network was then rasterized, resulting in cell values of 1 (containing a road) or 0 (not having a road). The origin and cell size of the satellite imagery were used to ensure compatible data sets. A "grow and shrink" or "dilation and erosion" algorithm was implemented in GRID. The first step was to grow the roads by five cells. The following figure is an example of this process. The original roads are shown in black, and areas added to the urban mask after the growth function are shown in gray.

Expand by five. Results of expanding the original street grid using the following ARC/INFO GRID command: newgrid = expand (gridname, 5, list, 1).

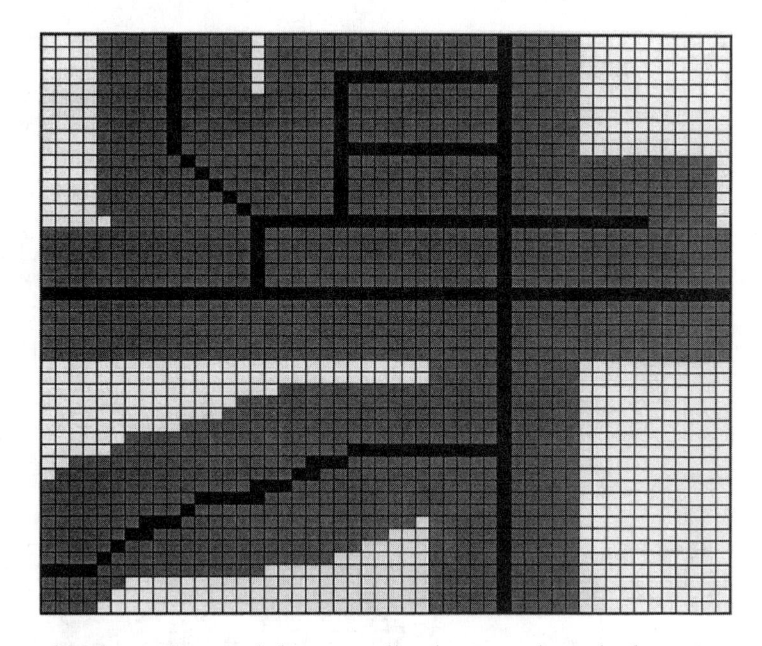

Shrinking the mask by six cells eliminated roads that were originally solitary. The resulting urban areas were then grown by three cells. This expanded the mask to include both sides of the urban roads. After each county was examined, it was joined into one large urban mask for the entire area. This full grid was imported into ERDAS as an image file.

Shrink by six.
Result of shrinking the expanded road network (previous figure) using the following command: newgrid = shrink (gridname, 6, list, 1).

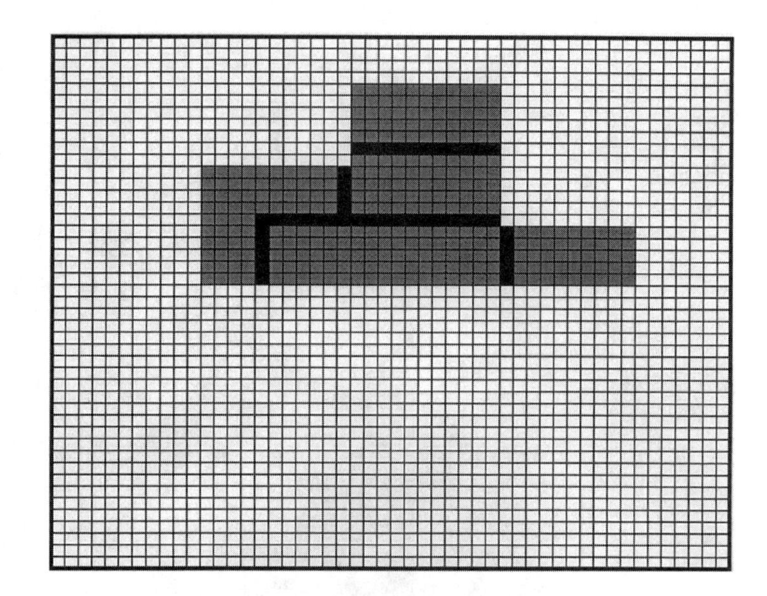

Expand by three.
The final expansion result using the following command: newgrid = expand (gridname, 3, list, 1).

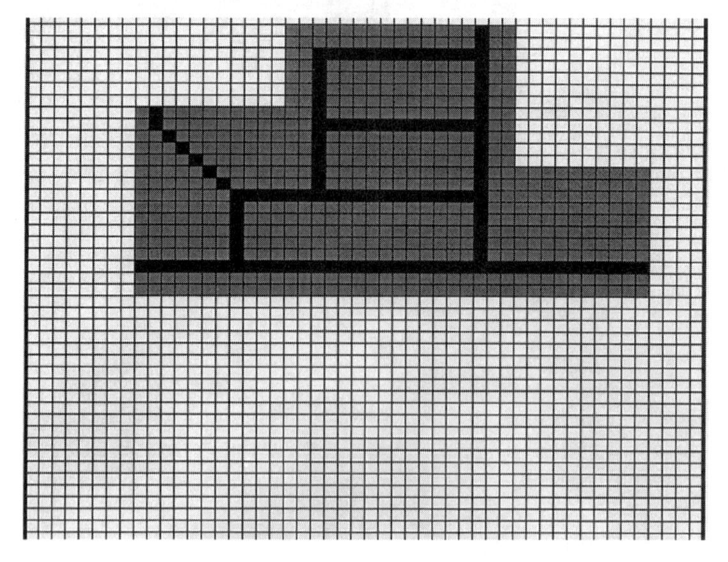

Results and Conclusions

The grow and shrink algorithm essentially erased solitary streets and converted more closely aligned streets into solid areas. The number of pixels by which the street expanded, as well as the number of pixels that were shrunk can be determined by the analysts, giving the user the freedom to choose how tight a street network has to be before it is labeled as developed.

Plate 36 shows an example of the urban mask over a satellite image. In the upper left corner of the image one can see the discrepancy between the location of some of the roads. These locational errors tend to be nonuniform, so geometric corrections would not help. However, the occasional misplacement of a road by eliminating solitary streets and grouping streets together into a common area seems to pose no serious problem.

One might note as well that some areas in the upper to middle right side of the image, which are apparently developed, are not contained in the urban mask. This provides an example of why the urban mask needs to be considered as a probabilistic mask indicating areas likely to be developed. The image of zeros and ones was used to mask the original image. It was done using the ERDAS commands Mask Function; one mask was applied to mask out the developed area, and another to mask out the undeveloped area. This resulted in the desired "a priori" separation of the image into developed and undeveloped areas.

Note that additional functions can be implemented by utilizing the TIGER vector data. For example, one can eliminate "holes" in the urban areas smaller than a desired threshold. Another possibility is to include buffers around the solitary roads if they are near large urban areas.

Benefits

The method seems to meet the goal of providing an objective way to label areas that are most likely developed. Using AMLs to produce the urban area mask allows one to keep explicit records on how the mask was constructed. The method itself is completely reproducible. TIGER data are available to users and the iterative expand-shrink algorithm can be adjusted to include larger or smaller areas into the developed area mask. The developed area mask does not guarantee that those areas are developed, but it helps provide additional information for areas typically difficult to classify with image data alone.

Combining Polygon and Raster Data in Cartography

Anthony D. Renaud, Maricopa County Department of Transportation, Phoenix, Arizona

Challenge

For years the Maricopa County Department of Transportation has produced an annual, general purpose highway map. The hand-drawn composition, with colored regions representing incorporated cities and some land use, concentrated less on detail than on covering a wide variety of subject matter. Terrain was represented by showing mountainous areas as primitive, pen-drawn "hairballs." When ARC/INFO GIS was chosen to reproduce these pen and ink mylars, terrain illustration was destined to be improved.

Data and Methodology

Digital elevation models (DEMs) were obtained, and, through the use of ESRI's GRID package, a suitable grayscale hillshade image was created. After some filtering to eliminate tile seams, hillshading experiments and some selective resampling yielded a grayscale image that showed the major landmarks and surrounding mountains without clutter in the flatter, populated areas of Greater Phoenix. The challenge now was to combine the image with all the polygon data without one masking the other.

Grayscale hillshade image created from DEM.

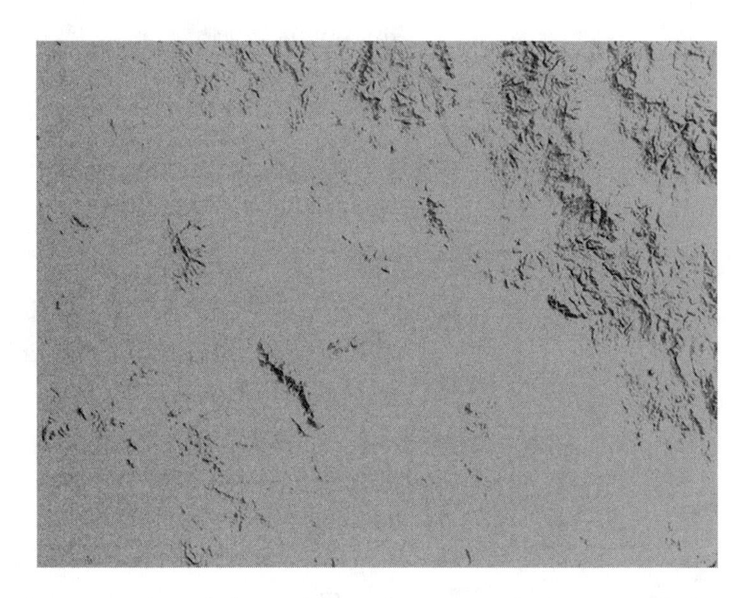

It was assumed that printing the grayscale image with the colored areas could be accomplished at the prepress stage, but a way was needed to display both types of data on the screen and produce variations of them on in-house raster plotters. Some research into how color is perceived and software capabilities eventually led to a very appealing background as versatile as any full-color image. The methodology has since been used to provide a variety of solutions for combining polygon and raster data.

First, the color polygons were converted to raster data via GRID and sampled to the exact resolution of the hillshade image. Information from the DESCRIBE command was used to duplicate the cell size and origin for the POLYGRID command. The color information was converted into a stack of grids using the HSV color model. Along with RGB and CMY(K), HSV is one of the three basic color models where H (hue) is the color described in degrees around a color wheel, S (saturation) is the amount of color from 0% (white) through pastel tints to rich tones (100%), and V (value) is the amount of light reaching the viewer (0 – 100%).

It was thought that by substituting the grayscale image for the original V component, the colors would appear translucent over a hillshaded background. The 0 to 255 range of hillshade grayscale values was compressed to a range of 0 to 100, and a composite was made using this new grid with the hue and saturation grids from the original colors. The results were less than aesthetically pleasing.

The carefully designed color scheme contained typical map colors, or pastels and light grays, and light tints of green and brown. The HSV composite turned the grays to white, the tans and browns to orange, and the pastels to a rich, mutually incompatible gloom. By replacing the original V component, the colors were unexpectedly altered. The data had been combined, but at the expense of design.

The solution lay in keeping the original saturation component and using the grayscale image to adjust it, rather than replace it. The grayscale values were compressed again and floated, this time to a range from 0 to 1.0. This grid was then simply multiplied with the original V component for the colors. Where the hillshade image was white, its cells would be 1 (unity), and multiplication did not change the original V values from the colors. As cells from the hillshade image approached 0, or black, multiplication reduced the V component of the original color proportionately toward black, passing on the hillshade effect.

Results and Conclusions

The desired effect had almost been achieved. The map was composed using this GRIDCOMPOSITE and overlaying the linework and text. The results on the screen were as pleasing as the results from the plotters. The printer, however, could extract only a black and white representation from our supposedly colorful image embedded in the output PostScript file. The solution lay in the color model.

ARCPLOT will not pass color information from an HSV grid composite to a PostScript file. However, it will produce a color image from an RGB composite. The grid stack was converted from HSV to RGB and the printing dilemma was resolved (see Plate 37). Twenty-four bit displays were used to accurately evaluate the aesthetic quality of the color scheme and plots of enlarged areas were identical to the mass-produced version. Since black is achieved through an equal absence of red, green, and blue in the RGB model, an eventual shortcut was developed by creating an RGB stack from the original colors and using the same hillshade "adjustment grid" to adjust each component equally.

Variations of this idea have been used to display polygon data over color images, such as aerial photography. Simple polygon coverages are converted to grids. The nominal areas are given a value of 1, while areas to be highlighted are given a value less than 1, depending on the intensity of the effect desired. This adjustment grid is multiplied with one or more of the original RGB components, and the effect is that of placing a translucent piece of cellophane over the photograph. For example, designated areas can be composed of grid cells whose values are made to be 0.75 and surrounded by cells whose values are 1.0. If this grid is multiplied with the green and blue components, the values of the target cells are reduced to 75% of their original value. When the image is composited, the desired area of the photograph will appear reddish. The results are much more readable than using undependable hatch patterns or a similar method to overlay polygons.

It should be noted that all adjustments thus far have reduced some color model component, asymptotically approaching zero. It is possible to make adjustments by increasing and decreasing a component, but this is problematic. Take the original example of the color polygons and the hillshade grid. Nominal or flat areas of the hillshade may be some value less than white. The grid can be stretched so that these values are unity (1) and the areas that do actually approach white (mountain highlights) are greater than 1. When this grid is multiplied with the value component, the resulting increase in original V value for some areas will produce the effect of lightening the colors of the sloped areas that are highlighted by the sun source. The effect is very dramatic, since some colors may already have a V component of 100 percent such as white; however, the results are inconsistent.

The important thing to remember is that a color scheme requires all three components to represent the true colors, regardless of the color model used. By substituting some unrelated grid for one or more components, the colors may be unpredictably changed. By creating an adjustment grid and using simple multiplication to proportionally reduce (or increase) a component, the new data are added and the original data are still present. In this way, polygon and raster data can be combined in endless varieties.

ASSESS: A System for Selecting Suitable Sites

Simon M. Veitch, National Resource Information Centre, Parkes ACT, Australia
Julie K. Bowyer, formerly of National Resource Information Centre, Parkes ACT, Australia

Challenge

A particular land use can be viewed as having a favorable or unfavorable impact depending on what and where it is. For example, if nearby park land were to be rezoned as a nature reserve, this land use would generally be accepted. If a rubbish dump were proposed for the same site, the attitude would generally be negative.

The impact of a proposed land use is shaped by how it changes the land and how a community perceives its impact. These perceptions vary because attitudes and circumstances change. For example, if a rubbish dump is located well away from a suburb but unusual winds carry odors to the suburb, residents' attitudes may change from initial general acceptance to objection.

The following acronyms summarize negative reactions to land use siting decisions:

- NIMBY - Not In My Back Yard
- BANANA - Build Absolutely Nothing Anywhere Near Anyone
- LULU - Locally Unacceptable Land Use

To make well-informed and acceptable decisions on where to locate land uses, we need to know the impact of the land use and what constitutes a suitable site. We also need to consider people's attitudes before proceeding to trade off between the ideal and the practical. This establishes the need for systems that enable interaction with information and provide the ability to review different scenarios.

A suitable site for a land use is typically defined by a list of selection criteria. The selection criteria refer to suitability characteristics, but relevant spatial information is needed to find the site. If a decision is to be fair and objective, then everyone's backyard should be considered initially, while applying the same selection criteria simultaneously everywhere.

The ASSESS method can be applied to small regions or entire continents with equal rigor. There are limitations imposed by the availability and quality of data sets. ASSESS's transparency and ease of use can assist by highlighting these information gaps and alerting users to the need for more data for certain considerations. ASSESS has been used for applications ranging from radioactive waste disposal, new agricultural and agro-forestry lands, and spacecraft landing sites, to marketing alternate termite deterrents in new housing developments.

As recently as the early 1990s, GIS software packages typically described spatial objects using either vector or raster approaches, but not both. This meant that in the vector environment, mapped areas could be maintained with their structured descriptive databases, but rasters were available only as backdrop images. The 1992 version 6 release of ARC/INFO enabled both grid and vector approaches to be used with attributes, and even provided a means to convert from one to the other while maintaining the attributes.

Before version 6, merging or intersecting many large polygon data sets was restrictive because the computational task quickly became enormous. In addition, where two or more data sets had near but not exactly coincident unit boundaries, the resulting data set contained "slivers" that were meaningless or unresolvable for the quality or scale of the data used. In contrast, a raster based data set with its simple square edges, but slightly decreased precision, could match others perfectly and enable rapid computation for many large databases, while still maintaining the scale of the original data set.

The ability to convert polygons to grids with attributes was crucial to the development of ASSESS as a site selection method. NRIC's continental databases were typically very large and primarily vector-based. Conversion to grids produced major size reductions and enabled the exact spatial matching of many data sets to each other. In this way each grid cell in a database could be directly compared with a corresponding grid cell in any other database. In addition to the descriptions for each cell, a numeric rating could be attached to represent its relative suitability. The rating could be recalculated for any or all data sets and a new outcome revealed.

Data and Methodology

For the radioactive waste repository, site selection criteria were developed by an agency independent of the ASSESS application team. The criteria were then reviewed by the team to identify issues and information needs. For example, the following criteria—"the repository should be located in an area of low rainfall and free from flooding, while possessing good surface drainage and stable geomorphology"—were assessed using the approach below:

- ❒ Identify the issues: (1) low rainfall; (2) no flooding; (3) good surface drainage; and (4) stable geomorphology.
- ❒ Approach professionals in relevant disciplines to advise on data, and data availability and appropriateness for the issues.
- ❒ Apply professional interpretations of the data to attach relative suitability ratings to the data set descriptive attributes.
- ❒ Obtain the data and convert to GIS format if in non-GIS form.
- ❒ Attach numeric ratings to the data representing suitability categories (e.g., rating 1 = suitable, rating 3 = intermediate, and 5 = unsuitable).
- ❒ Load the data set into ASSESS.

Typically, there were more issues for each criterion than GIS-based data. Where information existed as hardcopy maps, the maps were converted to vector GIS either by scanning (and converting the image to line work) or by manual digitizing. The data set was edited and labeled with descriptive attributes in the vector environment and converted to grid cells.

In this case, with a large, country-wide assessment, only data with a 1:5,000,000 scale or better was used. A 5x5 km grid was used as a conservative sampling cell (1mm on the map = 5km on the ground), thereby establishing the minimum modeling resolution. Each grid consisted of approximately 300,000 data cells.

The numerical rating approach converted the descriptive attributes of vector coverages to grid cell values representing relative suitability. These ratings of 1 (suitable) through 5 (unsuitable) were the grid cell values. The following table provides an example of the geology value assignment. The rating approach did not require that every theme had a complete set of ratings. For example, at the country-wide scale, some themes required only an indication of presence or absence of a feature, such as major fault lines. Thus, the use of only two values representing either suitability (1) or unsuitability (5) was appropriate.

Geological type	*Suitability category*	*Rating*
Fine-grained sediments	Suitable	1
Acid igneous, gneiss, medium-grained sediments	Mainly suitable	2
Meta-sediments, coarse-grained sediments, basic volcanics, coarse-grained sediments	Intermediate or indeterminate	3
Ultramafic and mafic intrusives	Mainly unsuitable	4
Limestone, dolomite	Unsuitable	5

The ASSESS interface was developed using AML to assemble a series of functional menus to frame an ARCPLOT display window (Plate 38). The primary purpose of the interface was to provide an interactive set of tools for viewing, modifying, and overlaying environmental suitability grids, so that the user could concentrate on the site selection issues instead of the GIS technology.

The sequence for a user to develop a scenario follows:

1. Click on the button with the name of the data set to display the grid (e.g., Regolith - large menu). A data set menu with slider bars and category descriptions appears as a smaller menu to the right.

2. Preview the suitability ratings for the attribute as appropriate in the small menu. At the bottom of each data set menu are Preview and Apply buttons. Preview takes the values from the slider settings and writes a color shade remap table to display the suitability ratings using traffic light colors, where red is unsuitable and green is suitable. The use of this color psychology renders an immediate understanding of the spatial distribution of suitability.

3. If required, change the preset ratings. The menu lists the attribute groupings adjacent to a slider bar with current suitability ratings. For changes, click-and-drag the slider to assign a new rating value. The slider bar has a zero option that will exclude the attribute group entirely. This recalculates the group to NODATA, which in the ARC/INFO GRID additive overlay (Add function) is propagated to any output grid irrespective of the other values added to it.

4. To invoke the new rating for a scenario, click the Apply button (next to the Preview button). This creates a temporary, reclassified new grid. The Apply button recalculates a new grid using the slider bar

settings to create a reclassification table (Reclass) of the original grid. The reclassified grid is used in any subsequent modeling for the current session.

5. View and modify any other information grids by repeating steps 1 through 3.

6. Weight the data set grids relative to each other. Each data set name button has a weighting slider bar to the right. The slider value ranges from 0 to 5 where 0 will exclude the information layer from the scenario and 5 will weight the layer five times more than a data set with a weight of 1.

7. Overlay the activated (non-zero weighted) grids. In GRID this is done by using the Add and Multiply arithmetic operators. Each of the activated grids is multiplied by its weighting and subsequent values are added together. As a default, ASSESS reallocates the resulting summed distribution of values to five suitability categories using a five-category equal-interval slice. The scenario is then displayed using the traffic light color map.

8. Reset the theme grids to their original versions and repeat steps 1 through 7 to produce an alternative scenario.

ASSESS also has button-activated, on-line documentation. Each information grid has an adjacent information (Info) button which, when activated, pops up text to explain the rationale behind the attribute categorization and a list of relevant selection criteria. There is also a README document under the i button (top left menu bar) to describe the functionality and modeling methodology of the interface. Reading to the right of the i button, the National button allows the user to change to different country-wide site selection applications and the Regional button changes to smaller regional siting studies. The Tools button activates ARC/INFO ARCPLOT and GRID commands such as IDENTIFY, CELLVALUE, MEASURE, RESELECT, LIST, and HISTOGRAM. Many of these functions direct queries to "behind-the-scenes" data sets (raster or vector) such as the following:

- Descriptive grids containing the full attribute descriptions of the original vector data sets.
- Polygon and line (vector drape) coverages.
- Point databases such as population centers, earthquake locations, mineral deposits, and water bores.

Results and Conclusions

For this assessment, continental-scale biophysical and socio-economic information was assembled from various sources. Examples of the information used to address the repository selection criteria are bedrock geology; surface cover geology; proximity to major fault lines; regolith terrain (weathered surficial materials and geomorphic setting); relief and landform; soil properties (permeability, water holding capacity); land use; rivers, streams, lakes, and swamps; hydrogeology (aquifer type, productivity and distribution); groundwater quality; dryness (precipitation/evaporation); present vegetation (growth form and foliage cover); vegetation density; rare or threatened plants; national heritage sites; land ownership; major transport routes (road, rail); population centers; and earthquake risk.

Once assembled, all the continental data sets were combined for a national view of suitability. Many different combinations and treatments of the data were attempted. After reviewing many scenarios, some regions (clusters of cells) repeatedly showed a higher level of relative suitability (lowest sum of rating values). As a result of the continental view, several regions were identified as potentially suitable for the disposal of low-level radioactive waste.

The second phase of the project identified eight regions for more detailed assessment. With a region identified, more detailed data sets could be obtained and assembled to modeling suitability. For each region the grid resolution was 250m. A series of scenarios was generated for each region with the result that some showed potential difficulties on certain criteria. Others confirmed that at this finer scale a suitable site could be located. Considering the regional assessments, the study aims to identify a single region for detailed field survey. The choice of one region will be based primarily on analyses of relative suitabilities, but it will also be sensitive to public commentary on the characteristics of each region as defined by ASSESS.

Benefits

ASSESS is a user-friendly graphical user interface (GUI) developed by the National Resource Information Centre (NRIC). It provides a versatile and powerful decision support tool for siting potentially contentious land uses such as Australia's low-level radioactive waste repository. ASSESS uses the Arc Macro Language (AML) of the ARC/INFO GIS to create a point-and-click user interface for accessing spatial information over areas ranging from an entire country to a small region. The system is fast, flexible, and easy to use, with a strong emphasis on providing on-line documentation to explain the modeling strategy and the rationale behind decisions on assignments of suitability ratings for a particular land use. The user can interactively make changes to the ratings to reflect differing viewpoints on land use suitabilities, and can immediately implement such changes to see the spatial consequences of such decisions.

The essence of ASSESS's speed and responsiveness is its use of the GRID module of ARC/INFO. Because the GRID module is raster based, it has the speed efficiencies of cell-based data and allows the contents of very large data sets to be reinterpreted, assigned new suitability ratings, and compared with dozens of other large data sets within minutes. In contrast, similar approaches using the vector modeling GIS create unduly complex geometries and typically take hours to achieve the same end.

ASSESS is currently designed to meet the requirements of the radioactive waste application, but it also satisfies many clients with other site selection issues. It is also an important educational tool used to train Australian and foreign students in spatial information systems and resource management techniques. ASSESS provides quick and simple demonstrations of the following:

- Availability, types, and limitations of biophysical and socio-economic resource data
- Spatial concepts
- Vector and raster representations of data
- GIS functionalities
- GIS applications
- Simple yet effective overlay modeling and analysis
- On-line documentation
- Menu interfaces
- Decision support and scenarios

The macros for ASSESS are currently customized for use within NRIC, but a generic AML with user assistance to generate third party applications is entirely feasible.

Shaded Relief Images: Visualizing Terrain Without Contour Lines

Jan Benson and Bob Greene, Facility for Information Management, Analysis and Display, Los Alamos National Laboratory

Challenge

Vector contour lines provide a topographic view of an area but can be impeded by slow computer response times with older versions of viewing software. Contour maps, especially those with small contour intervals, can require a long time to draw on the screen due to the density of their data files.

The Los Alamos National Laboratory property outside of Santa Fe, New Mexico, contains 46 sq mi of rugged mesas and canyons. The elevation varies from 1,650 to 2,400m (5,400 to 7,900 ft) above sea level, and 23 major canyons cross the area.

The laboratory is being mapped for the Environmental Restoration Project. The project's objective is to identify and remediate any sites where contaminants may have been released as a result of past operations. A phase of the project involves collecting samples which will be tested for analytes and compounds which may exist on the site.

Accurately mapping the landscape to find the locations of drainage networks is a necessary component in determining the sources and regional extent of contamination.

Data and Methodology

Los Alamos has stored in an ARC/INFO library 0.6m (2 ft) contour maps for the entire Laboratory property and 3m (10 ft) contours for the surrounding environs. These analytical data are archived in the Facility for Information, Management, Analysis, and Display (FIMAD). FIMAD also provides access to base layer data through HP UNIX ArcView.

Accessing these data and displaying them on the screen, however, takes too much time for practical use with older versions of ArcView. For instance, a 30m (100 ft) subset of the 0.6m (2 ft) contour map consumes 135 megabytes of disk space. An alternate means of finding drainages and viewing the contour lines had to be developed.

The general assumption was that the slow redraw problem resulted from too much data to be viewed in a reasonable period of time. The first attempt to resolve the problem was, therefore, to reduce the size of the data coverage.

A subset of the 30m (100 ft) contours was generated using ARC/INFO. The GENERALIZE command uses the Douglas-Peucker algorithm to weed vertices from the selected arcs. Technicians experimented with several distances in the generalize process, but the results were unsatisfactory for several reasons. First, the appearance of the contour lines became too angular. Secondly, with the vertices removed from selected arcs, some contour lines crossed, especially those on steep slopes where contours are located close together. This was unacceptable for the scope of this project. Lastly, and perhaps most importantly, the contour lines still took too long to draw on the screen. The complexity of the contour data was merely reflecting the complexity of the landscape. At this point, project technicians realized that further degrading of the data would serve no purpose. A means other than vector contour maps was needed to view the topographic landscape.

The next idea was to use a raster image as a backdrop to the base data. The reasoning was that if the image size could be made small enough, the software could display the view in a reasonable period of time. The raster image could be derived from a grid data set.

Digital elevation models (DEMs) at 0.3 and 1.2m (1 and 4 ft) had already been created from the Los Alamos contour data. These DEMs were stored in grid format. The procedure used the ARC/INFO HILLSHADE command to create the image from the grid data. A test area measuring approximately 3,636 x 2,424m (12,000 x 8,000 ft) was selected from the 1.2m DEM for the hillshading test.

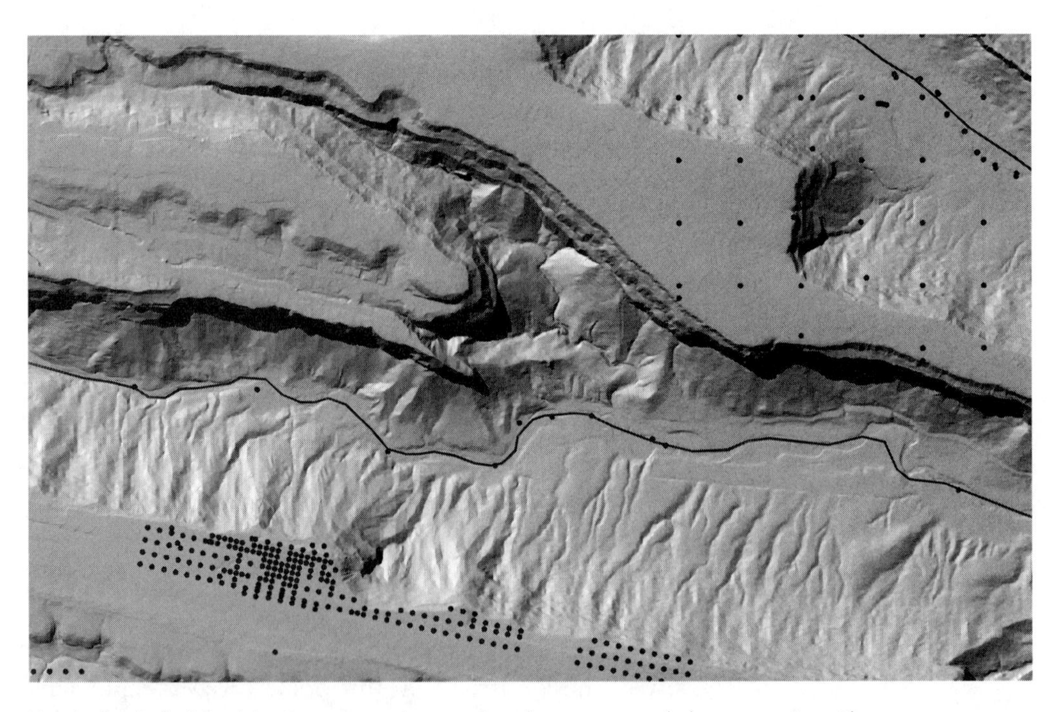

Shaded relief of Pueblo Canyon with sampling locations and drainage, Los Alamos, New Mexico.

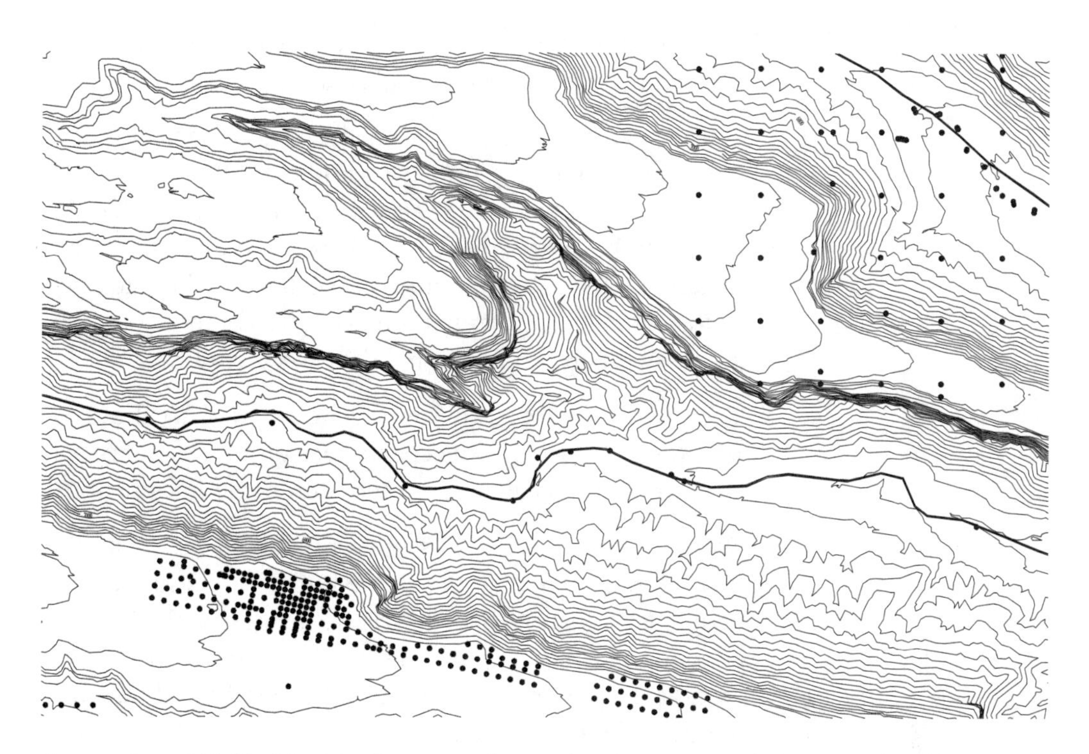

Contour lines (20 ft) of Pueblo Canyon with sampling locations and drainage, Los Alamos, New Mexico.

Hillshade, also referred to as shaded relief, uses elevation data and a hypothetical sun angle to cast a shadow over the DEM landscape. A grayscale image results with topographic relief appropriately highlighted. In this project, the shaded relief grid was created with a sun azimuth of 315° and an altitude of 45°, default parameters which provide extremely realistic relief shading. Grid cells in the shadows were assigned a grayscale value of zero, while the brightest areas were assigned values of 255. The HILLSHADE command also allows the user to specify local illumination angle and shadows.

The GridPaint command in the ARC/INFO grid module was utilized to view the shaded relief grid. The size of the test grid file was just under six megabytes. The results were startling. Canyon edges were well defined with mesa tops and canyon bottoms readily seen.

While creating an image from the grid, technicians also considered stretching the actual grid cell values over a full range of 256 gray shades. However, they also wanted the adjacent sheets to edgematch. To ensure analysis would include the edges of the grid, one cell was added to each of the four sides, except at the boundary of the laboratory property where no additional data were available.

The shaded relief grid was then converted to the TIFF image format, which was chosen because it has an associated word file containing its georeferencing information.

The technicians wrote a script program to automate the procedure for creating the shaded relief grid and the TIFF image. It took about a day to create images for the 71 grids covering the project area. The total disk space consumed by the 71 TIFF images and their associated word files was about 308 megabytes. The 0.6m topographic library consumes about 4.5 gigabytes of disk space.

An image catalog was created in ARC/INFO so that users could access the correct image easily. AVENUE, the programming language of ArcView, was used to create tools to access and display the shaded relief images. ArcView makes it easier for a user to pull up the image and display it quickly. The user can then click on the mouse to choose the desired image(s). The user can specify an area of many images and then change to view a smaller map area.

Results and Discussion

The shaded relief images proved even more valuable than anticipated. Clients appreciated the images because they provided a three-dimensional view of the landscape. The image allowed laypersons to understand topography more easily than they could with a contour map. Moreover, dense contour lines can result in clutter that makes it difficult to see other parts of the map.

The images were extremely sharp, providing excellent resolution of natural features such as mesa and canyon edges. Drainage patterns and networks were clearly visible and ultimately proved more accurate than the vector version, which was based on a 1:100,000 scale digital line graph edited to lie within the 3m contour lines.

Some of the vector data could be enhanced and updated with the new information. Several older features, such as roads and pits, that were missing from the vector data showed up on the images. In addition, some new geophysical details, such as geologic benches and fault traces, were evident in the images. The shaded relief images proved to be excellent base maps for plotting these geologic features.

The shaded relief images did, however, present a few drawbacks. The first concerns the nature of images. If an object of interest is in shadow, there is little that can be done to increase its visibility. Another drawback was that a user needed to consult the vector coverages if an actual elevation value or associated attribute were required.

Benefits

Large data files and associated hardware/software constraints can present obstacles; in this project they were instrumental in forcing experienced spatial data users to reconsider how data are depicted. The authors are now looking at putting data on a CD for PC and laptop computer use. A CD holds about 650 megabytes of data. The shaded relief images fit on a CD, but the contour data do not.

The shaded relief images can also easily be migrated across platforms, and since they consist of 3,636 x 2,424m tiles, the users can view pieces of the landscape without having to load one very large file. As technology changes, our perspectives are challenged to find better ways to depict the land surface.

Building Attribute Tables for Raster GIS Files with ARC/INFO

Zhenkui Ma and Roland L. Redmond, Montana Cooperative Wildlife Research Unit, The University of Montana

Challenge

The inability to set up attribute tables in a raster GIS environment effectively delayed the widespread use of GIS software for managing remotely sensed data and associated attributes. Most GIS packages were developed, at least initially, for use with vector files. GIS routines capable of handling and processing raster data in raster format are relatively new. They not only lack some of the functionality of vector based routines, but they tend to be less widely used. Thus, for a long time, processed images and any other raster data had to be converted to vector format before attributes could be assigned. With the introduction of functional, raster based modules, such as ARC/INFO's GRID, managing image data and their attributes for large geographic areas is finally practical using GIS software.

Data and Methodology

To classify TM imagery for a statewide land cover map of Montana (for GAP Analysis), a procedure was developed using the GRID module in ARC/INFO (version 6.0 and higher) to build attribute tables for raster GIS files. This procedure represents an important breakthrough for people using digital image classification methods to map large areas because it eliminates the vector conversion step that previously was required to build attribute tables. Each continuous area, or *raster polygon* (or *region* in GRID terminology), is treated as a single record in the GIS database. Associated biophysical and/or spectral attributes for each record can be stored more efficiently than in vector format and then be used in the GIS for supervised classifications. Other benefits to raster GIS processing include file sizes that can be roughly five times smaller than in vector format and file structures that both increase computational efficiency and lend themselves to standard GIS modeling techniques.

The six steps below describe building an attribute table for raster GIS files. The first step is performed using standard image processing software such as ERDAS IMAGINE. The remaining steps are carried out with ARC/INFO.

1. Classify digital imagery (output file named *CLASSIFY.GIS*).

2. Aggregate small regions to a user-specified minimum mapping unit (MMU). A rule and object based merging algorithm can be used to remove small pixel groups based on attribute similarity. Commercially available options include the ERDAS commands Clump or Sieve or ARC functions such as Nibble or Eliminate. None of these options, however, allows the user to control the aggregation process by comparing attribute values.

3. Input data (*CLASSIFY.GIS*) into ARC/INFO using the IMAGEGRID command in the ARC/INFO subroutine (output file named *CLASSGRID*). An example command line follows:

```
<ARC>:IMAGEGRID CLASSIFY.GIS CLASSGRID
```

4. Build attribute table using the REGIONGROUP command in the GRID subroutine. This creates an INFO file with the following three attributes: *value*, representing the region identifier; *count*, the number of pixels in each region; and *link*, the spectral class code from the original classified image (see following table). An example command line follows:

```
<GRID>:CLASSGRIDREG=REGIONGROUP
(CLASSGRID)
```

5. Collect additional attribute information for each region. Transfer attribute data for each region in *CLASSGRIDREG* by overlaying the file with other GRID files that contain the desired attributes. For example, to add the mean spectral value for TM channel 1 (TM1) to each region, a table file (named *TM1.TAB*) could be created as follows:

```
<GRID>:TM1.TAB=ZONALSTATS(CLASSGRIDREG,TM
1,MEAN)
```

6. Add items to the INFO file (the value attribute table). Use the RELATE and CALCULATE commands in the INFO subroutine to transfer the mean spectral value of TMl (rounded up to the nearest integer value) for each region from *TM1.TAB* to *CLASSGRIDREG.VAT.* Example command lines follow:

```
<ARC>:ADDITEM CLASSGRIDREG.VAT
CLASSGRIDREG.VAT TM1 2 4 B O LINK

<INFO>:SELECT CLASSGRIDREG.VAT

<INFO>:RELATE TM1.TAB BY VALUE

<INFO>:CALCULATE TM1=$1MEAN + 0.5
```

Additional steps similar to step 6 could be used to add other information about each region to the value attribute table (see table below). Once the attribute table is created, the CELLVALUE command in the GRID subroutine allows the user to interactively identify attributes for selected regions just as the IDENTIFY command does for vector polygons in the ARCPLOT subroutine.

An example of the structure of an attribute table created in the ARC/INFO GRID subroutine for classifying Landsat TM imagery

Column	Item Name	Width	Output	Type	N.DEC
1	Value	4	10	B	-
5	Count	4	10	B	-
9	Link	4	10	B	-
13	TM1	2	4	B	-
15	TM2	2	4	B	-
17	TM3	2	4	B	-
19	TM4	2	4	B	-
21	TM5	2	4	B	-
23	TM6	2	4	B	-
25	TM7	2	4	B	-
27	M_NDVI	2	6	B	-
29	Elevation	2	4	B	-
31	Slope	4	4	F	0
35	Aspect	2	4	B	-
37	Cover_Code	2	4	B	-

Acknowledgments

An earlier version of this paper was printed and distributed by the National GAP Analysis Research Project sponsored by the U.S. Fish and Wildlife Service. We thank Michael Jennings and J. Michael Scott for their encouragement and feedback and Melissa Hart for valuable comments and suggestions regarding this revision. Michael Blongewicz at ESRI was instrumental in getting us started with ARC/INFO. Our work has been funded by the U.S. Fish and Wildlife Service, the U.S. Forest Service, IBM Corporation, and the National Fish and Wildlife Foundation.

Water, Crops and Weather

Water, crops, and weather form a logical unit for purposes of this book. Agriculture was one of the preeminent stimuli for satellite sensing, in part because the technical requirements for a few broad bandwidth channels of data dictated using spectra related to vegetational reflectances. It was also recognized that crop assessments required frequent coverage. This, in turn, influenced orbital altitudes and swath widths to provide bimonthly data collection (cloud cover permitting). Water (too much or too little) and weather are the factors controlling health and vigor of crops. Without a fully integrated raster and vector GIS, it is unlikely that agricultural monitoring schemes are very efficient. Temporal changes in raster DNs by themselves can only assist some of the many needed applications. Likewise, vector GIS by itself lacks the necessary temporal detail to model the heart of agricultural applications. In addition to timely coverage of crop reflection, information systems are needed to describe agricultural practices from local to regional scales. Agribusiness and precision farming are phenomenal growth areas for GIS, but only if GISs employ all available data. The case studies described here focus heavily on local level irrigation and irrigation rights. They represent a snapshot of what has attracted considerable attention among regulatory and monitoring agencies. These case studies may represent the harbingers of new sectors of concern as GIS techniques gain acceptance.

Water resources management represents another huge area of interest. Indeed, water in all its states, from vapor to liquid to ice, represents challenges to land-based management operations and over time scales ranging from the geologic past to the present. Expansion of GIS applications to atmospheric moisture measurements and lightning detection are conceptually very exciting in that they attempt to model constantly moving attributes instead of the surface manifestations of those phenomena. In a contrasting way, maps of the paleo extent of glaciers produced with raster GIS may stimulate other approaches toward understanding Earth's past climates. There is no doubt that as raster and vector GIS techniques for water and atmospheric assessments become more widely used, each for their unique and joint contributions to problem solutions, global society will benefit from expanded applications that fit very specific user needs.

HYDROLOGY

Targeting Wetlands Restoration Areas

Richard G. Kempka, Ruth E. Spell, and Andrew T. Lewis, Pacific Meridian Resources
Frederic A. Reid, Ducks Unlimited, Inc.
Scott Flint and Kari Lewis, California Department of Fish and Game

Challenge

Alteration of hydrology has greatly affected the abundance and function of wetland systems throughout the United States. California's Central Valley is a classic example of an area affected by severe wetland degradation. U.S. Fish and Wildlife Service calculations show that of the original 5 million acres in the Central Valley of California, only 319,000 wetland acres (130,000 ha) remain. Satellite imagery analyses suggest there may be less than 260,000 acres (105,000 ha). The importance of the Central Valley to waterfowl habitat has made this a priority area for the North American Waterfowl Management Plan (NAWMP). At present, several federal, state, and private organizations are working to restore and manage wetlands for several species and provide buffers to human development. Numerous variables must be considered when selecting such locations for enhancement. Many of these habitat, hydrologic, and energetic variables are spatial in nature.

The goal of this study was to develop a GIS database of wetland and riparian areas for the Central Valley of California and San Francisco Bay. Specific objectives included developing image processing techniques to identify wetlands using multitemporal satellite imagery; establishing a baseline wetlands inventory for the Central Valley and San Francisco Bay area; and developing a GIS model to track wetland restoration efforts within a North American Waterfowl Management Plan focus area and identify priority areas for future wetlands restoration.

Data and Methodology

The project area for the Wetland and Riparian Inventory was defined initially by the boundaries of three TM scenes. Of the three scenes using the 300 ft contour as digitized from 1:250,000 USGS topographic maps, the valley was a subset. The project area included 9,977,907 acres (4,309,600 ha) and covered three key regions: Sacramento Valley, San Francisco Bay and the Sacramento and San Joaquin River Deltas, and North San Joaquin Valley.

Two seasons of satellite imagery from each of three path/row locations were required for this project. First, images from early mid-summer (6/28/93 and 7/7/93) were needed to identify wetland emergents during the growing season. Second, scenes from the previous winter (1/3/93) were required to determine the location and extent of winter flooding. Combining data from both the summer and winter images allowed flooded wetland and agriculture classes to be identified.

The National Wetlands Inventory is a U.S. Fish and Wildlife Service mapping program that delineates detailed wetland classes using manual interpretation of aerial photographs. The most recent NWI data for most of the Central Valley were derived from aerial photography collected in the late 1970s and early 1980s. For this project, the NWI data were used along with additional data sets to stratify areas of potential wetlands as a precursor to image classification.

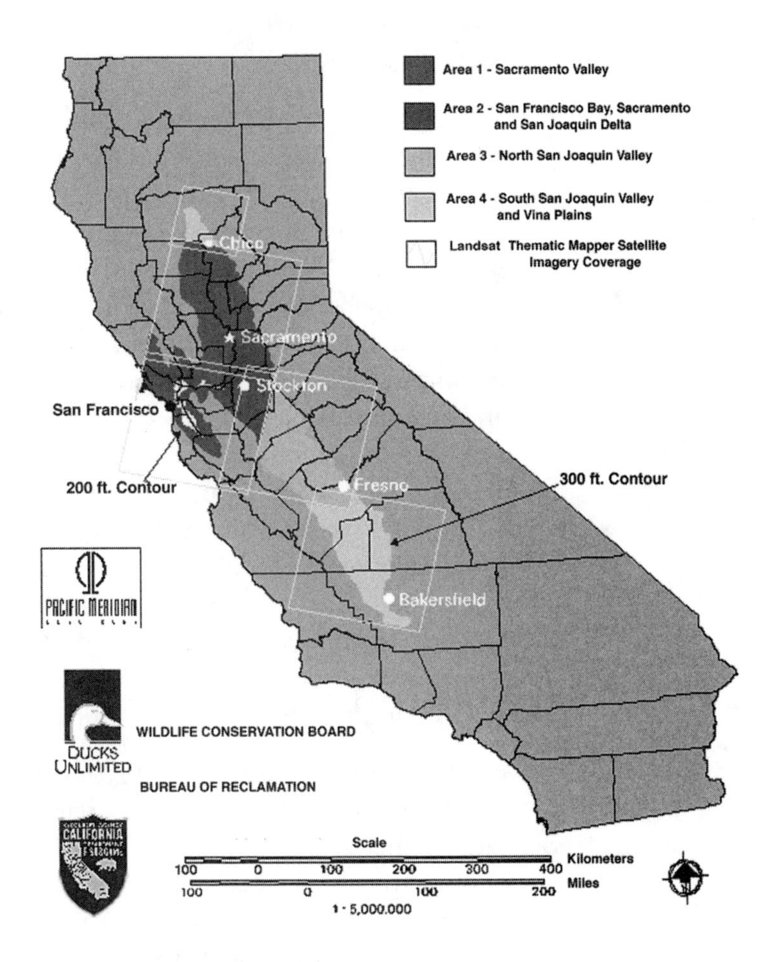

California Wetland and Riparian Inventory project area.

USGS 1:100,000 scale Digital Line Graphs (DLGs) were modified to include River Reach attributes. This data set was used to build a buffer around hydrographic features for delimiting riparian areas.

A digital coverage of agricultural lands was acquired from the California Department of Conservation Farmlands Mapping Program. These data were used along with the NWI data to build a mask for stratifying the imagery to reduce confusion between spectrally similar agriculture and wetland classes.

The classification scheme was modeled in part based on "Classification of Wetlands and Deepwater Habitats of the United States" (by Cowardin and others) and the NOAA C-CAP protocols. A riparian category was also added for woody vegetation found in the river flood plains after the classification system described in "Riparian Resources of the Central Valley and California Desert." Classes were organized under three categories: agriculture, wetlands, and uplands. The summer satellite scenes were initially stratified into these categories using a combination of ancillary data sets and same-year winter imagery classified to identify winter flooding. This stratification was performed to reduce confusion between spectrally similar classes (i.e., rice and emergent wetlands) and to refine the spectral signatures. The resulting strata were then classified into more detailed land cover classes, including permanent and seasonally flooded wetlands.

A combination of supervised and unsupervised classification techniques were used. Field data, aircraft reconnaissance, ancillary data, and 1:40,000 scale NAPP aerial photographs were used to identify training sites and label spectral clusters. Post-classification modeling was performed using ancillary data (NWI and hydrography) to further refine the classification. Specifically, NWI data were used to label the emergent wetlands as estuarine or palustrine. Non-persistent wetlands were identified using a combination of summer and winter imagery (i.e., emergent in summer, flooded in winter). Finally, a three-pixel buffer on 1:100,000 scale perennial streams and canals was used to stratify the woody class into riparian and non-riparian.

Results and Discussion

The GIS model was developed by Ducks Unlimited and Pacific Meridian Resources under a NASA EOCAP grant to provide assistance in waterfowl restoration decision making, evaluate current waterfowl habitat, and provide an educational tool for illustrating seasonal and between-year changes in waterfowl habitat in the Sacramento Valley. This model utilizes the vector and raster modeling capabilities of ARC/INFO and GRID with a user-friendly interface developed to guide the user through the specific model applications. The model consists of four applications: rice phenology cycle; change detection; waterfowl energetics model; and wetlands restoration site analysis model. Only the latter two are presented.

The energetics model was developed to assist resource managers in evaluating how well the Central Valley Habitat Joint Venture wetland restoration/enhancement goals are being met in the Sacramento Valley under various scenarios. In the model, the acreages of various waterfowl habitats from the Wetland and Riparian Inventory are input along with user-defined population and energetics parameters to determine how closely the habitat goals are being met. Equations established by the Central Valley Habitat Joint Venture (CVHJV) are used to calculate the energy requirements of the user-defined target waterfowl population. Next, the amount of food available to waterfowl is calculated using habitat acreages from the 1993 Wetland and Riparian Inventory along with user-defined assumptions about how much food is available to, and consumed per day by, waterfowl in each habitat type. The results of these calculations are then presented in terms of target use days, actual use days available under the given user inputs, and the percentage of target use days that were supported under the specified scenarios. The following illustration is a conceptual diagram of the energetics model.

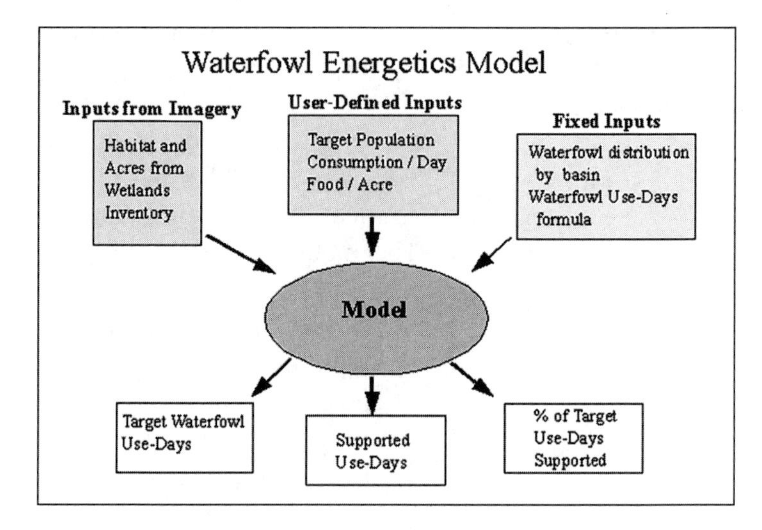

Conceptual diagram of the waterfowl energetics model.

The site analysis model was designed to assist waterfowl resource managers in evaluating potential locations for restoring wetlands and targeting agricultural enhancement efforts. Three data layers are considered in the analysis: distance to refuges; distance to wetlands; and distance to water delivery. The features in these three GIS data layers can be coded in terms of their value in meeting site criteria. Various GIS data layers can then be combined to create a composite output map showing a site suitability value for each grid cell in the map. Plate 39 is an example of the model's output.

Combined use of multidate satellite imagery and existing digital ancillary data sets has proven to be a useful technique for producing a baseline inventory. Initial review indicates that the energetics model is useful for tracking joint venture accomplishments throughout the Central Valley of California. Multiple GIS layers can be applied for similar regional landscape analysis to other Joint Venture regions throughout North America. Multidate analysis of critical agricultural lands, such as rice, can reveal changes in landscape flooding strategies. The energetics model is being used to reevaluate waterfowl habitat goals. The siting model allows for quantitative assessment of land parcels for locating potential wetland restoration/enhancement projects. The model functions effectively in a GIS environment but additional data layers need to be added to improve its utility for wetlands conservation purposes. Additional variables such as soils, parcel ownership, water districts, and others make the siting model more robust. This model will be reprogrammed to run on a desktop ArcView system to allow greater accessibility by waterfowl biologists and wetland ecologists.

Water Resources Management Applications

Thomas H. C. Lo, South Florida Water Management District, West Palm Beach, Florida

Challenge

The South Florida Water Management District is responsible for planning, facilities management, and regulating water resources for a 17,930 sq mi (46,439 sq km) area with a population approaching 6 million. Satellite image data have been incorporated into the District GIS, which has been built upon ARC/INFO and, to a lesser extent, ERDAS, to cope with the demands of these functions. In order to maintain a timely information database, the District has been acquiring SPOT data covering the entire District every year since 1988.

Traditionally, remote sensing data were stored in quarter-inch or 8mm tapes. It took a fair amount of effort for users to get to the tape storage area, manually sort through the tapes, find the space on their hard disk drives, and load the tapes for use. These extra steps tend to discourage users. To make them user friendly, all current data have been put on-line since early 1994, and users have been able to access the data effortlessly across the network. Both panchromatic and multispectral 1995 SPOT data in SPOTview format and a 1994/93 Landsat TM statewide mosaic in full resolution are on-line along with older data sets.

Users need the most current data to update GIS coverages or to map study areas. The District's remote sensing program does its best to speed the process between date of data acquisition and availability as enhanced georeferenced data. The current turnaround time is about three months. The value of remote sensing data depreciates rapidly after a year for many of the District's applications.

Raster image data are used routinely and indispensably in the following GIS applications:

- *General Reference.* Satellite data are mosaicked to produce digital and hardcopy maps of the District. These image maps provide the most current general overview of South Florida and are critical references for planners and decision makers. Depicted in Plate 40 is an image map of south Florida composed of 29 scenes of 1994 SPOT multispectral data.
- *GIS Coverage Update.* Satellite data are used extensively to update the vector data layers, including land use/cover, transportation, and others. One example is the use of SPOT data for reconstruction of 1990 land cover scenarios for hydrologic modeling for that specific period of time.
- *Water Permit Checking.* Regulatory staff use SPOT data for day-to-day water permit evaluation and compliance checking.

- *Change Analysis.* Ecological change analyses are routinely performed by overlaying two or more sets of satellite data taken at different points in time. The results provide useful pictures of long-term ecological changes on a local or regional scale.
- *Base Map.* Image data are used for base mapping. All map features are mapped into the spatial framework of the base map. The data used in this application are mostly rectified scanned aerial photos or orthophotos characterized by fine spatial resolution and good positional accuracy.

Data and Methodology

Described below are applications using SPOT panchromatic imagery as a backdrop and as a hardcopy image map for evaluating new water permits, improving the accuracy of the existing permit database, and compliance monitoring.

Permits are required for water use and for managing and storing water within the District's boundaries. Water use permits are issued to applicants for urban supply and/or irrigation. Surface water permits are issued to applicants who want to control, impound, or obstruct surface waters.

To apply for a permit, an applicant is required to submit several documents including a map of the project location. Until the late 1980s, the project boundary was hand-drawn on USGS quad maps at 1:24,000 after the project was permitted. Each permit was color coded to refer to another sheet of information containing the project permit number, name, and a short description. These hand-colored quad maps were used on a daily basis by District staff and the public to locate permits.

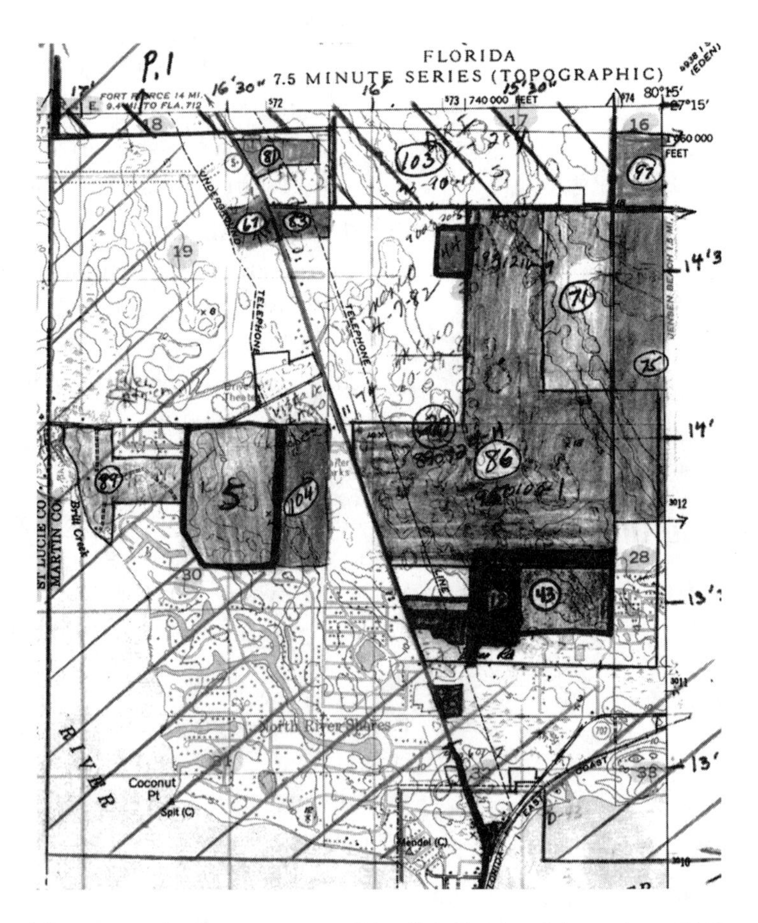

Portion of well-used paper map.

These quad maps were originally digitized into AutoCAD drawings and every new permit issued was also digitized from the applicant's map into AutoCAD. In 1991 the AutoCAD drawings were converted into the District's ARC/INFO GIS. Once the GIS was set up, each newly issued permit was digitized from the source map submitted by the permittee on a digitizing tablet directly into ARC/INFO. The permit coverages contain only one attribute, the permit number. All other data related to a permit are stored in the Oracle database. The permit number provides the necessary link from the GIS to Oracle.

Due to variations in quality of the source maps, digitizing on a tablet was cumbersome and time-consuming. Many source maps had poor to no calibration points for georeferencing. Even if the calibration was possible, the actual line work of some of these maps was of poor quality.

In 1994, panchromatic SPOT data became available online and answered most of the tablet digitizing problems experienced by the regulatory staff. Most of the permits are digitized on screen using SPOT data as the base map to georeference source maps to permit coverage. Not only does this help the GIS permit data to become more accurate, it also saves enormous amounts of time. Staff members save about 60 percent of their time using the images to digitize the permits on screen.

To save even more time, a graphical user interface menu has been developed by regulatory staff based on ESRI's Arc Macro Language (AML). The menu allows the user to easily edit coverages without having to know and type all of the command syntax in ARCEDIT. It also contains built-in safeguards. The menu, for example, will not allow the user to build a point coverage as a polygon coverage. By entering section, township, and range values, it controls the display of images and other associated GIS layers in the ARCEDIT environment. Once mapping, editing, and evaluation have been completed, the new permit map is stored in a temporary database to be transferred to the master GIS when the application is approved.

Once the permit boundaries are in the GIS system, they are plotted over the panchromatic SPOT image in the 1:24,000 USGS quadrangle format. These paper maps are used by District staff and the public who cannot view the data online. By viewing both of these layers of information together, the water permit boundary line work—which was originally converted from AutoCad before the images came on-line—can be corrected to match the image. Horizontal positional accuracy of permit boundaries mapped in this manner is estimated at about 30 ft.

The third regulatory use of images is compliance monitoring. Remote sensing data are employed to identify potential problem areas, monitor compliance progress, and as evidence in court. The District is charged with monitoring permit compliance from Orlando to Key West, a region too large for traditional surveillance methods to be effective. Worse, a large portion of lands are inaccessible by road. The regulatory staff have to rely upon costly transportation means such as helicopters. In addition, more information is required for the field personnel to know which activities on the ground were permitted and which occurred before environmental laws were passed. By checking with current and historical SPOT image data associated with permit information, field personnel are able to identify potential violation areas for further investigation, thereby eliminating excessive and expensive field time.

For large environmental impact projects, such as wetland mitigation, post-permit compliance reports are filed by permittee to inform the District of the progress of wetland mitigation which includes wetland restoration and/or creation. By evaluating a time sequence of satellite images in an interactive mode or in hardcopy map form, District staff are able to verify progress and allow them to work with the permittee to correct anticipated problems.

When an investigator finds signs of illegal canal digging or other land-altered development, the District must prove that the activities took place after wetland protection laws were passed. By comparing newly acquired images, regulatory staff can easily and quickly identify changes and compare those changes to a list of legal activities in the area. An illustration of a classic court case is provided in the following figures. Dikes and ditches were detected in an isolated wetland area. Although the company who owned the property insisted that the activity had taken place before wetland laws were passed, current and archival SPOT imagery provided an important piece of evidence that the claim was false. When the imagery revealed the truth, the defendants settled out of court.

February 1988 SPOT scene shows area with few drainage ditches.

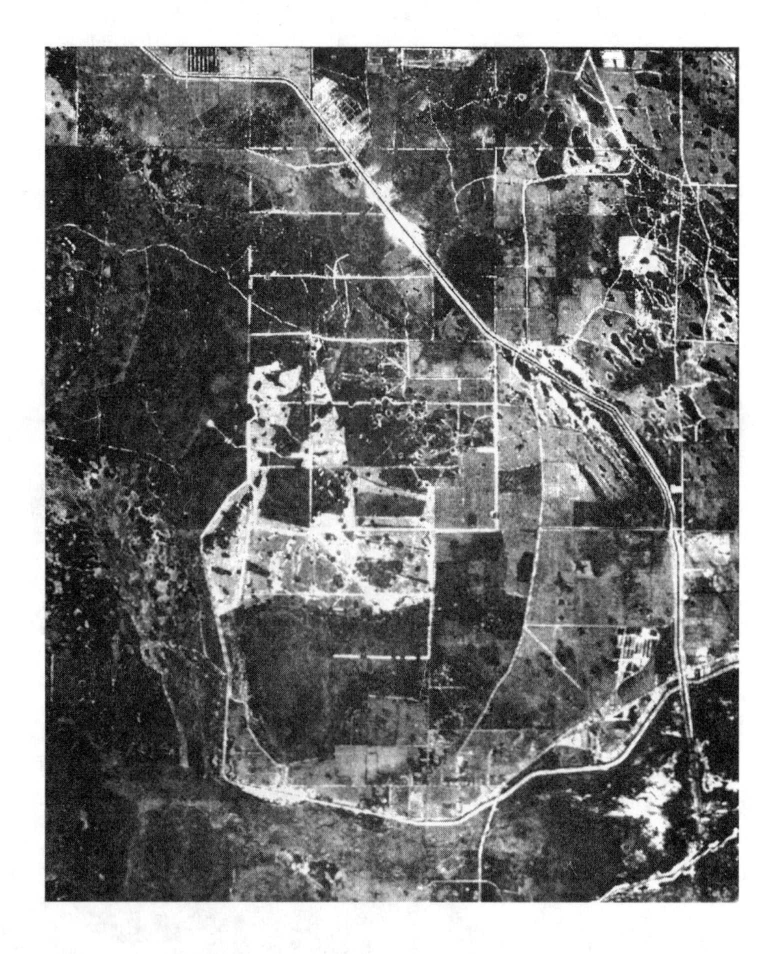

December 1989 scene shows new ditch system (white linear features) installed to drain wetlands in violation of Florida law.

Results and Conclusion

The District has succeeded in integrating image data into the GIS over the past two years. However, as the volume of image data increases at geometric rates, led by the USGS orthophoto program and the new generation of high resolution satellites, the District will be hard-pressed to accommodate them in an on-line, seamless manner. Hardware infrastructure upgrades and vast improvement of image handling software to store, process, display, and plot such a large amount of data in a fast, interactive, and user friendly manner will be necessary.

Acknowledgments

The author acknowledges the following regulatory staff members: Cecilia Conrad, who created the USGS quadrangle image map series and provided help with this manuscript; Carlos Piccirillo, who wrote the AML interface; and Al Harkabus, who provided compliance information.

Mapping Glaciers with SPOT Imagery and GIS

Andrew G. Klein and Bryan L. Isacks, Department of Geological Sciences and Institute for the Study of the Continents, Cornell University

Challenge

Glaciers in the central Andes have been retreating at an accelerated pace over the past century. Researchers are mapping the modern extent of glaciers and their limits during the Last Glacial Maximum (prior to 20,000 years ago) to gain insight into the climatic changes that occurred in the region and that influence glaciers and other Earth processes. The inaccessibility and size of the mountainous central Andes necessitate the use of satellite imagery for glacial mapping activities. Using medium-resolution imagery, interpreters face many challenges trying to identify small-scale geomorphic features that indicate past glacial activity.

A primary objective of NASA's Earth Observing System (EOS) program is to monitor the Earth's atmosphere, oceans, and continents to understand the causes and potential consequences of climatic changes on the Earth and its environment. Past and present glaciers offer significant insight into climatic variations shaping current environmental conditions.

Researchers in the Department of Geological Sciences at Cornell University have formed one of the 29 Interdisciplinary Science investigation teams funded by NASA to develop applications and models using EOS data. The Cornell study, entitled "Climate, Erosion, and Tectonics in Mountain Systems," focuses on modern climate hydrologic and glacial processes operating in major mountain belts such as the Andes and the Himalayas.

Glaciers serve as excellent measuring sticks of climatic change because they shrink and expand with variations in temperature, precipitation, and radiation. Glacial activity leaves behind geomorphic features in the landscape that can be mapped, allowing researchers to reconstruct the glacial extent during the past.

Comparing the former extent of glaciers with present size yields clues to climatic conditions existing at the time the geomorphic features formed. For instance, by mapping the size and positions of moraines left behind by a melted glacier in a low-elevation mountain area, scientists can estimate the areal extent and volume of such paleo-glaciers. Climatic conditions on modern glaciers are used as an analog for understanding former larger glaciers in the same area.

Mapping has shown that today's glaciers in the central Andes seldom extend below 4,500m, whereas 18,000 years ago they existed in mountain areas as low as 3,200m. A logical conclusion is that the current climate at the lower elevation in the Andes during the past was similar to conditions found only at much higher elevations today.

Satellite based glacier studies also have practical applications. Satellite imagery combined with aerial photography or ground studies can be used to map the rate at which modern glaciers are retreating. By understanding the factors that influence glacial retreat, researchers may be able to predict how long modern glaciers will survive under present climatic conditions.

In the project study area of Bolivia and Peru, for instance, the depletion of Andean glaciers would have a devastating impact on fresh water supply. Runoff from nearly 2,000 glaciers high in the Andes supplies millions of people with water for drinking and generation of electric power. At current rates of recession, many of these glaciers could be gone within 100 years.

Data and Methodology

Cornell researchers are examining present and former alpine glaciers in the Andes and Himalayas. A recently completed project focused on a section of the central Andes stretching from 15 to 22° south latitude, an approximately 700 x 700 km area.

SPOT panchromatic satellite images with 10m resolution were used to supplement lower resolution Landsat imagery to define and map landscape evidence of Pleistocene glaciation. Researchers used ER Mapper image processing software to link display windows so that SPOT and Landsat imagery of the same geographic region could be viewed simultaneously. Vector maps outlining the extent of the Pleistocene glaciers were created in the image processing software and ported directly to ARC/INFO for further analysis.

Twenty-two archived Landsat images were used to create a mosaic as a regional backdrop of the entire study area. SPOT images were acquired for selected areas where evidence of glaciation was known to exist. Paper topographic maps and aerial photographs of some parts of the study area were also available.

As an alpine glacier flows down a mountain, it carves a wide valley in its upper reaches. Along its lower reaches, material bulldozed from above is deposited along the front and sides of the glacier as terminal and lateral moraines, respectively. When the glacier retreats, these moraines remain as narrow ridges. In satellite imagery with sufficient spatial resolution, moraines can be identified by their linear shape and location along the sides and mouths of mountain valleys.

The 30-meter Landsat Thematic Mapper imagery served as the base map providing an overview of the study area so that researchers would know where to focus detailed mapping efforts with the SPOT imagery. Much of the area is a semi-arid plateau, and the moraines are visible on Landsat TM imagery. However, in a critical study area in the eastern cordillera of the Andes, broad U-shaped valleys—classic evidence of former glaciation—extend down extremely steep vegetated slopes. While these valleys were visible in the lower-resolution imagery, the higher resolution SPOT imagery was needed to detect the moraines within the valleys. These moraines were mapped to determine the extent of the retreated glaciers (Plate 41).

Working in ER Mapper, the researchers first created a regional mosaic of the Landsat scenes in a Transverse Mercator Projection. Ground control points were selected from topographic maps using points such as road or stream intersections easily located on both the map and the satellite image. GCP selection is extremely easy and straightforward in ER Mapper. The image and ground coordinates of each GCP are entered into the image processing software which assesses the root mean square error associated with the point and its effect on the polynomial warp. Errant GCPs can be identified and corrected before warping the image.

ER Mapper offers a feature called "dynamic algorithm compilation" which allows the user to work with numerous images on screen at once without saving the images to an intermediate file. In creating a mosaic of this size, the feature saved time and disk space because multiple images could be accessed and manipulated interactively as if they were one image from the same file.

An important aspect of mosaicking multispectral imagery acquired on different dates is radiometrically calibrating the scenes so that colors are constant throughout the images. This removes differences in the scenes that result from seasonal variations in sun angle, time dependent changes in the sensor, and atmospheric changes. Radiometric corrections are easy to implement as formulas within ER Mapper. Again, dynamic algorithm compilation allows these corrections to be applied to the original images without creating an intermediate image for each scene in the mosaic.

SPOT panchromatic images were also loaded into ER Mapper and registered with the mosaicked scenes. Registration of the two data types was extremely important because this allowed the researchers to take advantage of an ER Mapper tool called Geolink, which accelerated the on-screen mapping process.

With Geolink, the researchers displayed both the Landsat and SPOT scenes simultaneously on the screen in two Geolinked windows. As the researchers scrolled through the Landsat scene and zoomed in on a possible glacial valley, the exact same area was displayed in the SPOT window. This allowed for easy feature comparison between SPOT and TM imagery.

As each valley was identified in the imagery, the researchers conducted the mapping on screen using the ER Mapper Annotation Overlay. This tool kit allowed the scientist to draw vector lines and polygons over the satellite image for marking the locations of the moraines. The vector mapping process was as simple as drawing lines on the screen with the mouse.

Using the terminal and lateral moraines as geographic guides, the researcher joined the moraine vectors and extended them around the divides surrounding each glacial valley. This created a polygon that encompassed the entire estimated areal extent of the paleo-glacier.

The vector map was then saved in ER Mapper as an ARC vector file so that it could be ported directly to ARC/INFO for further analysis.

Results and Discussion

In the Bolivia project, most of the GIS analysis was carried out visually, but other studies used automated GIS procedures. Both are described to illustrate the flexibility of the GIS.

After the vector maps were edited in ARC/INFO, they were printed at the same scale as the topographic maps. The minimum and maximum altitudes for each paleo-glacier were determined from the Bolivian 1:50,000 scale paper topographic maps.

In a similar mapping project in the Himalayas, topographic data were available as a digital raster image. They were loaded into ARC/INFO as a GRID. ARC/INFO was used to automatically determine the minimum and maximum elevations for each paleo-glacier, thereby saving staff time.

To correlate elevation of paleo-glaciers with modern glaciers, an equilibrium line was calculated for each paleo-glacier. The equilibrium line is the elevation at which ice accumulation equals ablation on a yearly basis. Based on research with modern glaciers in the region, this altitude was estimated to be 45 percent of the distance between the minimum and maximum elevation for each paleo-glacier. The equilibrium line served as a reference point to compare paleo with modern glacier conditions.

Researchers created contour maps of the equilibrium line elevations of both modern and paleo-glaciers for the entire region. These contour maps were gridded into raster images in ARC/INFO. The paleo-surface was subtracted from the modern surface to calculate how much lower the former glaciers extended from their present positions. In the central Andes, approximately 20,000 years ago glaciers reached elevations 0.5 to 1.2 km lower than at present (Plate 42). This expansion was due to lower temperatures and/or increased precipitation.

Mapping moraines with satellite imagery and GIS allowed Cornell researchers to reconstruct the climatic conditions which existed in the central Andes during the Last Glacial Maximum approximately 20,000 years ago. The study concluded that in the study area the average temperatures then were about 5 to 8° C colder than at present.

The western portions (and perhaps all) of the central Andes had more precipitation during the Last Glacial Maximum than at present. Future phases of this project will focus on studying modern glacial processes to better understand climatic controls on the region's glaciers. This understanding will improve ability to reconstruct past climates based on the geomorphic mapping.

Identifying Efficient and Accurate Methods for Conducting the Irrigated Water Use Inventory, Estancia Basin, New Mexico

Douglas J. Paulson, Earth Data Analysis Center, University of New Mexico

Challenge

Irrigation of field crops accounts for the bulk of water withdrawn or depleted in New Mexico. Estimates are that 77 percent of surface water depletions and 86 percent of groundwater depletions are related to crop irrigation. Management of New Mexico's water resources is the responsibility of the State Engineer Office (SEO), and is accomplished through regulation of declared water basins which may consist of either surface or underground waters. For purposes of sound water management, the SEO must be able to determine the acreage of irrigated cropland associated with any particular declared basin with a high degree of accuracy and confidence.

The study area is the Estancia Basin, a topographically and hydrologically closed region covering more than 2,200 sq mi (5,700 sq km) of central New Mexico, approximately 40 mi (64 km) east of Albuquerque. The Estancia Underground Water Basin was declared in 1950 and expanded in 1975, and currently covers 1,724 sq mi (about 4,450 sq km). Annual rainfall averages 12 to 14 inches (300-350 mm) over most of the basin. Economic activity is dominated by ranching and irrigated farming. Primary crops are alfalfa, corn, wheat, beans, irrigated pasture, and sorghum.

The purpose of the study was to determine the best methods, in terms of correctness and time efficiency, for conducting a periodic Irrigated Water Use Inventory in the Estancia Basin. The final product had to provide a means to inventory irrigated fields (active and inactive), as well as for periodic monitoring and measuring actively irrigated acreage. Completion of the project required integration of satellite imagery, image processing techniques, GIS and GPS technologies, and ground truthing.

For inventory purposes, a decision was made to utilize a GIS polygon coverage of identified irrigated croplands within the basin. The polygon coverage contains boundaries of irrigated fields identified through this pilot study and is to be expanded in the future as additional fields are identified. Various satellite imagery and image processing techniques were evaluated as to their capability of identifying actively irrigated cropland and automatically estimating irrigated acreage. GPS technology was used to accurately record selected field boundaries in order to assess the precision of automated acreage estimations. At the time of GPS data acquisition, Earth Data Analysis Center (EDAC) personnel also conducted ground truthing of crop type and irrigation status with the resulting information used to assist in image processing techniques.

Software used included ERDAS IMAGINE 8.2 for image processing and ARC/INFO 7.0.2 for GIS coverage generation. GPS data were collected and manipulated using Trimble hardware and software.

Preliminary analysis began with the acquisition of imagery covering the Estancia Basin, which included both SPOT 10m Panchromatic and Landsat 30m multispectral images. Beginning with the Landsat TM data, various image processing techniques were examined in an attempt to highlight active agricultural fields. First, a false color composite

(FCC) of TM bands 4 (red), 3 (green), and 2 (blue) was generated, resulting in active agriculture appearing as bright red. This FCC was used for plotting, screen display, and to select representative fields for GPS data collection and ground truthing. Fields exhibiting representative high, low, and uncertain spectral responses were included in this selection. Next, statistical signatures of the image were found using the ISODATA (iterative self-organizing data analysis technique) clustering method. These signatures were used to classify the image using a maximum likelihood rule. The resulting image contained one class that definitely indicated active agriculture and several marginal classes representing both active agriculture and other healthy vegetation. A principal components analysis was also performed on the Landsat image in an attempt to create a component image highlighting only agricultural fields. While the vegetation response was divided between two components, neither was a unique agricultural component and much data noise was also enhanced.

Alternative image enhancement techniques in the form of band ratios were also attempted on the Landsat data with the objective of isolating only agricultural vegetation. One technique was the NDVI (normalized difference vegetation index) which ratios biomass response in the near infrared against biomass absorption in the visible red (TM4-TM3)/(TM4+TM3). While the NDVI has been found to be a reliable indicator in many vegetation studies, its brightest responses were not limited to agricultural areas but included other healthy vegetation as well as marginal vegetation along field boundaries. Other ratios considered were TM4/TM5 to contrast plant type response in near infrared to response in mid-infrared, and TM5/TM7 as a moisture indicator to distinguish well-watered healthy crops from healthy crops that are not well-watered. When the three ratios were combined into a single file they highlighted agricultural fields very well but again the brightest response was not limited to agricultural fields. With the additional processing time needed to create the ratios, the resulting information was no greater than from the original FCC.

The final image processing technique investigated was to resample the Landsat TM data at 10m and merge it with the SPOT 10m data. This technique combined the spectral resolution of Landsat with the spatial resolution of SPOT, resulting in a much more accurate image. (See Plate 43.)

GPS accuracy is increased by taking multiple position fixes at a single point, then correcting and averaging the data. Point data at a minimum 180 fixes per point were collected for 29 representative fields throughout the basin. To further increase acreage accuracies, actual crop corners rather than fence boundaries were recorded. Fields irrigated with center-pivot systems were gathered with one point at the pivot and another at the crop boundary. After being corrected and averaged, the GPS point data were transferred to ARC/INFO for polygon creation and acreage calculation. The resulting polygon coverage was regarded as the standard when assessing acreage results of other field delineation methods.

Eight different methods of field delineation and the resulting acreage measurements were compared to the GPS-derived polygon data. These methods, involving both manual and on-screen digitizing, are summarized below.

Manual digitizing

- ❒ Hardcopy plot of SPOT/TM merged color image

On-screen digitizing

- ❒ SPOT/TM merged color image
- ❒ Landsat TM principal components

- ❒ NDVI of Landsat data
- ❒ Combined ratios of Landsat data
- ❒ Merged image of classified TM with SPOT
- ❒ Landsat TM false color image
- ❒ SPOT 10m Panchromatic image

Results and Conclusions

Effectiveness of the various methods was evaluated by both the accuracy of the acreage calculations and time required for completion. Manually digitizing from a hard-copy plot resulted in acceptable acreage accuracy but was time-consuming both in terms of the actual digitizing and plot preparation. Error may also be introduced due to the multiple steps involved and the inability to see the digitized polygons overlaid on the image. Of the on-screen digitizing techniques, the most time- and accuracy-effective were using the SPOT/TM merged image or the classified Landsat TM resampled to 10m.

This project highlighted the advantages to be gained by an integration of geographic information technologies. While each is effective in its own right, satellite imagery, image processing, GIS, and GPS are all enhanced when used in combination.

Benefits

Methods used here were evaluated for both the accuracy of acreage calculations and time required for their creation. Digitizing field polygons, whether on screen or from hard-copy plots, was time-consuming, but both methods resulted in acceptable acreage accuracy. Hardcopy plots are more difficult because they do not allow one to see the digitized polygons overlaid on the image and because they require more time and expense to produce and register. Moreover, the additional steps involved open more opportunity for error. Of the on-screen digitizing techniques, the most effective in terms of time and accuracy were achieved by using the SPOT/TM merged image or the classified Landsat TM resampled to 10m.

AGRICULTURE

Polygon Mode Filter for Remotely Sensed Agricultural Land Cover Data

Casson Stallings and Siamak Khorram, Computer Graphics Center, North Carolina State University

Rodney L. Huffman, Department of Biological and Agricultural Engineering, North Carolina State University, Raleigh, North Carolina

Challenge

GIS interfaces to water quality models are valuable tools to estimate the effects of alternate agricultural management practices over large areas. Methods are being developed to integrate into a GIS the many types of data required for water quality modeling. The crop grown every year is an important theme for water quality modeling in agricultural areas. Management practices applied to a field are highly dependent on which crop is growing.

In recent times the Farm Services Agency (FSA) has used intensive, low-altitude aerial photography interpreted by hand to determine the crop type on individual fields. The advantage of satellite data is that they can be obtained and processed for large areas relatively cheaply. One of the problems with satellite data, however, is the "salt-and-pepper" or speckle in land cover classifications occurring within agricultural fields and other regions that are known to consist of a single cover type. This within-field heterogeneity creates conceptual problems in applying water quality models. The work described in this case study demonstrates how knowledge of agricultural field boundaries can be applied in a polygon mode (or object) filter to altered land cover data so that they contain only one crop type per agricultural field. The process increases the classification accuracy of the land cover data by 15 percent.

The study site for this work was the 2,044-ha Herrings Marsh Run Watershed in Duplin County, North Carolina. This watershed in the Coastal Plain physiographic region is about half agricultural land; the main crops are soybeans, corn, cotton, tobacco, and hay. The watershed supplies water to the Northeast Cape Fear River which has been classified as support threatened by the NC Department of Education, Health, and Natural Resources.

Data and Methodology

Building GIS model interfaces around commonly available data is advantageous because it allows the interfaces to be used in modeling the largest possible area with the least effort. The GLEAMS (Groundwater Loading Effects of Agricultural Management Systems) interface for which the data were prepared assumes that Natural Resources Conservation Service (NRCS) soil series data are used. It also makes use of United States Geological Survey (USGS) 1:24,000 digital elevation models. The field boundaries were digitized from 1:7000 FSA aerial photography flown every five years. The reference cropping data for the fields were obtained by the Cooperative Extension Service via farmer surveys.

The reference cropping data and field boundaries were joined to form the reference data for crops. Agricultural fields were classified as containing bare soil based on visual examination of the image after a Tassled Cap transformation. Spatial and tabular data were taken from other sources to complete the reference data. Upland forest polygons were identified on the FSA aerial photography. The USGS 7.5' quadrangle was used in conjunction with the imagery to identify water polygons. The National Wetlands Inventory data were used to complete the reference polygons with the addition of wetland types. The reference polygons were rasterized to match the Landsat Thematic Mapper (TM) imagery and transferred from ARC/INFO to ERDAS.

Image Processing and Accuracy Assessment

Landsat Thematic Mapper data were obtained for July 1990, and ERDAS software was used to process the imagery. Steps in the procedure are listed below.

- The image was rectified to NC State Plane coordinates.
- Reference polygons to be used for training sites were defined randomly for each class.
- The region growing algorithm (SEED) was used to define the exact pixels for most training classes.
- Additional training classes were defined for the water classes by hand delineation and for the wetland forest classes by performing a cluster analysis on the pixels within the set of randomly selected reference polygons.
- A maximum likelihood classifier was applied to the image.
- The resulting classes were merged into the five desired classes (water, forest, soil, open, and agriculture).

The classified and training site grids were transferred to ARC/INFO for further processing. The accuracy assessment was carried out in ARC/INFO and S-Plus (Statistical Sciences, 1993, Seattle, Washington). An error matrix was derived using the reference data to identify the true classes. All cells for which there were reference data, except those used to develop training classes, were sampled for the error matrix. The overall accuracy, user's and producer's accuracy, and the Kappa coefficient were calculated for the classified data.

Accuracy estimates for the original land cover classification

	Producer's Accuracy	User's Accuracy	Overall Accuracy
Water	58.1	99.0	75.4
Forest	84.1	98.3	Kappa
Open	56.2	19.9	0.632
Bare soil	69.4	70.0	
Agriculture	66.2	71.2	

Results and Conclusions

Initially, remotely sensed data are classified in a grid format. It was necessary to convert this format to a vector-based format to be compatible with the water quality interfaces being developed. A logical filter is often applied to classified data before vectorization. Usually, a mode or majority filter with a square 3x3 or 5x5 kernel is used. A 3x3 mode filter changes the class of each cell based on its class and the classes of the eight cells surrounding it. The central cell is changed to reflect the dominant class within these nine cells. An example application of a 3x3 mode filter is shown in the following illustration. The typical rationale for applying a filter is to reduce the resulting number of polygons. These filters also have the effect of reducing the visual "noise" and often increase the classification accuracy. Drawbacks of rectangular mode filters are that they can eliminate small and linear features from the land cover data as in the next illustration.

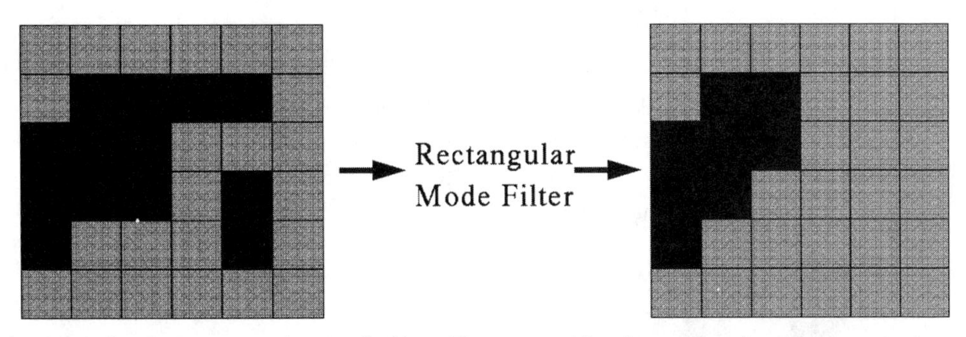

Application of a 3x3 rectangular mode filter. The original land cover grid containing two cover types is on the left. The resulting grid is on the right.

In cases where the boundaries, but not the crop type, of the agricultural fields are known, an alternate filter method is possible. This is referred to as a polygon mode filter, or sometimes as an object filter. Rather than using a constant-shape kernel, this filter looks at all of the cells in a single agricultural field. This method requires data describing the extent of each agricultural field or homogenous region. Application of the polygon mode filter creates a new grid. The dominant land cover class from the classified data is determined for each region known to be homogenous. Then each homogenous region is assigned the corresponding dominant class. Areas that are not identified as being homogeneous (i.e., areas outside a reference polygon) are left undefined.

The cover type of the whole field is changed to the dominant or mode class within the field. An example application of this filter is shown in the following figure. It uses the same initial cover type data as the previous illustration. This method has two advantages over filters with constant-shape kernels: it looks only at cells that are known to be of a homogeneous class, and it can use many more cells to determine the dominant or mode class (a 3x3 kernel is limited to only nine cells). As already stated, the drawback of this filter is that the boundaries of the homogeneous regions must be known in advance. Also, as seen in the next illustration, the cover type of areas outside of the homogeneous regions are not defined after application of the filter.

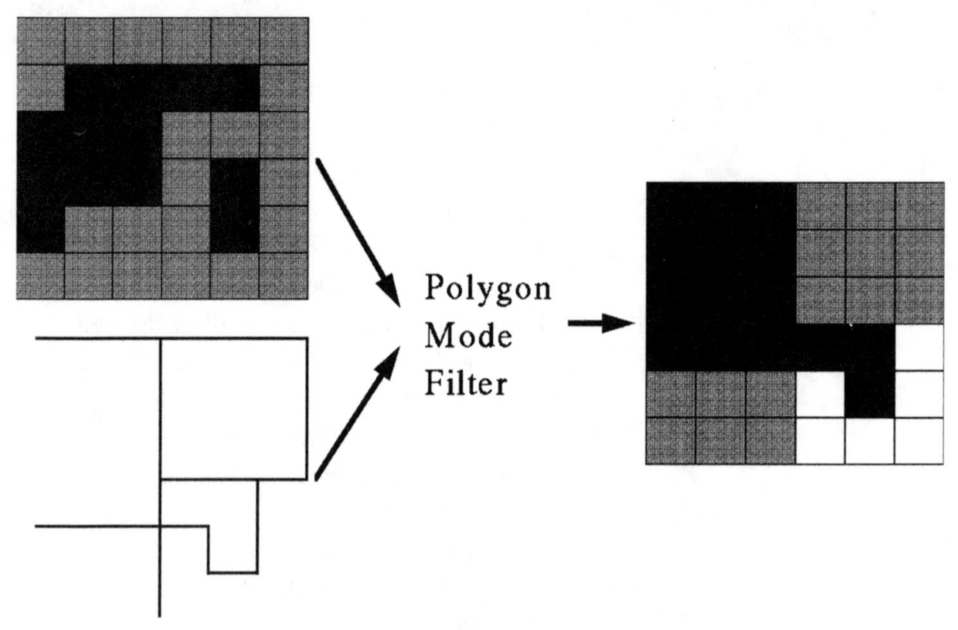

Application of the polygon mode filter. The top left grid is the original land cover and contains two classes. The bottom left data describe four homogeneous regions. The result is the grid on the right. Note that cells not part of a homogeneous region are left unclassified (unshaded cells) in the resulting grid.

The polygon mode filter was implemented in ARC/INFO using the GRID module. The homogeneous regions were determined using the gridded reference polygon data. It should be emphasized that only shape and location information were taken from the reference data; the class information was not used in the filter. There was no requirement that more than half of the pixels be in a single class before being reclassified. Prior to filtering, the homogeneous regions contained up to five classes. After the filtering, each region contained exactly one class. The accuracy of the filtered grid was quantified using the same methods described previously.

Accuracy estimates for the land cover classification after application of the polygon mode filter

	Producer's Accuracy	User's Accuracy	Overall Accuracy
Water	66.5	100.0	89.4
Forest	91.2	99.5	Kappa
Open	84.8	47.7	0.838
Bare soil	90.2	86.8	
Agriculture	87.4	87.8	

The driving force of this project, decreasing within-field heterogeneity, was accomplished via the polygon mode filter. In the process, the accuracy of the classified data according to each measure evaluated was increased. The filter increased the overall accuracy by 15 percent and the Kappa coefficient by 0.21, both substantial improvements. The class-specific accuracies as measured by the user's and producer's accuracies were improved by up to 28 percent.

Benefits

The polygon mode filter is a beneficial technique for post-processing land cover grids and is superior in some respects to rectangular mode filters. The method is applicable to data derived from many high resolution multi-spectral sensors (e.g., SPOT, AVIRIS). However, it requires additional information (compared to the rectangular filters) which may not always be available, especially for large areas. It may also only be applicable in limited situations, such as where boundaries remain constant, but the land cover changes with time. If the boundaries change from year to year, then they will have to be digitized every year and the efficiency of the technique is lost. Agriculture and forestry are the most obvious applications.

SPOT, ARC/INFO, and GRID: A New Concept in Agro-environmental Monitoring

Paul de Fraipont, Stephen Clandillon, and Dominique Esnault, DDAF du Bas-Rhin, Strasbourg, France
Didier Georgieff, Jean-Daniel Hennemann, and Alain Lefeuvre, SERTIT, Illkirch, France

Challenge

The central plain of Alsace, France (between Strasbourg and Colmar), is a resting and feeding zone for birds migrating through the Rhine corridor. This area also serves as a shelter for various sedentary bird populations. Previous studies based on remote sensing and other techniques have demonstrated the disappearance of Ried prairies in Alsace. Because of the high rate of prairie to corn tillage over the last two decades, European national and regional authorities have set up farmer conservation contracts to protect the environment (e.g., prairie upkeep, and livestock burden and nitrate fertilizer limitations).

In partnership with SERTIT, the DDAF du Bas-Rhin—a local government agency in charge of agriculture, environment, land management, and forestry in Alsace—is responsible for monitoring implementation of policies established within the framework of the European Union Common Agriculture Policy (PAC). The principal policy objective is environmental and groundwater quality conservation in the central plain.

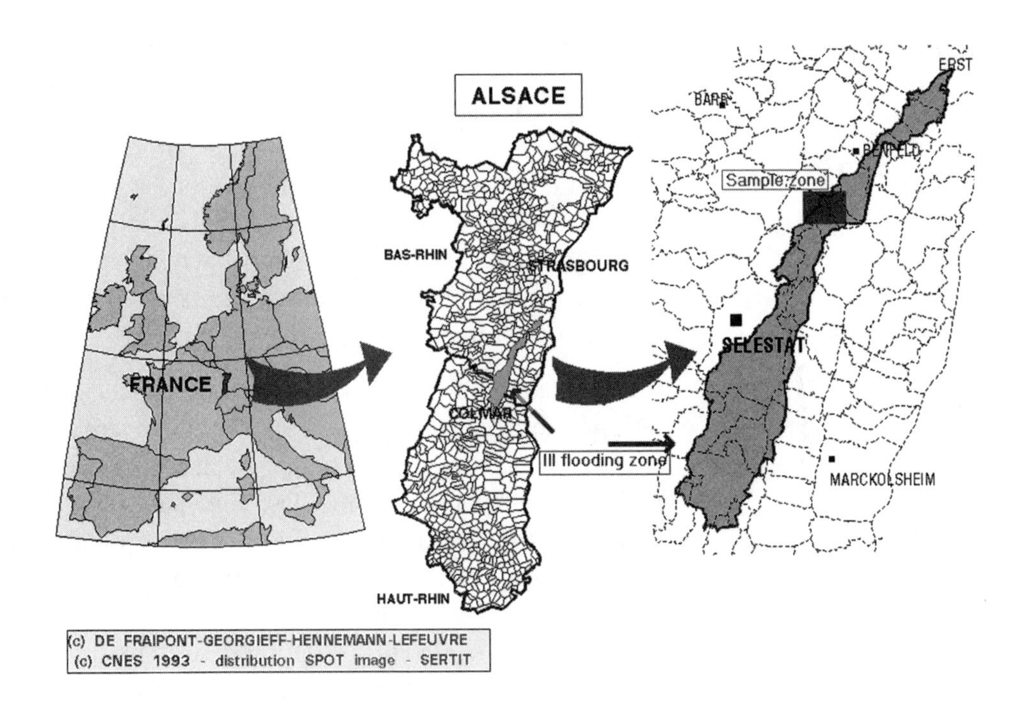

Study area.

Linking land use (monitored via SPOT and Landsat images) and fauna and flora environmental data to the area land parcel reference map was the first step in building the policy implementation monitoring system. The remote sensing data and GIS employed in this system are also useful for environmental impact assessment of socio-economic policies and regional planning. The table below lists primary remote sensing data sources.

Year	Prairie surface	Remote sensing data
1978	12,130 ha	Landsat MSS, 8 April 1978
1984	6,100 ha	Landsat TM, 17 October 1984
1989	2,870 ha	Landsat TM, 6 March 1989

Data and Methodology

The study area of 130 sq km encompasses 20,000 land parcels. The GIS contains data on all land parcels, including classification into 350 prairie grassland convention dossiers related to 3,000 declared prairie parcels under four different contract modes. With the use of SPOT imagery, a parcel-based classification algorithm has been developed. The algorithm classifies agricultural activity for the purpose of evaluating funding distribution along with optimizing environmental returns.

The GIS database includes a vectorized, parcel data layer with prairie declarations and a SPOT XS (November 1993) georeferenced image. The prairie declarations are related to the parcel layer via unique IDs. The spatial and spectral resolution and geometric precision of SPOT data are particularly well adapted to the observation of biophysical land use. The SPOT View data were directly managed through the ARC/INFO GRID module. The processing developed during this procedure is summarized in the following illustrations.

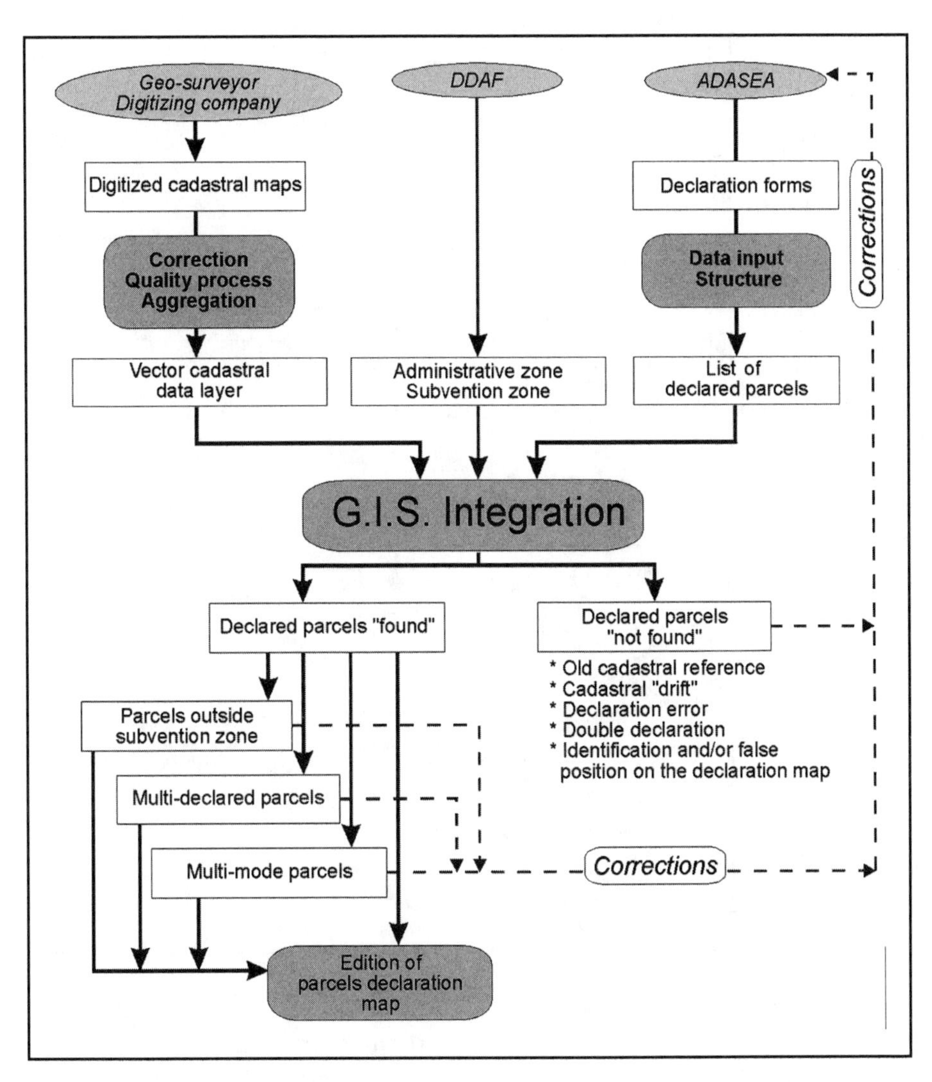

Classification by parcel.

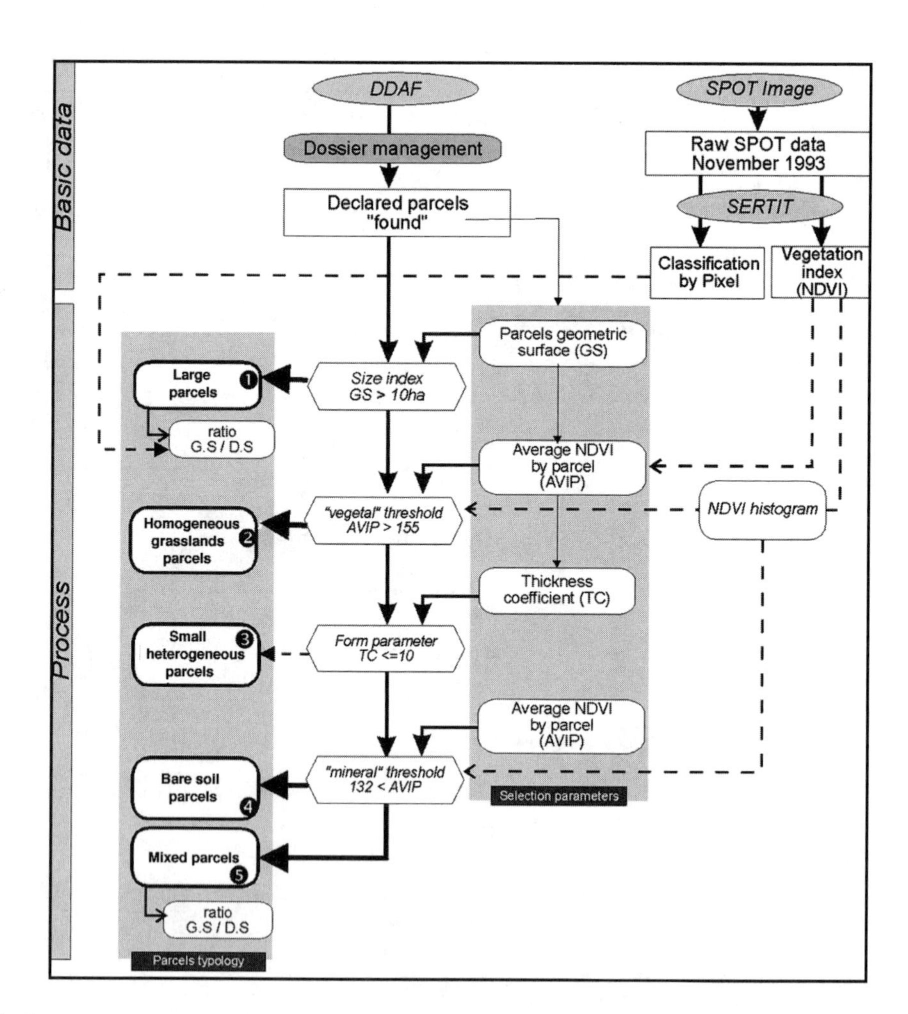

Parcels and declaration.

Because prairie parcels are considered permanently vegetated, a declared parcel can be monitored by a normalized difference vegetation index (NDVI) after inclusion in the declared parcels data layer. Two modes are observed on the NDVI histogram leading to a mineral/vegetal threshold.

In order to perform this parcel-based spatial analysis, a selection procedure is carried out based on parcel typology in relation to size, form, and spectral homogeneity. In this fashion, each parcel's radiometric homogeneity and declaration status are taken into account.

First, the vector cadastre layer is transformed to raster, and the NDVI index is then employed using average value per parcel. A sophisticated algorithm is constructed with the use of several GRID module tools: map algebra, vector to raster conversion, SPOT data integration, parcel analysis, spatial analysis, and graphic output. Hardware and software used in this process include a SUN SPARC 10, ARC/INFO, ERDAS IMAGINE, and the GRID module. Results of this procedure appear in the previous illustration.

The prairie contracts are managed with the SQL database OPEN ACCESS II, and then integrated into ARC/INFO in ASCII format. The vector parcel layer is managed with the use of ARC/INFO and the analyses performed through the GRID module and its map algebra subset. AML was employed to create a user friendly graphical user interface.

Results and Conclusions

Results of the above procedure are summarized in the table below.

Category	Number	Surface 96
Large parcels	27	12.6
Homogeneous grasslands	2,372	72.3
Small parcels	216	3,5
Bare soil parcels	24	1.0
Mixed parcels	251	9.9
Parcels outside flood zone	32	0.7
Parcels under size threshold	2	—
Total	2,924	100.0

Almost 85 percent of the declared parcels are homogeneous grasslands. The 10 percent of "mixed parcels" are categorized via a different relationship based on vegetal ratios versus declared ratios. Only 1 percent are "bare soil parcels." Remaining parcels are not included in the analysis due to small size.

The largest declared parcels (under 10 ha), which represent a small declared surface area, illustrate the limits of an administrative cadastre reference base in declaration control. The averaged NDVI makes little sense in the context of very narrow parcels (except when adjacent to other declared parcels) because of the 20-sq m pixel size (3.5 percent of declared prairie surface). An overview of the process appears in the illustrations below and in Plates 44 and 45.

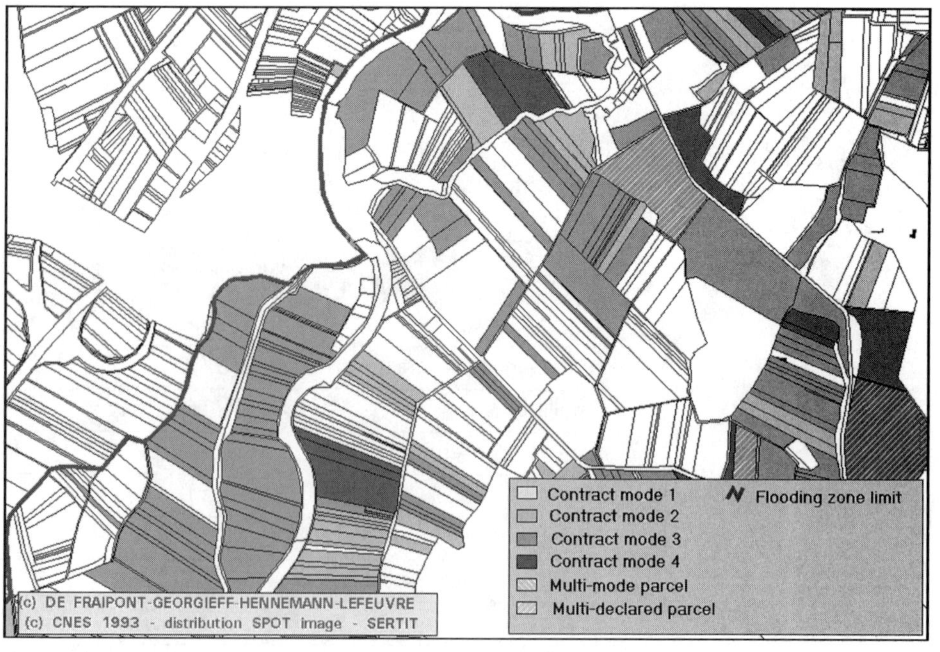

Declared parcels.

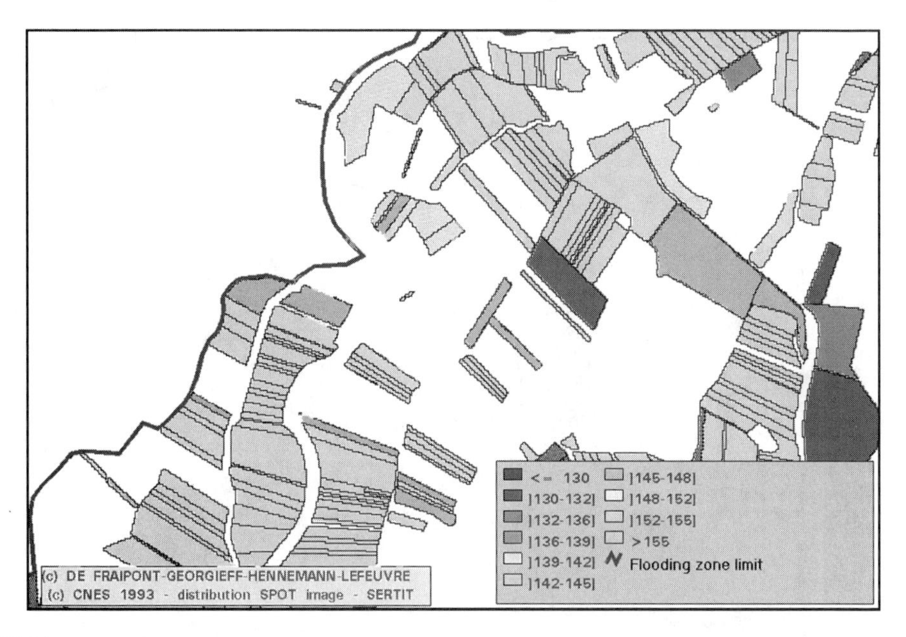

Average NDVI by parcel.

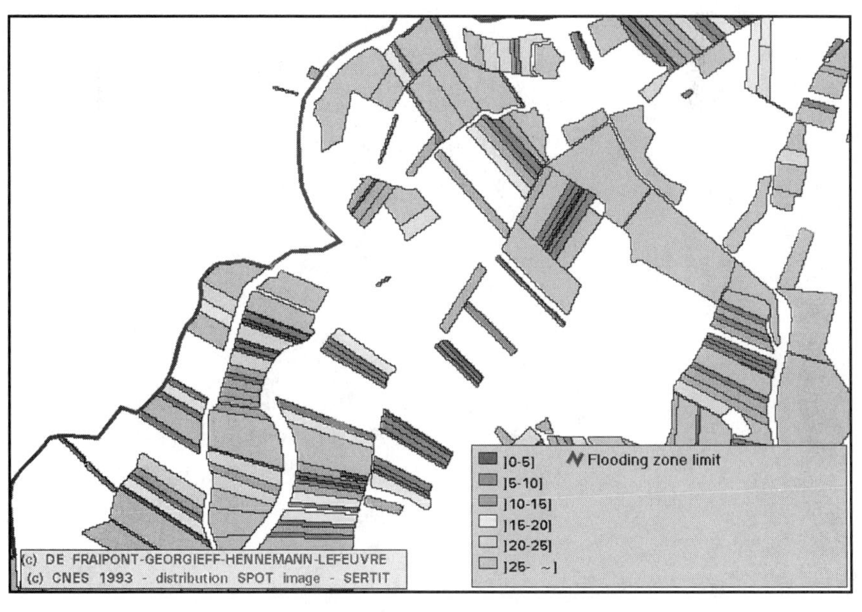

Thickness index.

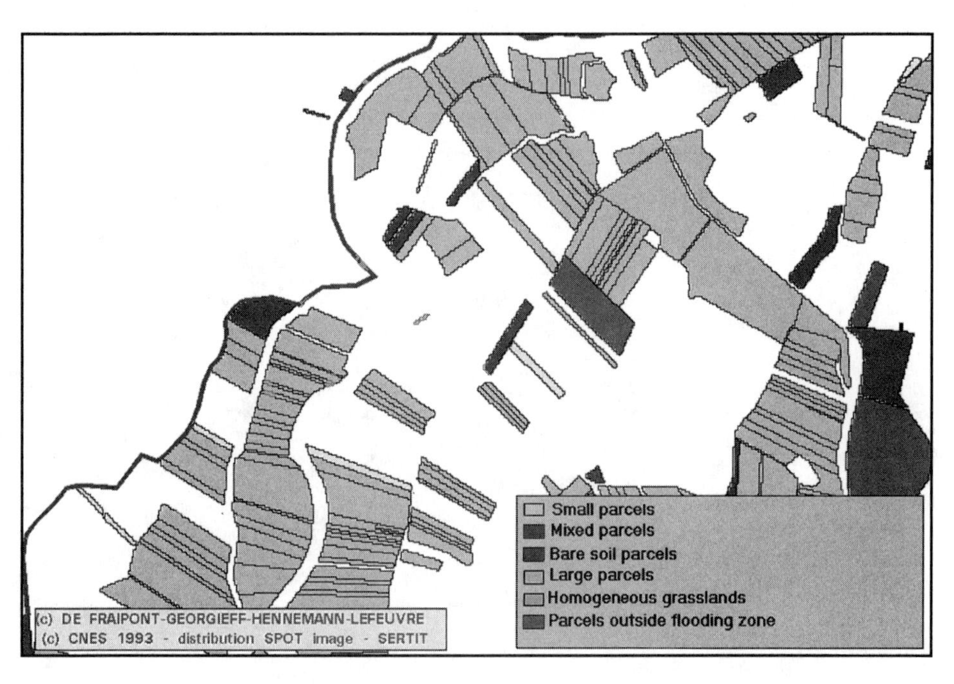

Parcels classification.

The spatial analysis of the data layers using 1:45,000 hard-copy output proved very useful in the parcel typological classification. For ground control purposes, 1:5,000 parcel maps and various parcel lists were output from ARC/INFO. This procedure (both experimental and operational) is efficient for monitoring land use practices even in instances of small parcels (project average is 0.0075 sq km). In this application, combining the radiometric-based indices with parcel homogeneity analysis is more efficient than a parcel-based analysis after land cover classification despite the larger data volume. The map algebra tools provided in the GRID module and the new SPOT image data and media render the process accessible to an administrative service lacking expertise in digital image processing.

Benefits

Another aspect of this study confirms that after automating the procedure via the AML, the entire process takes 16 hours from receiving the SPOT XS data on CD-ROM to statistics, lists, and cartographic hardcopy output related to the monitoring of the declared prairie grasslands. Moreover, an integrated, common approach to cadastral update and satellite data exploitation with other organizations in a regional database exchange would greatly improve the administrative process and data management efficiency, and therefore, further reduce minor discrepancies and project costs. The first phase of the project cost $US 300/sq km, including the initial elaboration of the parcel data layer (85 percent of total cost). At an annual update cost of approximately $US 60/sq km, the process described here is obviously cheaper and more accurate than conventional methods.

Vegetative Index for Characterizing Drought Patterns

James Rowland, Andrew Nadeau, John Brock, Robert Klaver, and Donald Moore, EROS Data Center
John E. Lewis, Department of Geography, McGill University

Challenge

In 1994, nearly 11 million people in the greater Horn of Africa (Burundi, Djibouti, Eritrea, Ethiopia, Kenya, Rwanda, Somalia, Sudan, Tanzania, and Uganda) were adversely affected by drought. Per capita food production declined in this region by more than 16 percent during the 1980-1993 period. Shortage of moisture for rain-fed agriculture has negative impacts on crop production, and if the moisture deficit is extensive both temporally and spatially, food production in this region is seriously jeopardized.

Satellite data processed into Normalized Difference Vegetation Indices (NDVI) are both timely and spatially comprehensive and can serve as an indicator of regional drought patterns. This is particularly useful because of the sparse distribution of rain gauges and the scarcity of timely and reliable rainfall data for the region. Ultimately, NDVI may be used as a planning tool for forecasting agricultural production.

Data and Methodology

The U.S. Geological Survey's EROS Data Center was requested by the U.S. Agency for International Development (USAID) to analyze factors affecting food security in the Greater Horn of Africa. The analysis was part of a larger agreement with USAID's Famine Early Warning (FEWS) Project. As part of that effort it is important to describe the historical and regional variability of NDVI and its relationship to drought. The NDVI is computed from surface reflectance, obtained from the National Oceanic and Atmospheric Administration's (NOAA) Advanced Very High Resolution Radiometer (AVHRR), as a ratio of the difference between the near infrared (NIR) and visible red (VIS) radiation. The formula is summarized below.

$$NDVI = \frac{NIR - VIS}{NIR + VIS}$$

Changes in internal leaf structure, vegetation chlorophyll content, and spatial density of vegetation cover—all of which influence the spectral reflectance—produce variations in NDVI. For example, an increase in chlorophyll pigment results in a decrease in visible reflectance and an increase in the near infrared reflectance. Consequently, the ratio of VIS and NIR produces a higher NDVI value. However, if the vegetation is under stress, a decrease in chlorophyll density and photosynthetic capacity results in lower NDVI values. The NDVI values range between -1 and +1, where negative values usually represent clouds, water, and other unvegetated surfaces, and positive values, especially those greater than +0.15, represent vegetated surfaces.

The NDVI data from 1982 to 1993 for Africa, resampled to 7.6km resolution from the original 4km resolution coverage, were obtained from Global Inventory Monitoring and Modeling Studies (GIMMS) at the National Aeronautics and Space Administration-Goddard Space Flight Center (NASA-GSFC). To minimize the effect of cloud and atmospheric contamination that systematically attenuate the value of NDVI, 10-day (dekad) temporal composites of NDVI were developed by choosing the maximum NDVI value for each individual pixel location. The maximum value composite selects the "greenest" value, which represents the least contaminated value for each dekad period.

Results and Discussion

In the past decade, NDVI has been shown to be an effective method for drought detection in certain regions of the world, and can provide a qualitative evaluation of the impact of moisture deficit upon vegetation. Furthermore, research indicates that clear relationships exist between vegetative primary productivity, vegetation phenology, and NDVI. The mean NDVI pattern for the greater Horn of Africa region is shown in the next illustration. This image map represents the mean of 12 annual maximum NDVI values, where annual maximum NDVI generally represents greenness at the height of the growing season.

The process of selecting annual maximum NDVI for each pixel disregards the seasonal timing of the NDVI values. Darker tones represent high NDVI values (>0.61). As shades become lighter, NDVI decreases to values of 0.01 - 0.10, representing regions of sparse or no surface vegetation. High annual NDVI values normally occur for southwest Sudan, Uganda, and throughout most of Tanzania; these high values are also found in the Kenyan and Ethiopian highlands. Substantial portions of Somalia, as well as northeast and eastern Ethiopia, are areas of very low annual maximum NDVI. Along with these regions of marginal rain-fed agriculture, lower NDVI values also exist in a crescent-shaped pattern from southeast Sudan, southeasterly to eastern Kenya, and southward to southern Kenya. Finally, the strong latitudinal gradient encountered in the Sahelian region is evident with the sharp changes in NDVI occurring through the central section of Sudan.

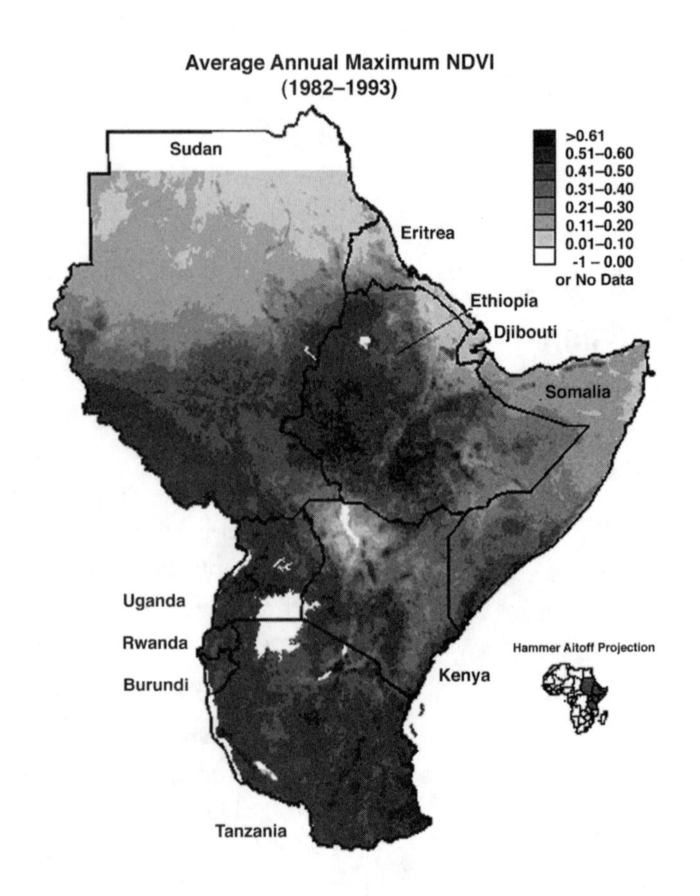

Average of 12 annual maximum NDVI (1982-1993). This image map represents the 12-year mean of the highest annual values of NDVI for each pixel in the region.

A measure of the variance in NDVI, derived from the 12 annual maximum NDVI values for each pixel, is portrayed by a coefficient of variability map. A gray area represents low variability, and includes regions of small relative change regardless of whether it is characterized by substantial or sparse vegetation cover. By inference, these regions experience sufficient rainfall or are consistently arid. With few exceptions, these regions of low variability coincide with either the highest or lowest NDVI areas in the previous illustration.

Regions of high variability in NDVI depict regions that are either highly variable in precipitation regime, or extremely sensitive to slight changes in pedologic and/or anthropogenic variables. The highest variability of NDVI (>11 percent) is found in eastern Sudan and northern Kenya, with secondary regions of high variability (6 to 10 percent) covering a substantial portion of the greater Horn region. Central Sudan, eastern Ethiopia, and a crescent-shaped region from southern Sudan to eastern Kenya to southern Kenya are regions of higher variation in moisture availability, which ultimately has an impact on food production. These same regions show variability ranging from 20 to 40 percent of annual rainfall variability (data prior to 1985), with striking geographic coincidence to the variability pattern for NDVI between 1982 and 1993. Furthermore, even 10 percent variability in NDVI has potential consequences for regional food production.

The NDVI is also sensitive to interannual variability of rainfall in areas where mean annual amounts are less than 1,200 mm (about 5 in). In such areas, year-to-year climatic variability will tend to have greater socio-economic impact than changes in the mean climatic condition. Next, there is evidence that interannual variability of precipitation in the greater Horn region has increased over the past 50 years. However, a longer NDVI time series may be required before a conclusive statistical statement can be made concerning the role of NDVI in appraising interannual variability.

Another map depicted the year of the lowest annual maximum NDVI for the period 1982 to 1993. Shades of light green represented 1983-84, the drought years in Sahelian Africa. The same dry years also appear in western Kenya (a major agricultural production region) for lowest annual NDVI, corresponding to dry years for that area. Another extensive pattern of minimum annual NDVI values for this geographic region coincides with the major drought for the 1991-92 growing season, which had disastrous effects in Kenya and northern Tanzania.

NDVI temporal patterns for two specific locations are representative of the 1984 and 1991-92 droughts: En Nahud in Sudan and Marsabit in Kenya. In these areas, the significance of NDVI related to wet and dry periods becomes more apparent. For each location, NDVI values were extracted for a 100km radius for each dekad from 1982-93. From each spatial sample, the median was calc ulated, and then the mean NDVI for each dekad was computed. The next illustrations display the average dekad NDVI with individual dry and/or wet years. These graphs emphasize the relationships among greenness, drought, and precipitation. The NDVI curves for the drought year of 1984 and the "good year" of 1988 (the highest rainfall in 20 years) in central Sudan are compared with the average NDVI values for 1982-93. The rainy season for En Nahud is generally from June (dekad 18) to September (dekad 25). Likewise, the agricultural drought of 1991-92 in Kenya is well illustrated by comparing these years with the 12-year average. The bimodal distribution of precipitation is well represented in the bimodal distribution of NDVI for Marsabit, with the long rains occurring in the first half of the year and the short rains primarily in November (dekads 31-33).

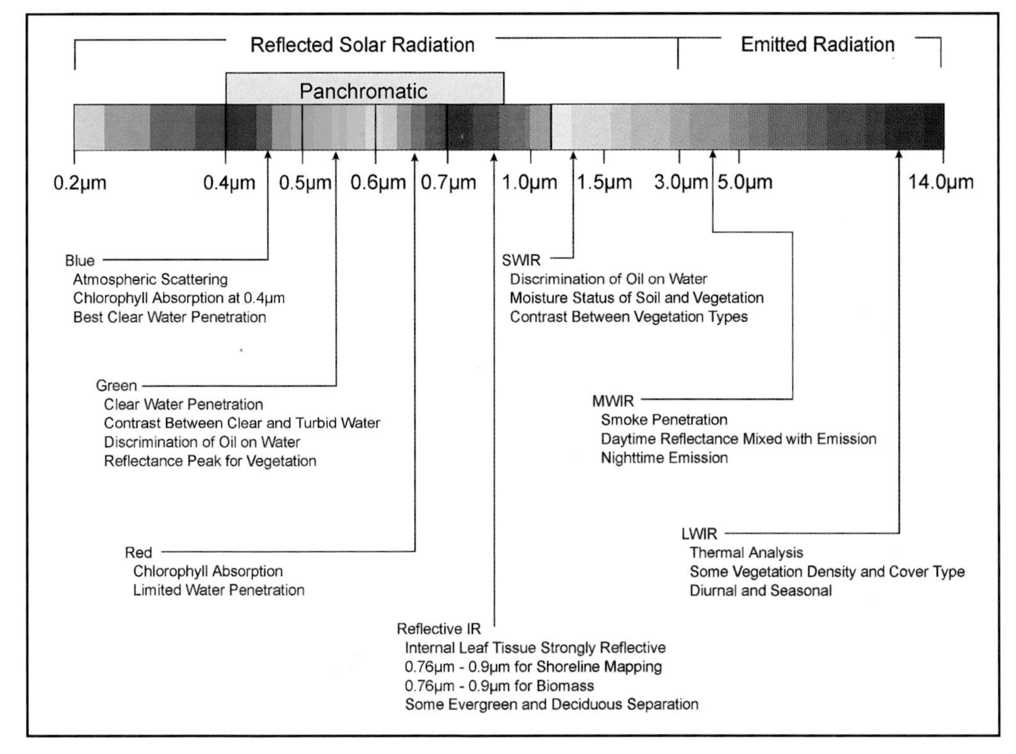

Plate 1.
Commonly used regions of the EM spectrum between the ultraviolet (0.2µm) and long wavelength infrared (14µm). *(Image Formation and Raster Characteristics)*

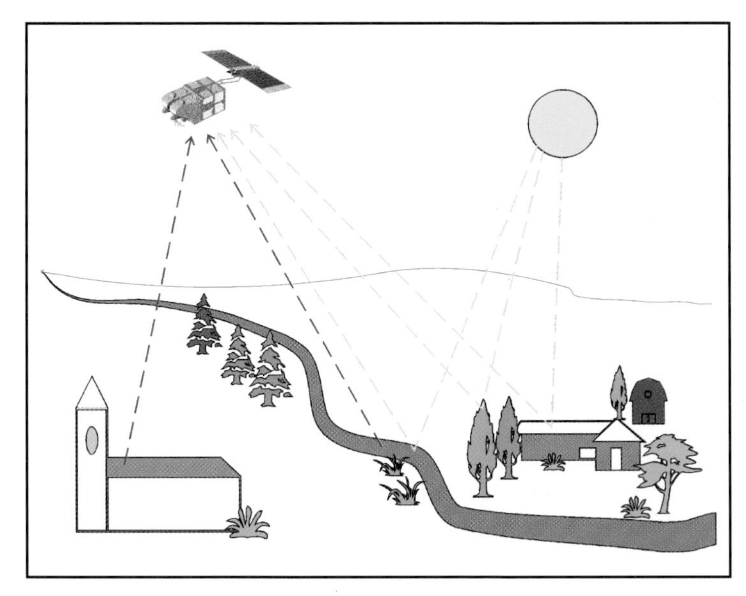

Plate 2.
Satellites collect reflected (yellow) and emitted (red) energy from surface features. *(Image Formation and Raster Characteristics)*

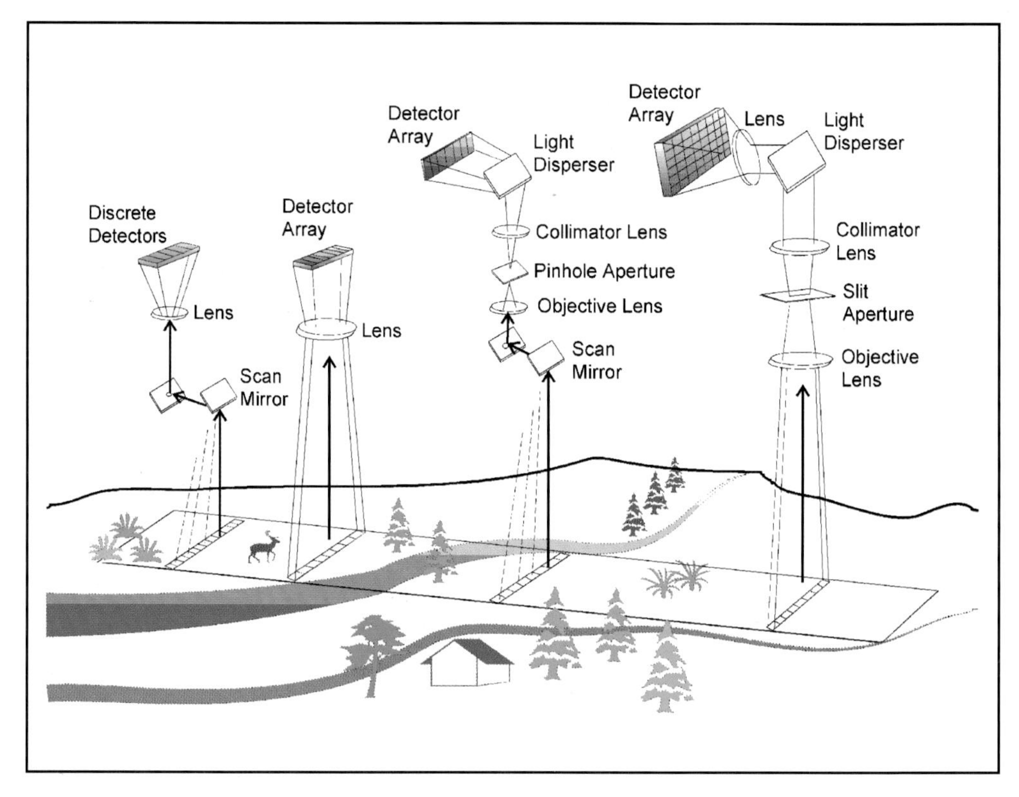

Plate 3.
Four design strategies for scanners and imaging spectrometers.
(Image Formation and Raster Characteristics)

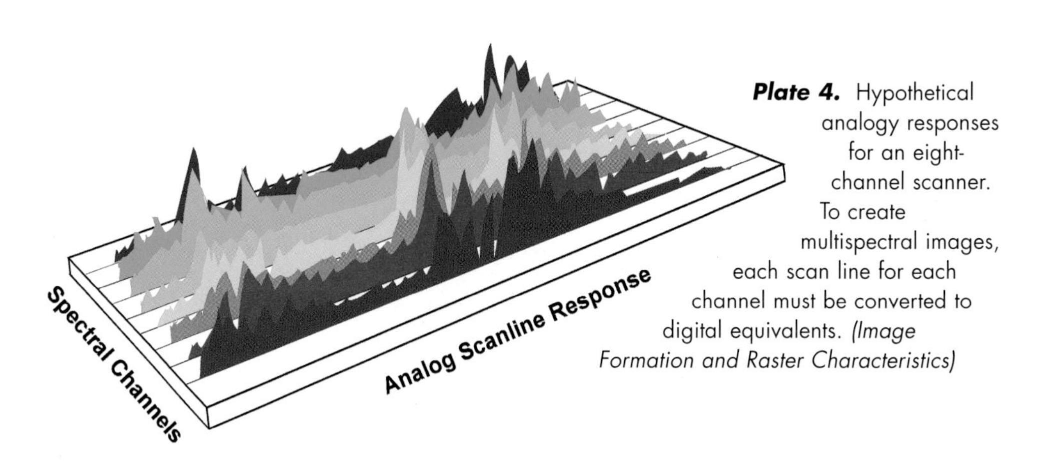

Plate 4. Hypothetical analogy responses for an eight-channel scanner. To create multispectral images, each scan line for each channel must be converted to digital equivalents. *(Image Formation and Raster Characteristics)*

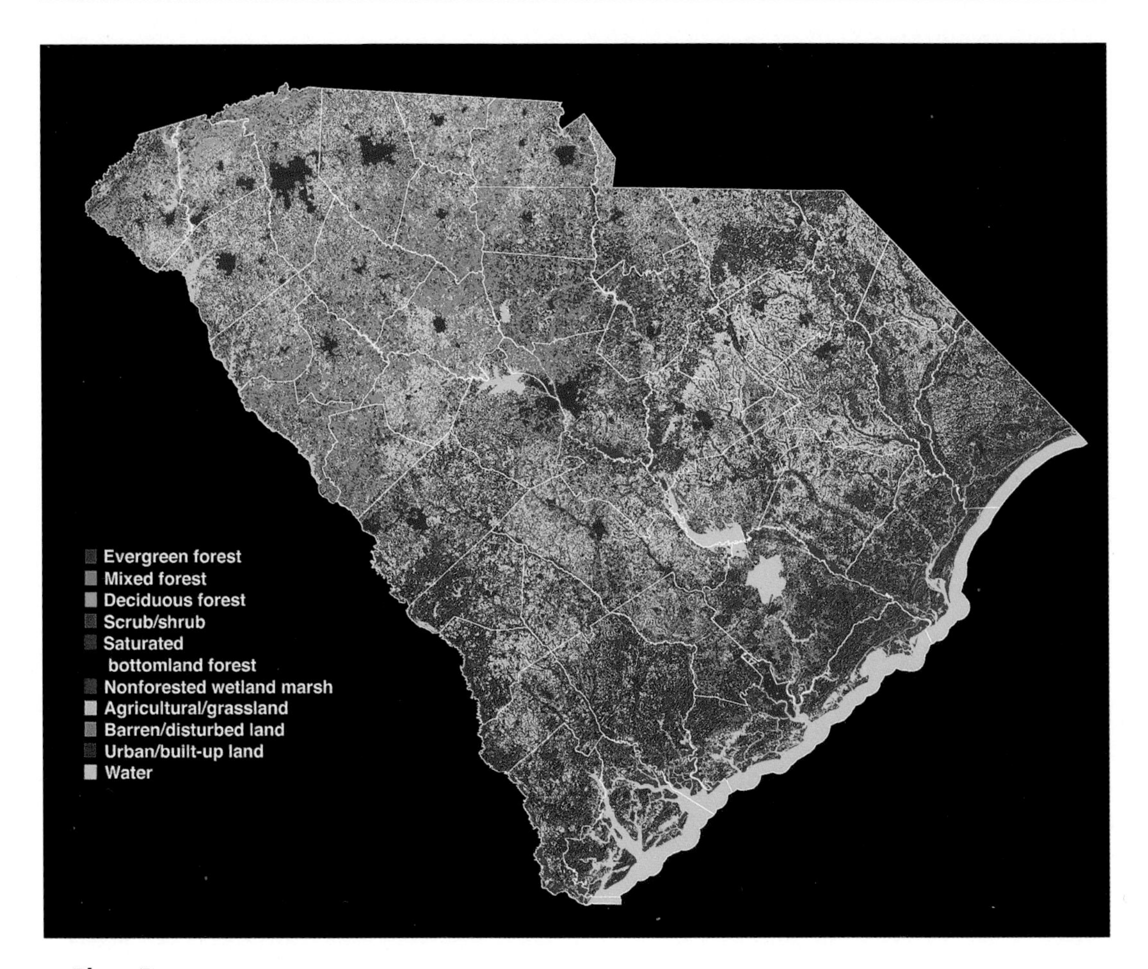

Plate 5.
Land cover classification of South Carolina created from SPOT statewide satellite image coverage (approximately 46 SPOT images). Produced by South Carolina Land Resources Conservation Commission; image processed with ERDAS software and ARC/INFO for GIS work. (SPOT Image Corporation CNES ©1993) *(Image Formation and Raster Characteristics)*

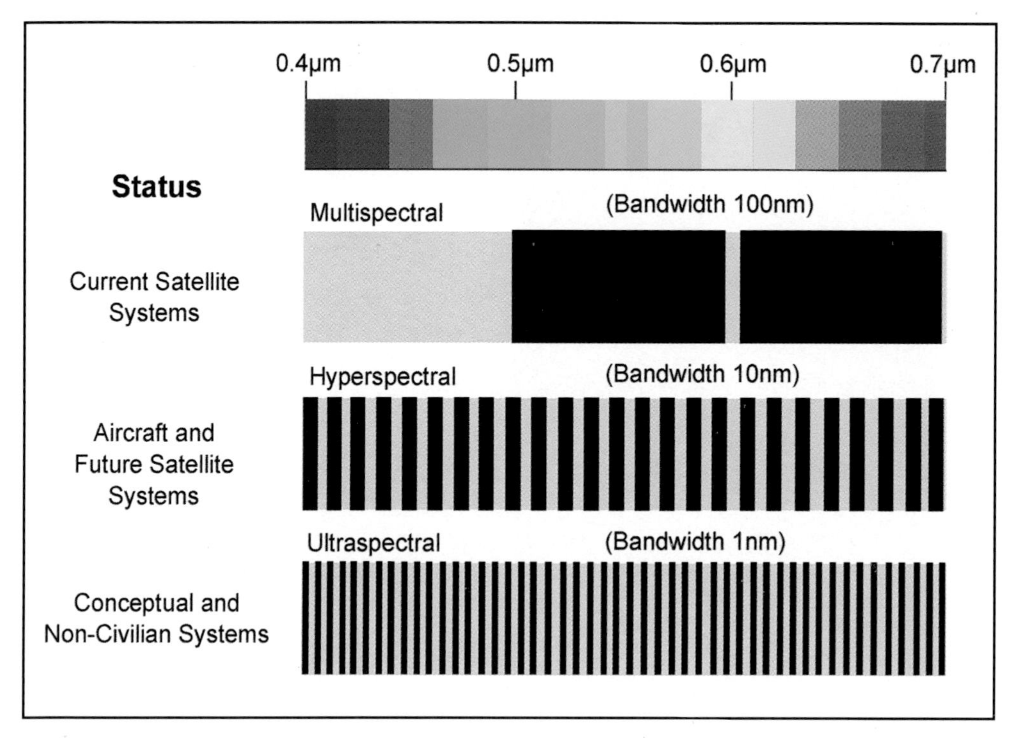

Plate 6.
Bandwidths of current satellite spectral channels (top) are broad by comparison to anticipated future systems (bottom two rows), but they enable wide swath widths, moderately fine spatial resolutions, and twice monthly revisit calendars. *(Image Formation and Raster Characteristics)*

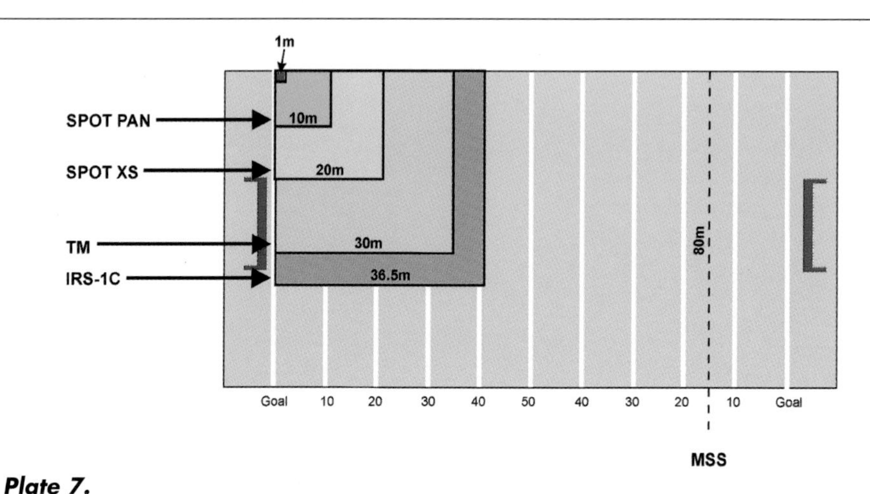

Plate 7.
Typical pixel sizes from current satellite scanner systems. The SPOT Image sensors obtain data with GSDs of 10m and 20m. The Landsat series has collected 30m and 80m data. *(Image Formation and Raster Characteristics)*

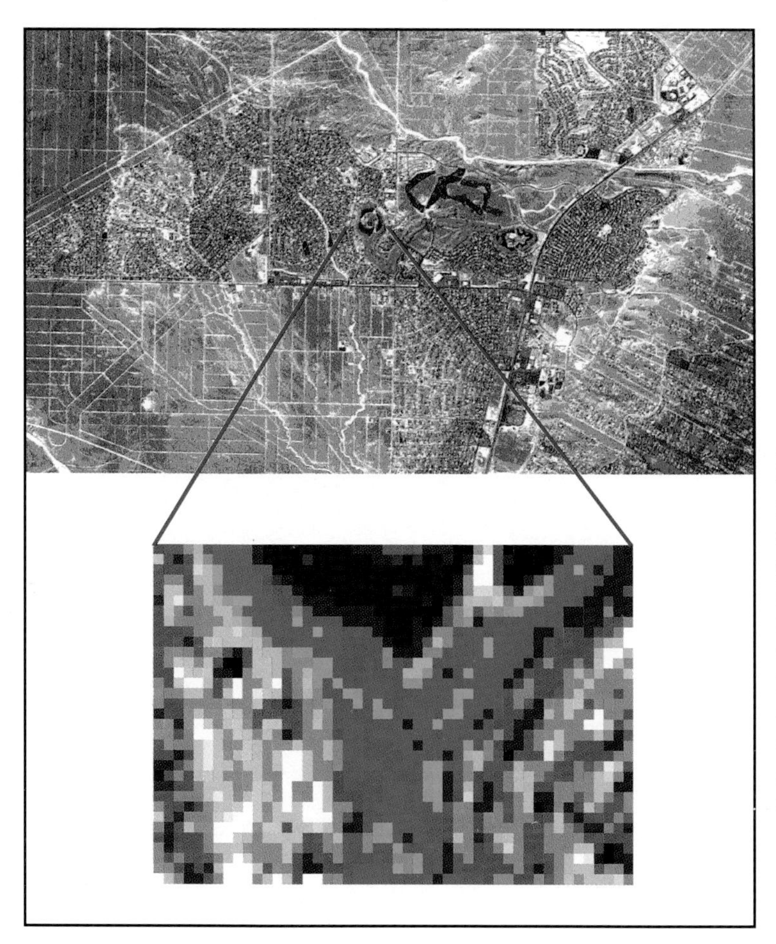

Plate 8.
Enlargement of part of a processed SPOT Image scene showing pixel colors. Pixel colors having nearly identical reflectance values in the three channels form homogeneous regions while those on the edges of regions show gradations between regions. *(Image Formation and Raster Characteristics)*

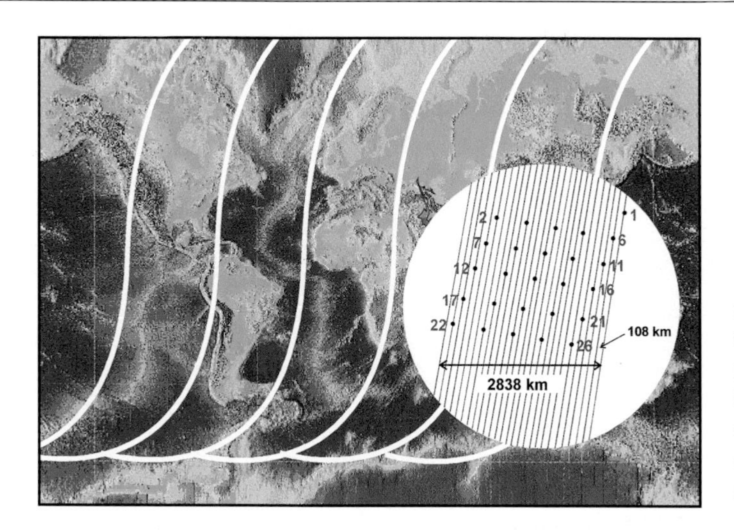

Plate 9.
Orbital progression (swathing pattern) for SPOT. *(Data Collection Systems, Formats, and Products)*

Plate 10.
This highly detailed image map of Seattle is a SPOT Image 10m panchromatic and 20m multispectral merge. The true color image is visually enhanced and GIS-ready. (SPOT Image Corporation CNES ©1993) *(Data Collection Systems, Formats, and Products)*

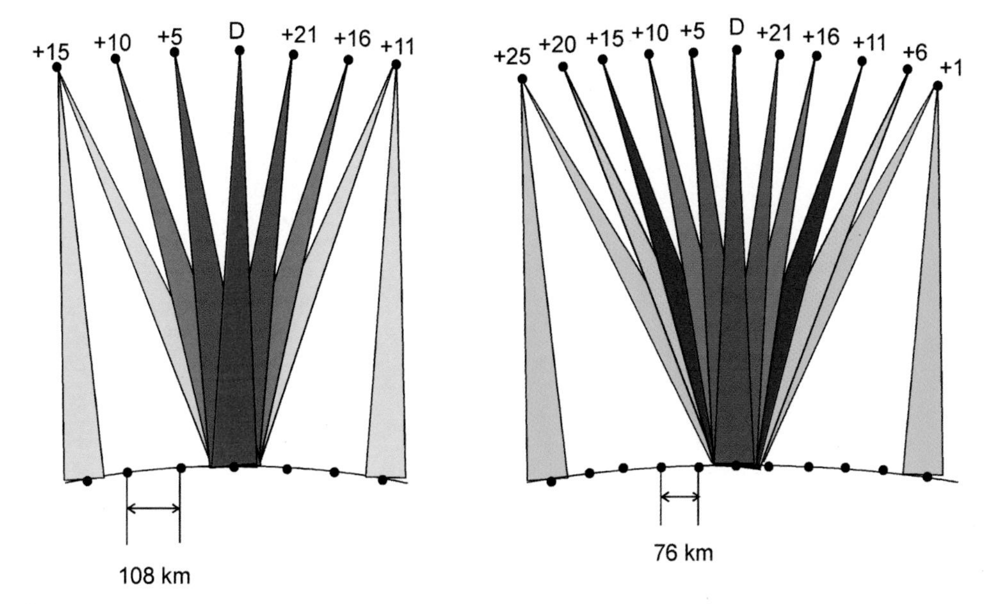

Plate 11.
SPOT revisit sequence at Equator (left). SPOT revisit sequence at 45° latitude (right).
(Data Collection Systems, Formats, and Products)

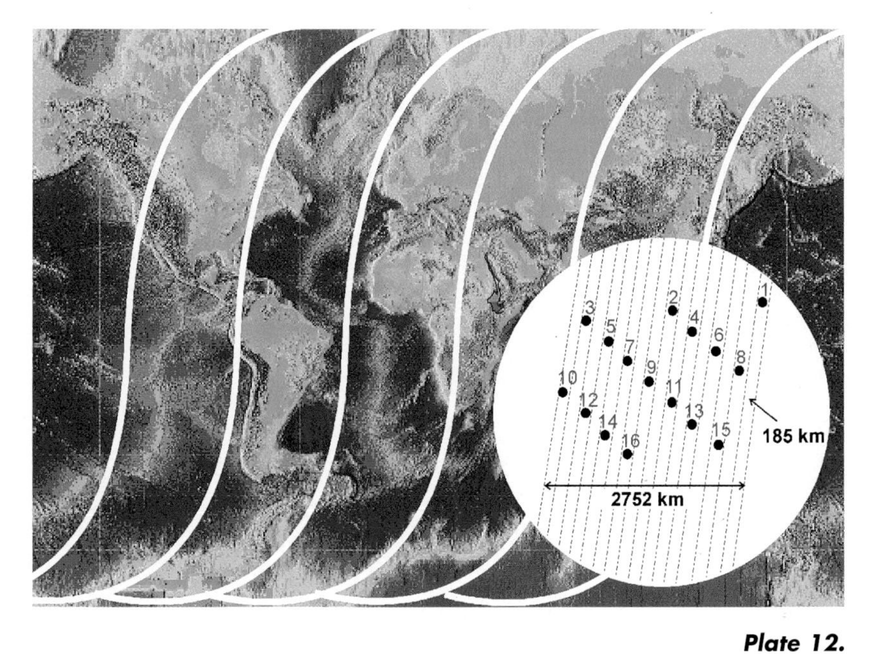

Plate 12.
Orbital swathing pattern for Landsat -4, -5.
(Data Collection Systems, Formats, and Products)

Plate 13.
TM image of Albuquerque, New Mexico. (Courtesy, Earth Data Analysis Center, University of New Mexico) *(Data Collection Systems, Formats, and Products)*

Plate 14.
First global image produced from AVHRR data. *(Data Collection Systems, Formats, and Products)*

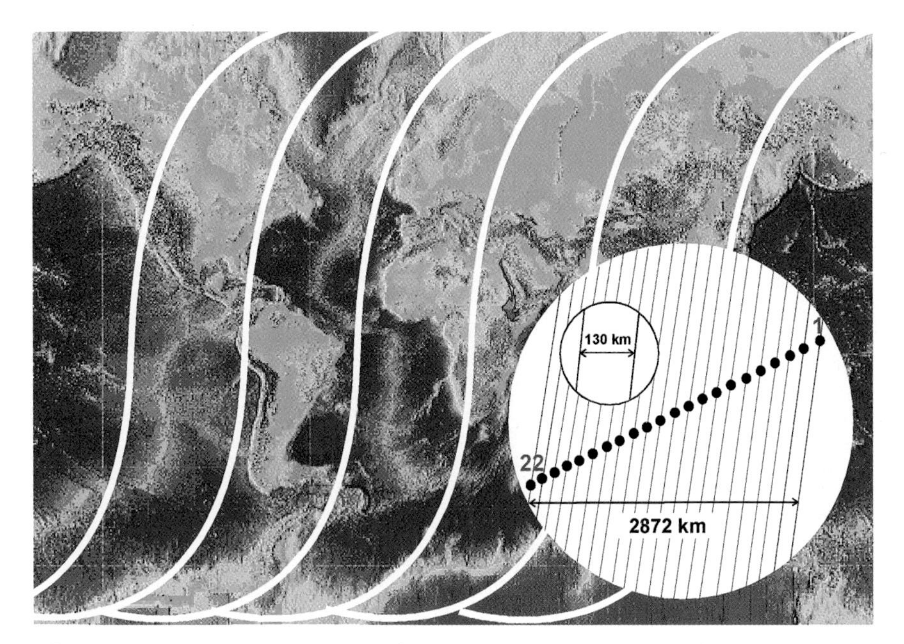

Plate 15. Orbital swathing pattern for IRS-1A, -1B. *(Data Collection Systems, Formats, and Products)*

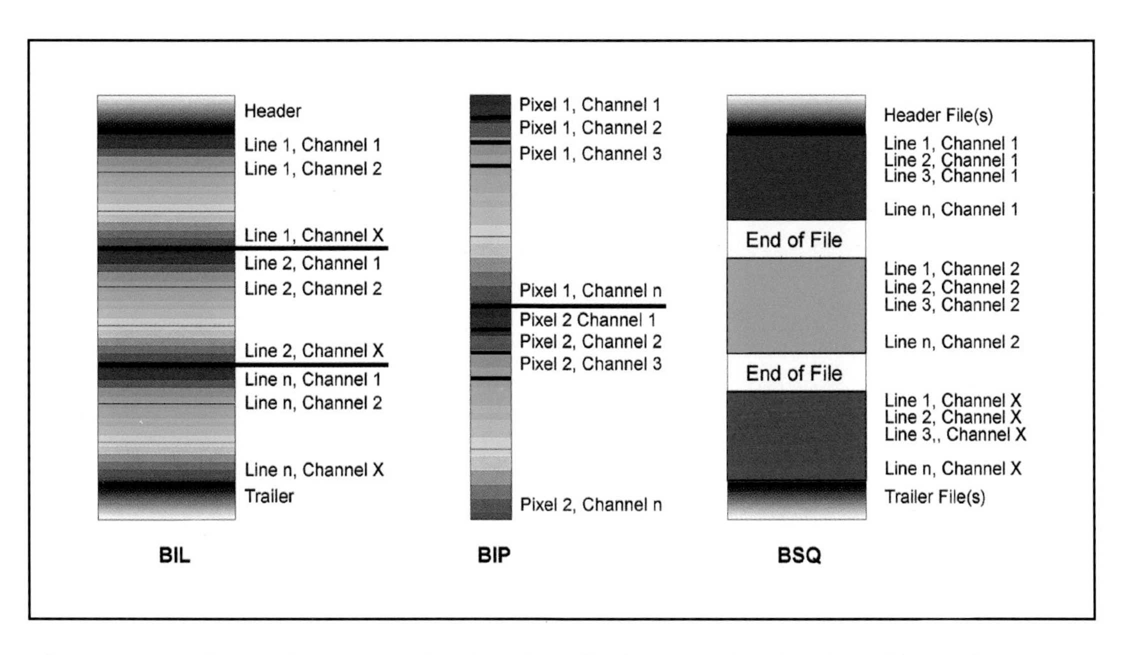

Plate 16. Digital image formats: BIL = band interleaved by line; BIP = band interleaved by pixel; BSQ = band sequential. *(Data Collection Systems, Formats, and Products)*

Plate 17.
A SPOTView image map showing the Kremlin, Red Square, and other features. The SPOTView is a merge of SPOT imagery and GPS control points taken from several bridges crossing the Volga River. (SPOT Image Corporation, CNES ©1993) *(Image Display and Processing for GIS)*

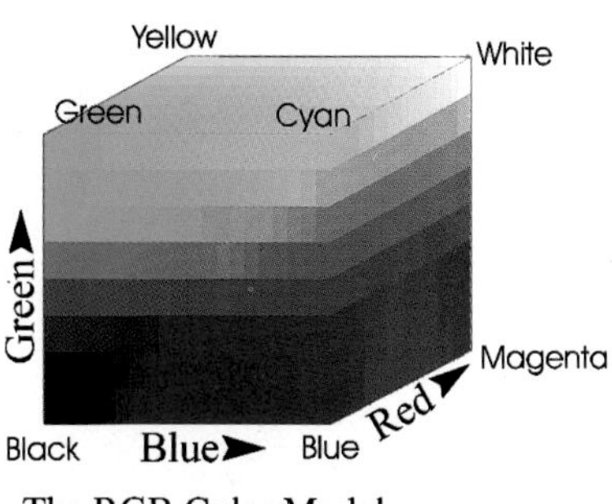

Plate 18.
The RGB color model. (Courtesy, Tony Renaud, Maricopa County of Transportation) *(Image Display and Processing for GIS)*

CATEGORY	GREEN CHANNEL	RED CHANNEL	INFRARED CHANNEL	COMBINATION
VEGETATION				
Coniferous				
Deciduous				
WATER				
Clear				
Sediment				
BARE SOIL				
CLOUD				

Plate 19.
Grayscale responses and equivalent yellow, magenta, and cyan intensities forming false color composites for typical terrain categories. *(Image Display and Processing for GIS)*

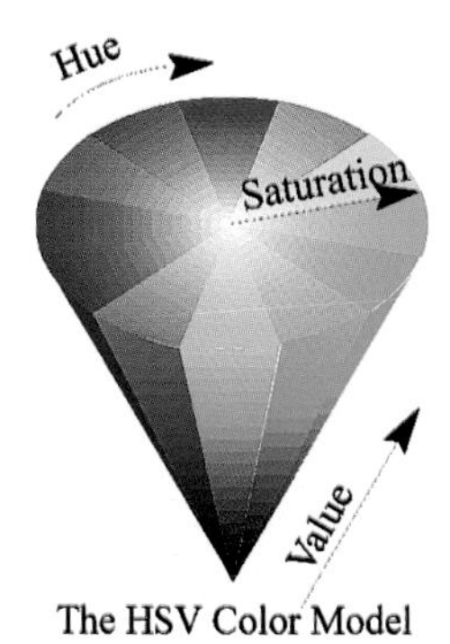

Plate 20.
The HSV color model. (Courtesy, Tony Renaud, Maricopa County of Transportation) *(Image Display and Processing for GIS)*

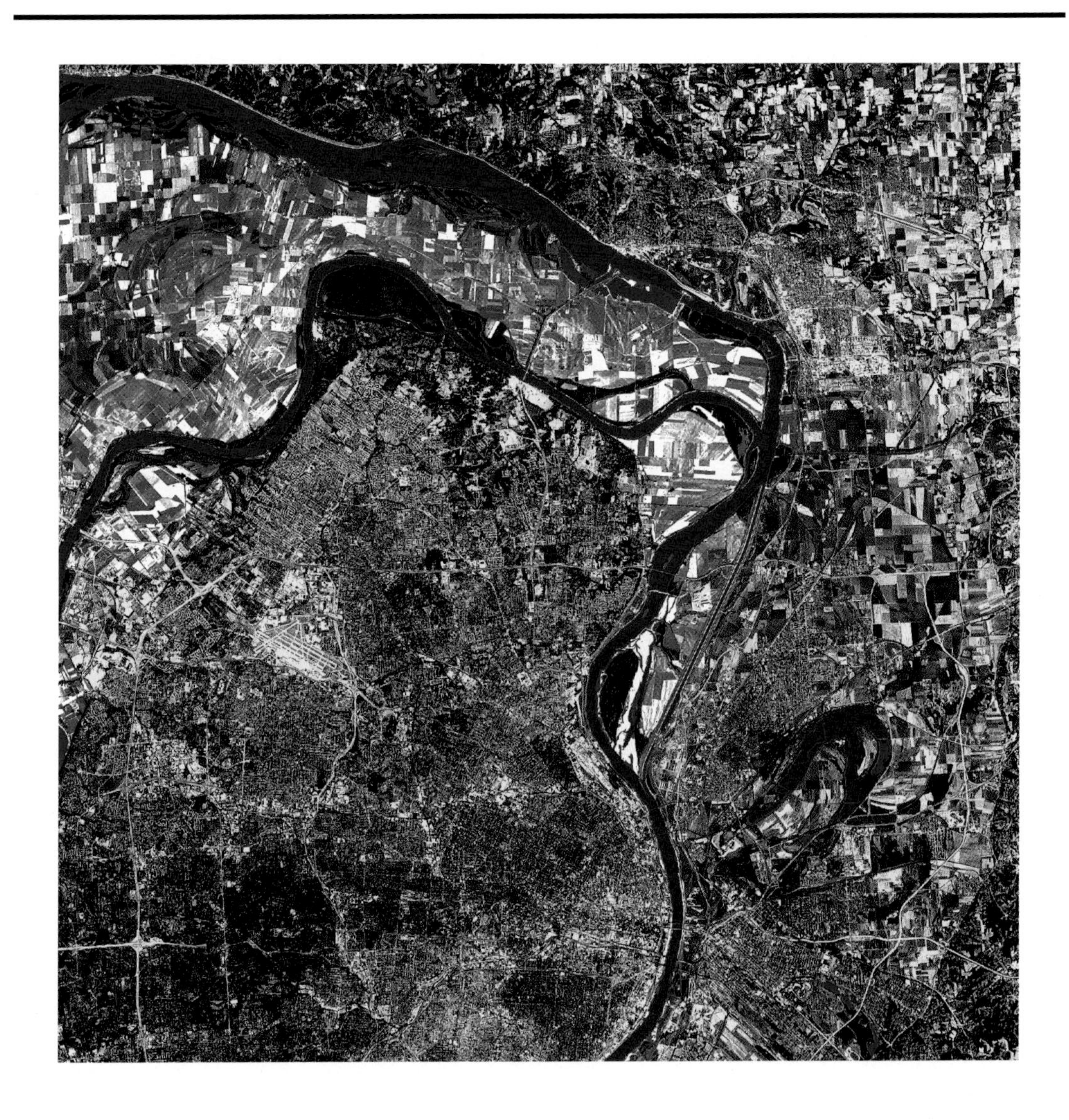

Plate 21.
Change detection showing pre-flood and flood stages of the 1993 Mississippi River flood. Normal river courses shown in dark blue, and flood stage in light blue. Processing reveals land use patterns beneath the flood waters, providing records of damage important to insurance companies and regional planners, among others. (SPOT Image Corporation, CNES ©1993) *(Image Display and Processing for GIS)*

Plate 22.
Perspective view looking southwest from mainland China across The New Territories toward Hong Kong. Created exclusively from SPOT satellite imagery, this digital elevation model (DEM) was used for wireless communication system planning. (Courtesy, SPOT Image Corporation, CNES) *(Image Display and Processing for GIS)*

Plate 23.
SPOT imagery is often used to verify, update, and correct vector locations. In this plate, digital line graph (DLG) vectors are overlaid on a SPOTView of Bee Ridge, FL (corresponds to the USGS 7.5 minute quadrangle map). Available DLGs are outdated and only show a small portion of roads (yellow) and hydrography (blue) which have been recently developed. (SPOT Image Corporation, CNES ©1992) *(Image Display and Processing for GIS)*

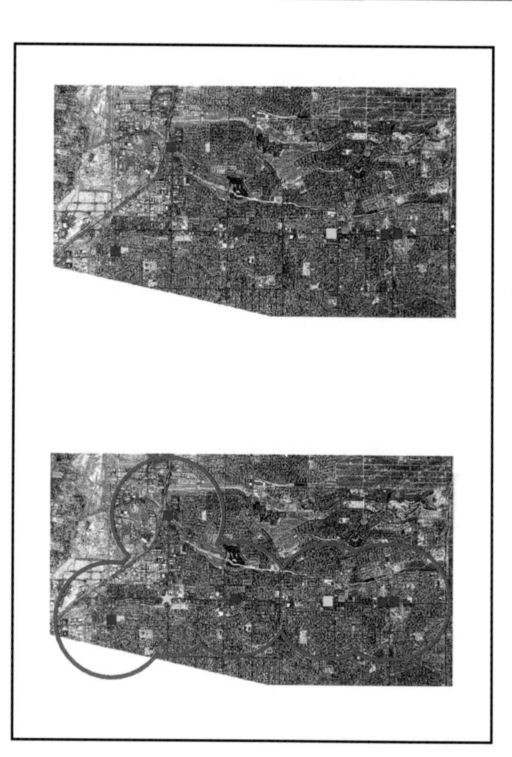

Plate 24.
GIS-generated distribution of copy and duplicating shops superimposed on a SPOT image of northeastern Albuquerque. The yellow square is an existing Company X franchise and the blue squares are competing copy and duplicating shops. A site for a new Company X franchise in northeast Albuquerque appears as a yellow star. The market potential of the site can be evaluated by accessing land use and street patterns. Other factors within the theoretical range of the shop can be delimited by a red buffer. *(Evaluating Franchise Locations)*

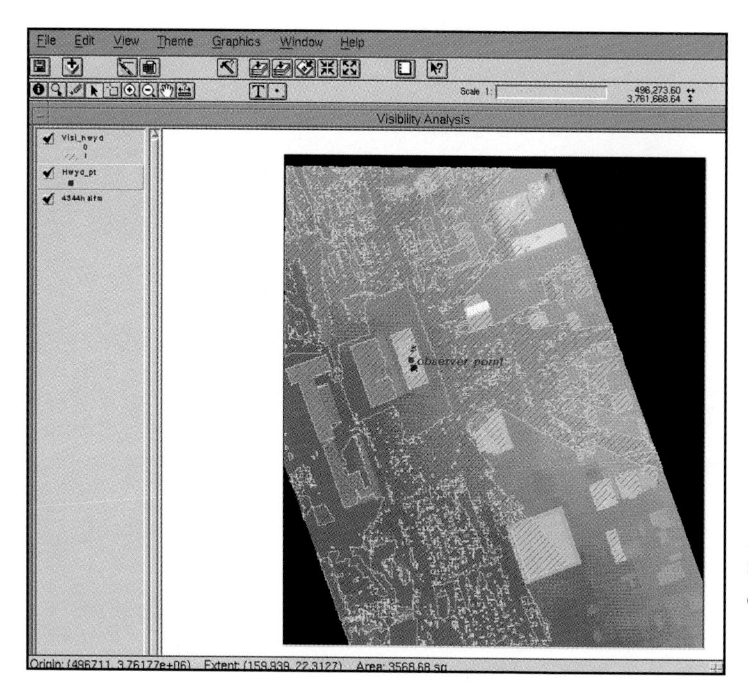

Plate 25.
Results of an intervisibility analysis. *(Cellular Phone Transceiver Site Selection)*

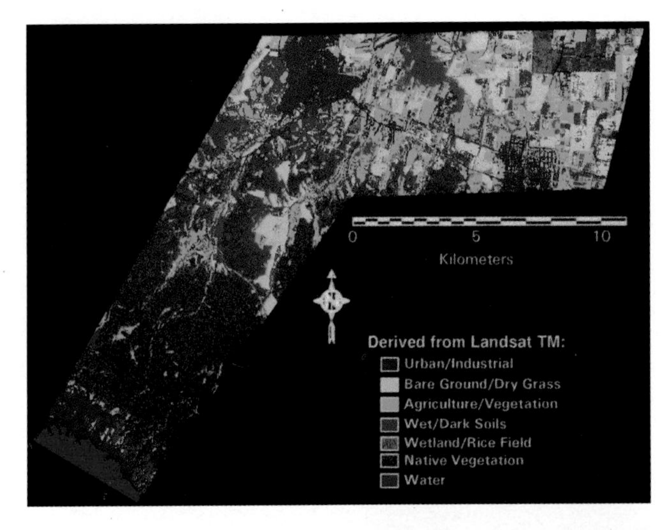

Plate 26.
Color composite of TM bands 3, 5, and 4 as blue, green, and red, respectively. The bands distinguish key land use classes used as input for a supervised classification. *(Finding a Least-cost Path for Pipeline Siting)*

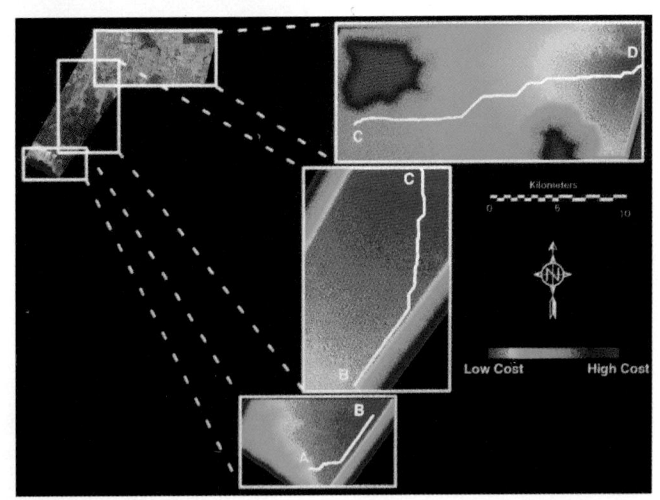

Plate 27.
Cumulative cost surface map. *(Finding a Least-cost Path for Pipeline Siting)*

Plate 28.
Terrain analysis, Los Angeles County. Digital elevation models (DEMs) were processed and overlaid with satellite imagery in ERDAS IMAGINE. This analysis is used to assist in identifying landmark terrain features for visual quality evaluation of transportation alternatives. *(Choosing Efficient, Cost-effective Transportation Routes)*

Plate 29.
Land use, San Francisco Bay area. To ensure that the majority of workers and residential populations could be served by a proposed transportation facility, while simultaneously minimizing displacements and relocations, land use types and densities were analyzed and mapped with satellite imagery in ERDAS IMAGINE. *(Choosing Efficient, Cost-effective Transportation Routes)*

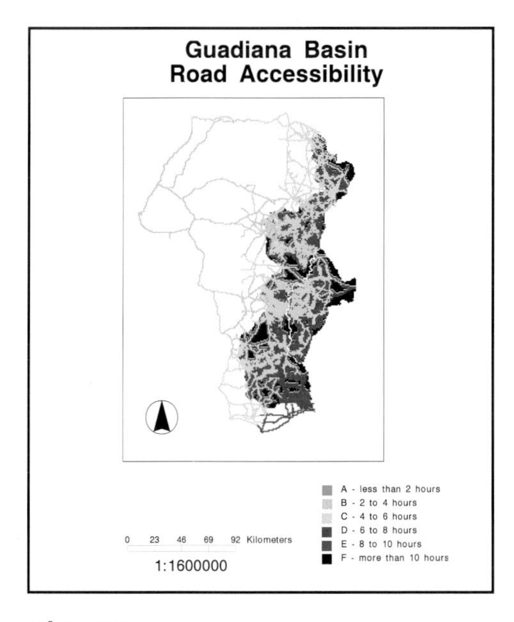

Plate 30.
Map used to calculate travel time to villages for cuisine of interest to tourists. *(Assessing Tourism Potential: From Words to Numbers)*

Plate 31.
Areas with tourism potential of over 50 percent. *(Assessing Tourism Potential: From Words to Numbers)*

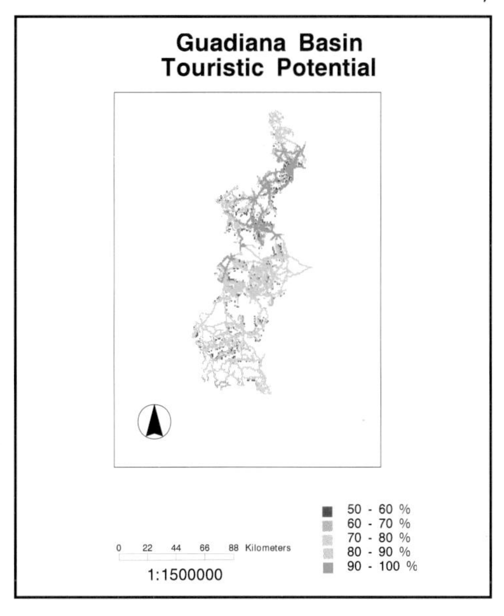

Plate 32.
On-screen (manual) updating 1990 land use coverage (yellow) to 1995 conditions (green) based on interpretation of SPOT Pan-XS merged and high pass filtered image. *(Editing and Updating Vector GIS Layers Using Remotely Sensed Data)*

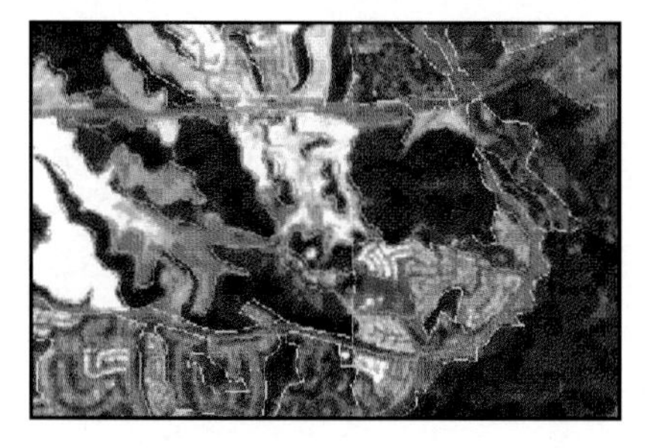

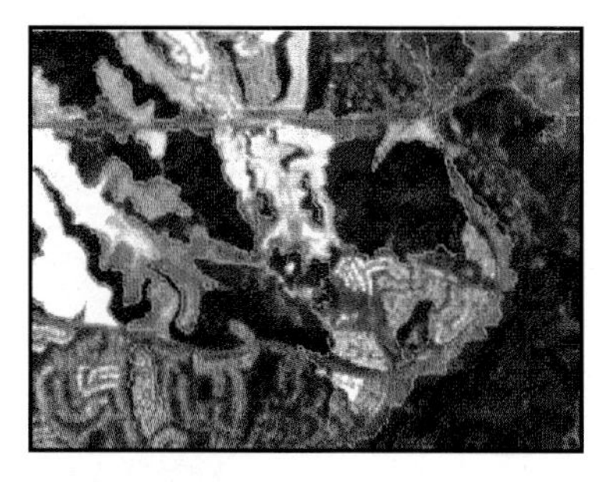

Plate 33.
Automatic generation of 1995 land use vectors based on classification of SPOT multitemporal panchromatic difference image and raster-to-vector conversion. *(Editing and Updating Vector GIS Layers Using Remotely Sensed Data)*

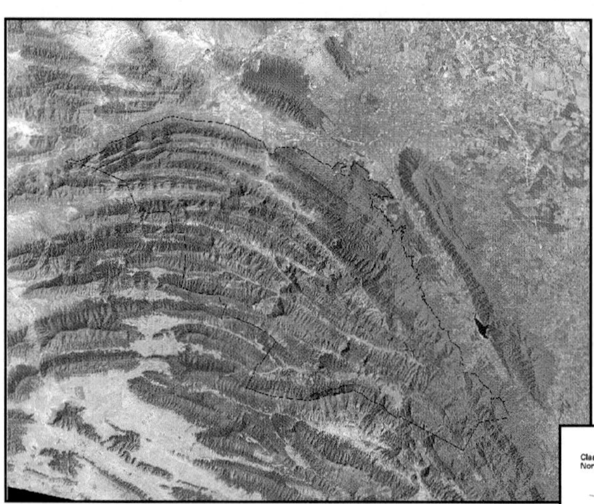

Plate 34.
Landsat Thematic Mapper image of Cumbres de Monterrey National Park. TM-4 appears in red, TM-5 in green, and TM-3 in blue. *(Modeling Vegetation Distribution in Mountainous Terrain)*

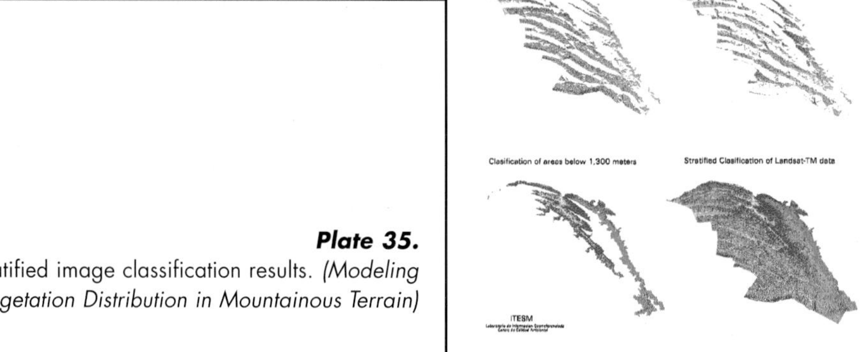

Plate 35.
Stratified image classification results. *(Modeling Vegetation Distribution in Mountainous Terrain)*

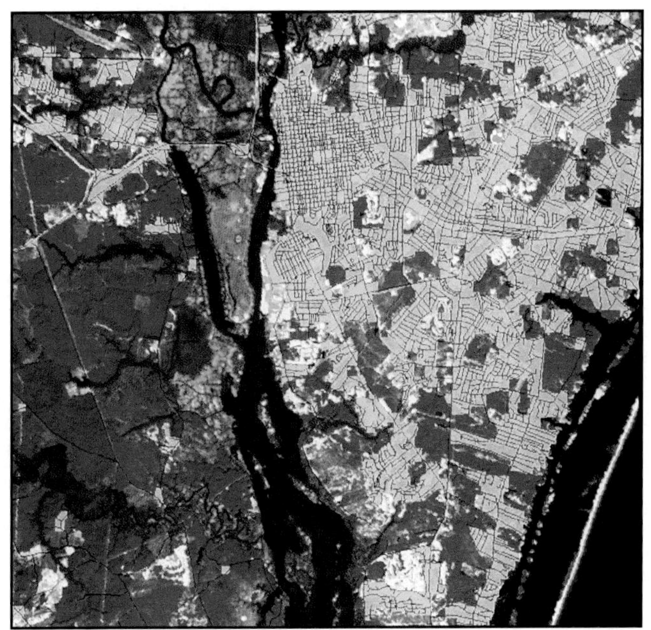

Plate 36.
Urban mask on satellite image, Wilmington, North Carolina, and surrounding areas. *(An Urban Mask Raster Image for Vector Street Files)*

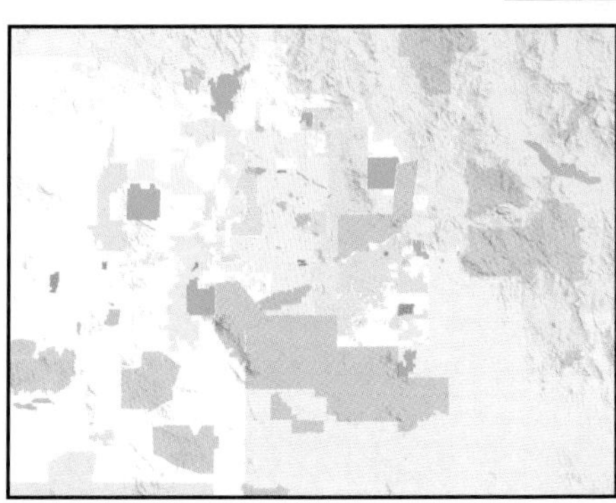

Plate 37.
Polygon data superimposed onto a DEM grayscale image. *(Combining Polygon and Raster Data in Cartography)*

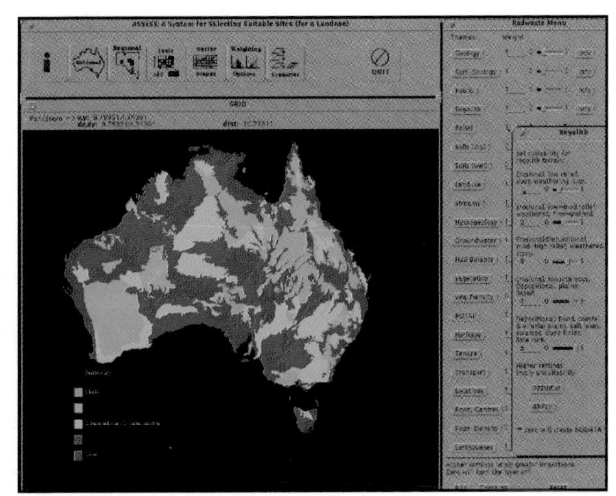

Plate 38.
ASSESS interface developed to assemble series of functional menus to frame an ARCPLOT display window. *(ASSESS: A System for Selecting Suitable Sites)*

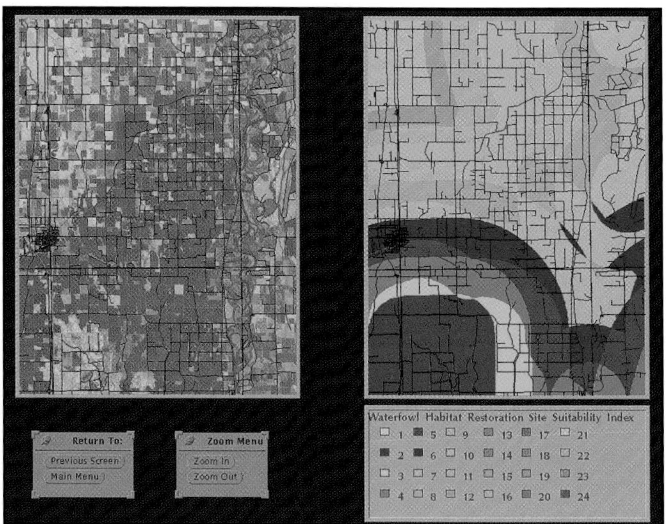

Plate 39.
Output from site analysis model. *(Targeting Wetlands Restoration Areas)*

Plate 40.
Image map of south Florida composed of 29 scenes of 1994 SPOT multispectral data. *(Water Resources Management Applications)*

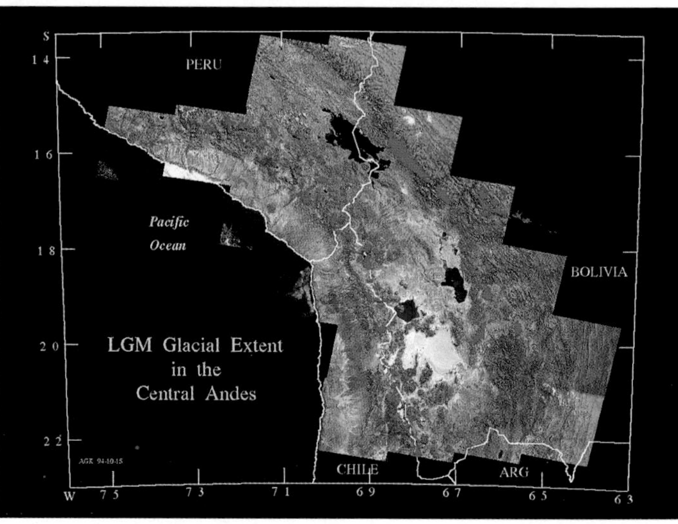

Plate 41.
At maximum extent during the Last Glacial Maximum, paleo-glaciers (shown in magenta) covered approximately 20 to 25 percent more area than at present and extended to much lower elevations. *(Mapping Glaciers with SPOT Imagery and GIS)*

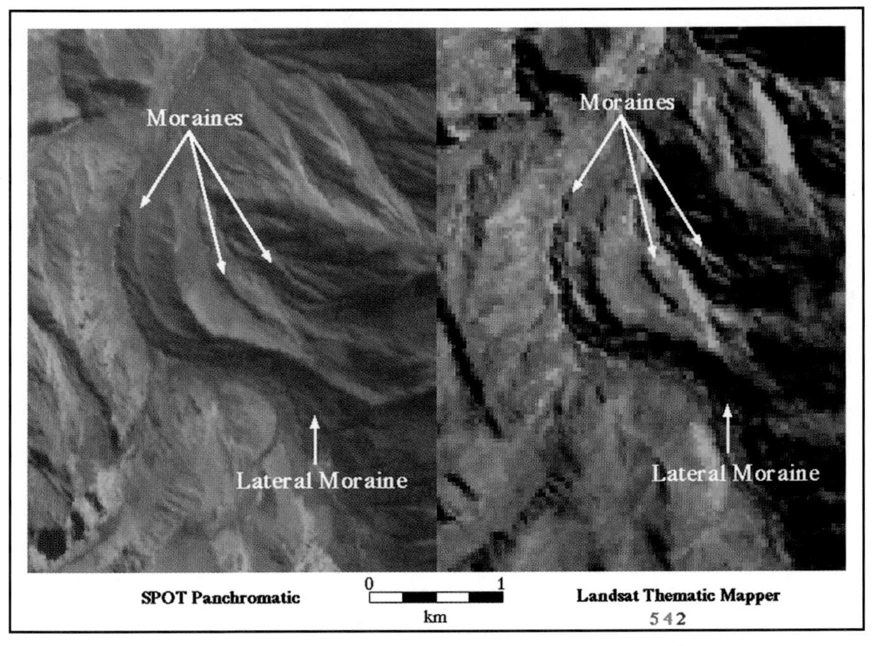

Plate 42.
In the rugged eastern Andes of Bolivia, the 10m resolution of SPOT panchromatic imagery (left) enables recognition of moraines that are barely visible at the lower (28.5m) resolution of Landsat TM (right). The images are centered at S 16° 27' latitude, W 67° 47' longitude. *(Mapping Glaciers with SPOT Imagery and GIS)*

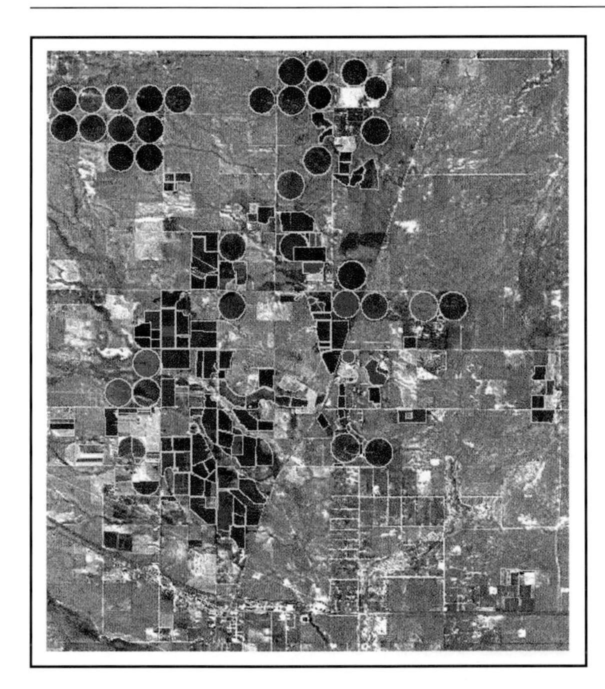

Plate 43.
SPOT/TM merged image with vector overlay of irrigated field boundaries in the northern Estancia Basin. *(Identifying Efficient and Accurate Methods for Conducting the Irrigated Water Use Inventory, Estancia Basin, New Mexico)*

Plate 44.
NDVI index. *(SPOT, ARC/INFO, and GRID: A New Concept in Agro-environmental Monitoring)*

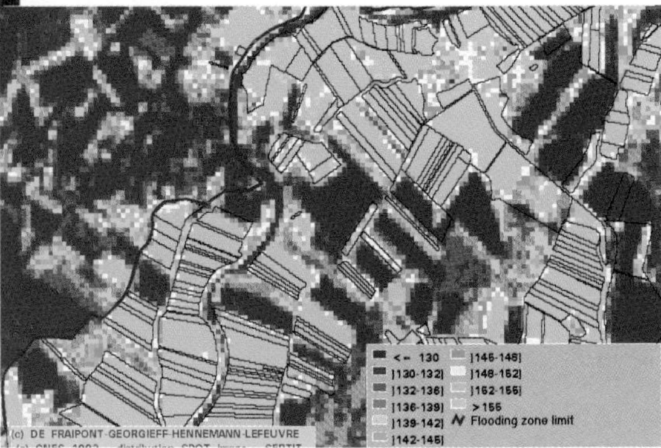

Plate 45.
SPOT XS, November 1993. *(SPOT, ARC/INFO, and GRID: A New Concept in Agro-environmental Monitoring)*

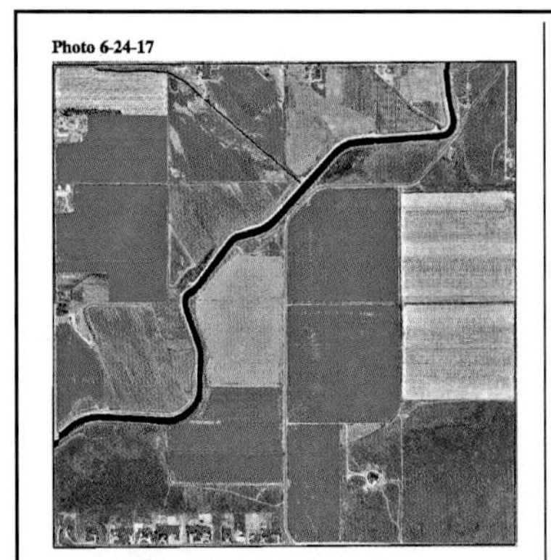

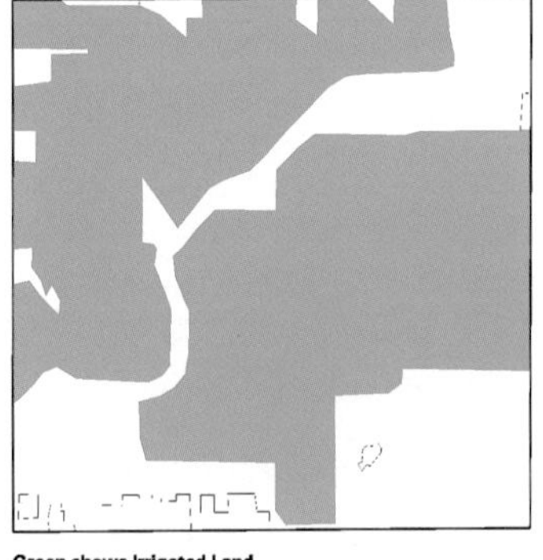

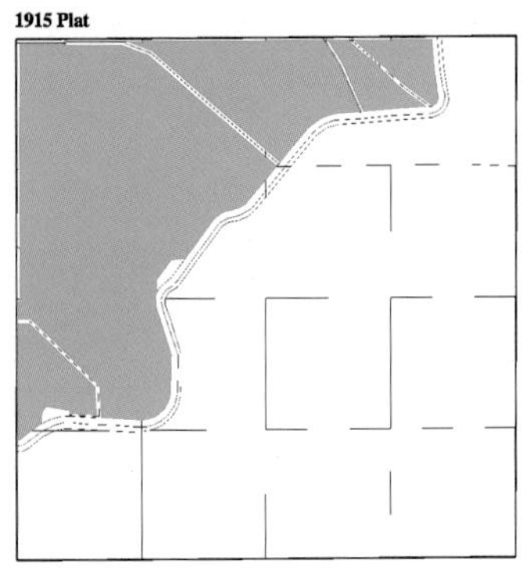

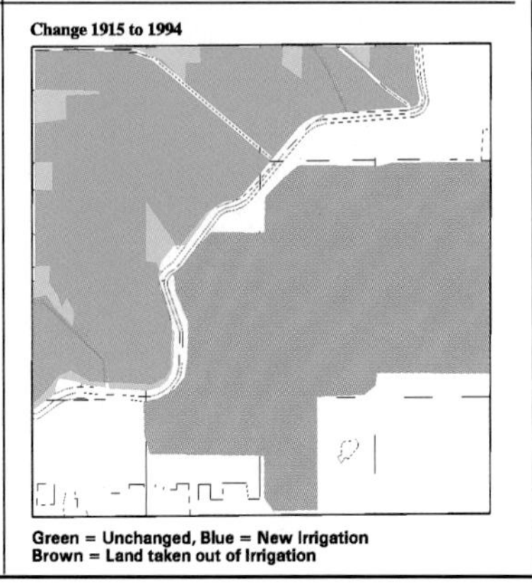

Plate 46.
Example of four data sets generated for each registered section. At lower right, white indicates land area that has never been under irrigation. *(Mapping Long-term Change in Irrigated Land)*

Plate 47.
Winter index differentials for Scotland. *(Integrating Raster and Vector GIS with Climate Data for Winter Road Maintenance)*

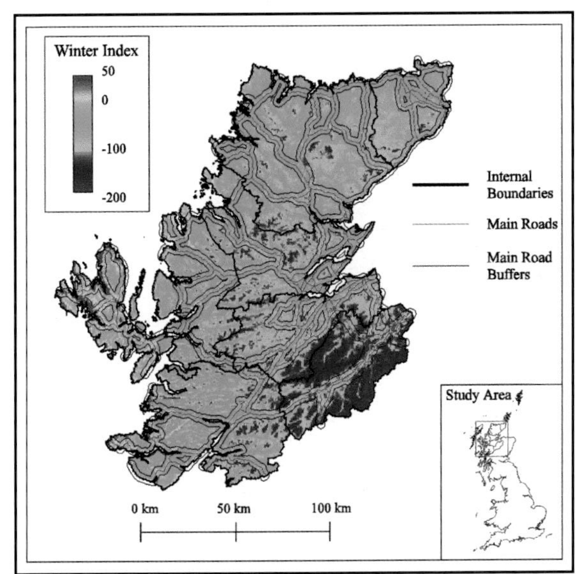

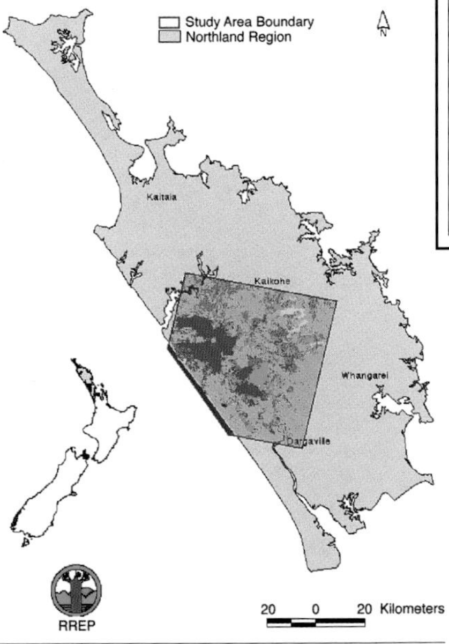

Plate 48.
Study area for Regional Resource Evaluation Project (Northland region, New Zealand). *(Landscape Structure as Input to Ecological Planning in Northland, New Zealand)*

Plate 49.
Continuum map based on factor weights that stress remoteness from population and access. *(Mapping the Wilderness Continuum Using Raster GIS)*

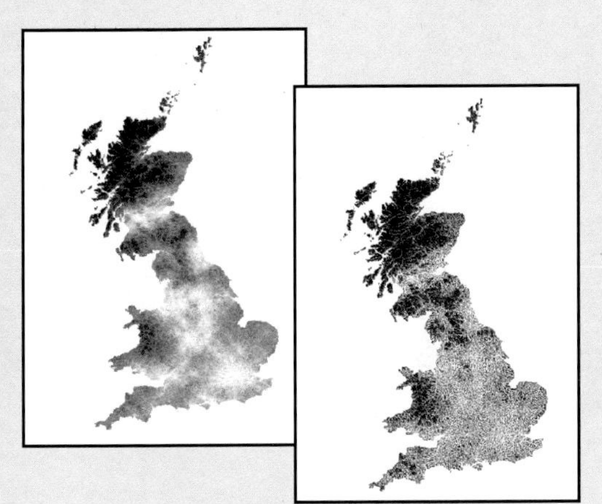

Plate 50.
Continuum map based on factor weights that stress apparent and biophysical naturalness. *(Mapping the Wilderness Continuum Using Raster GIS)*

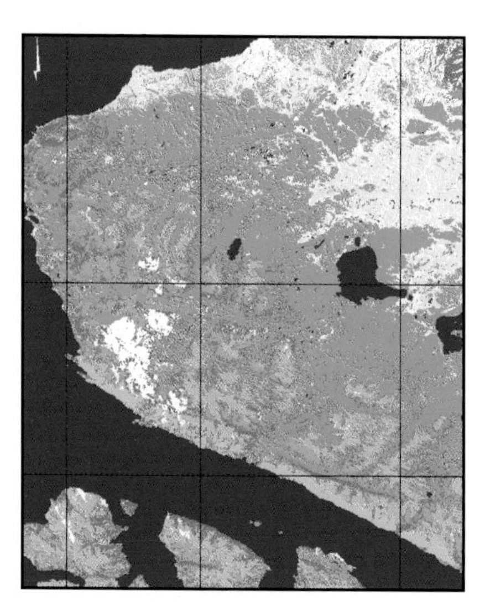

Plate 51.
Thematic classification containing 15 classes for eastern area of Tierra del Fuego. *(Technologies in a Regional Forestry and Biodiversity Project: Tierra del Fuego, Chile)*

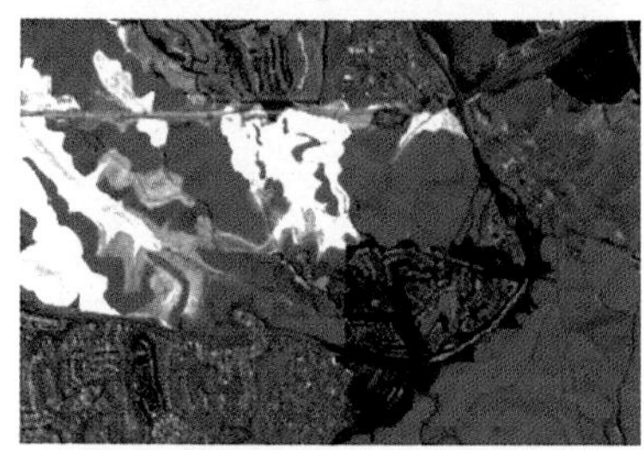

Plate 52.
Image change analysis products based on multitemporal satellite image data. (a) Multidate overlay image derived from summer 1990 and 1995 SPOT-Pan data (10m); (b) Differenced image derived from summer 1990 and 1995 SPOT-Pan data (10m); (c) Post-classification change identified map derived from summer 1990 LANDSAT-TM (resampled to 20m) and 1995 SPOT-XS (20m); (d) Transition sequence identified map derived from summer 1990 LANDSAT-TM (resampled to 20m) and 1995 SPOT-XS (20m). *(Enhancement, Identification, and Quantification of Land Cover Change)*

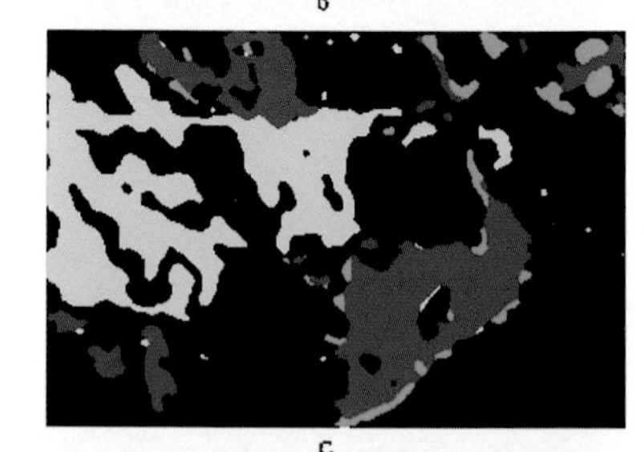

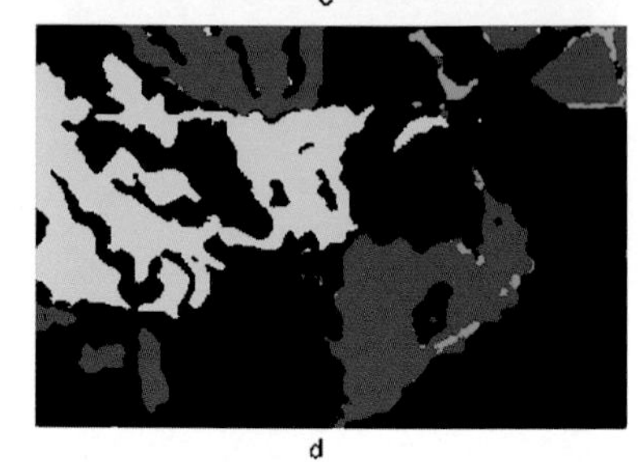

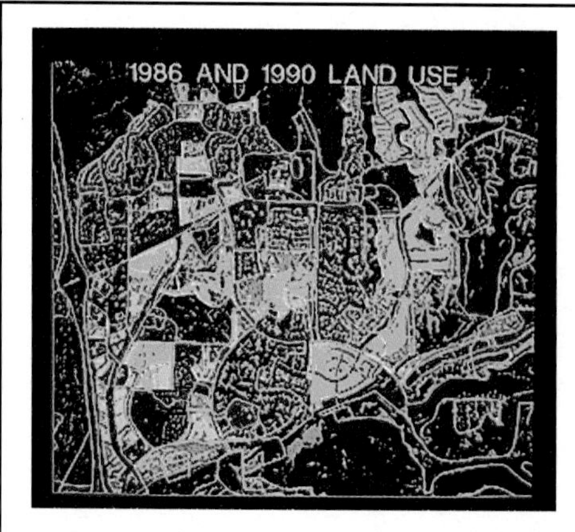

Plate 53.
Composite image display for the 1986 land use vectors in blue and the updated 1990 vectors in yellow. *(Updating Vector Land Use Inventories Using Multi-date Satellite Imagery)*

Tijuana River Watershed

U.S.

Mexico

Plate 54.
Tijuana River Watershed area. *(The U.S. Mexico Border GIS: The Tijuana River Watershed Project)*

Plate 55.
Post-integration vegetation polygons. *(The U.S. Mexico Border GIS: The Tijuana River Watershed Project)*

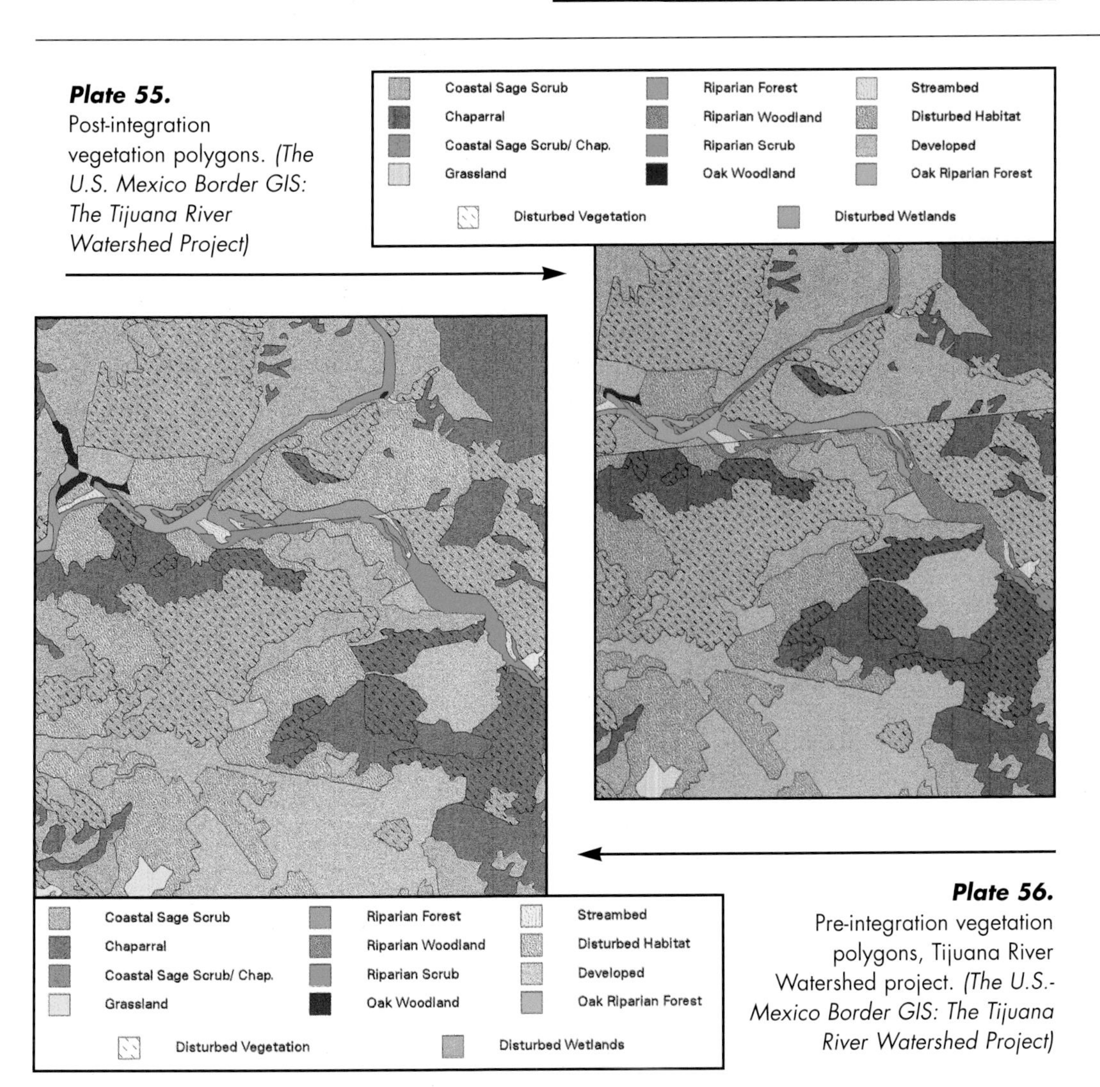

Plate 56.
Pre-integration vegetation polygons, Tijuana River Watershed project. *(The U.S.-Mexico Border GIS: The Tijuana River Watershed Project)*

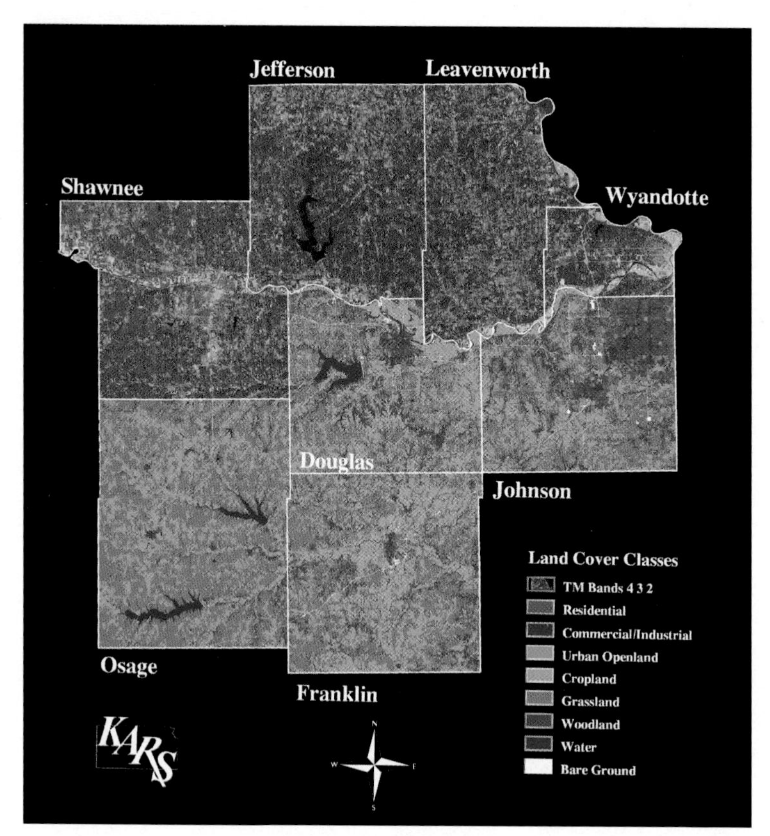

Plate 57.
Eight-county area of northeast Kansas. *(Development of a Land Use/Land Cover Map)*

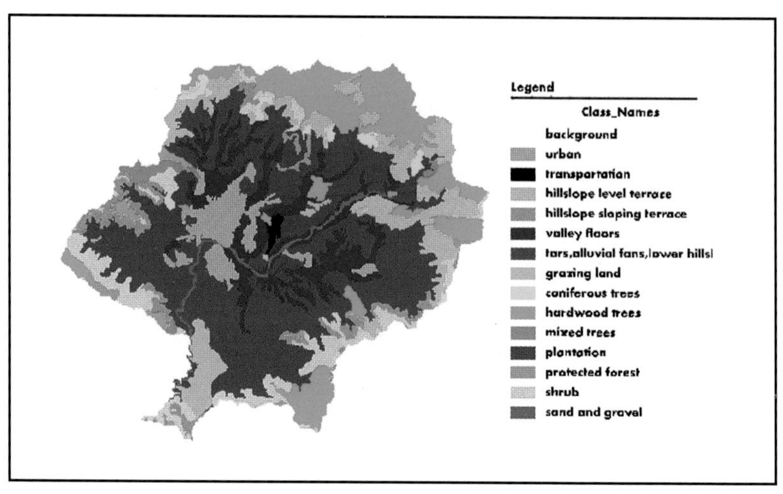

Plate 58.
Example of LRMP land use layer. *(Measuring and Modeling Urban Growth in Kathmandu)*

Plate 59. Layer overlaid with existing land use map as a means of identifying land that may be consumed by future growth. *(Measuring and Modeling Urban Growth in Kathmandu)*

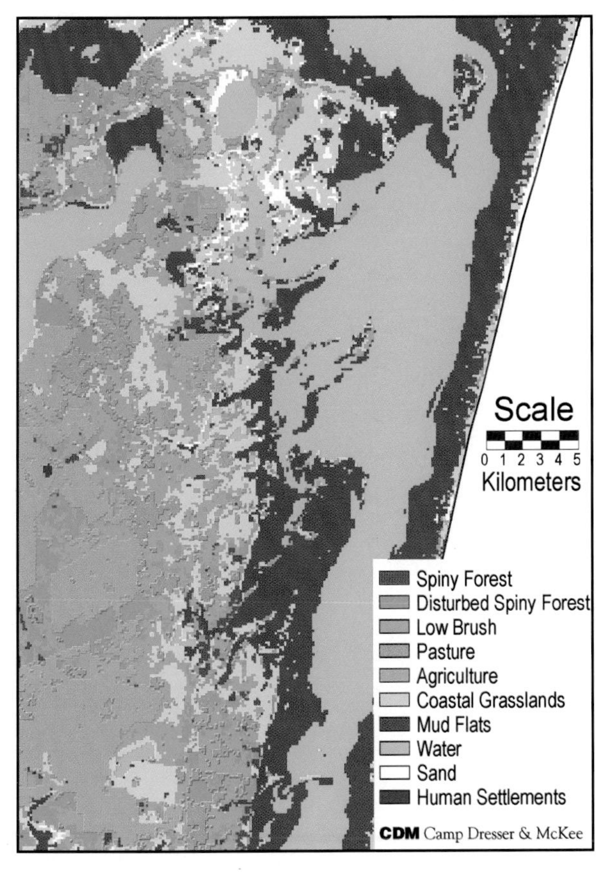

Plate 60. Land use classification for the San Fernando region, Tamaulipas, Mexico. *(Merged Raster and Vector Data for Regional Environmental and Land Use Planning: Tamaulipas, Mexico)*

Plate 61.
A surface showing U234 concentration with blue representing lower values and red representing higher values (top). A classified image with yellow and gold representing mine and mill tailings and cyan representing a high clay, high vegetation area that correlates with the area of high radionuclide concentration (bottom).
(Detecting Contamination From Uranium Mines)

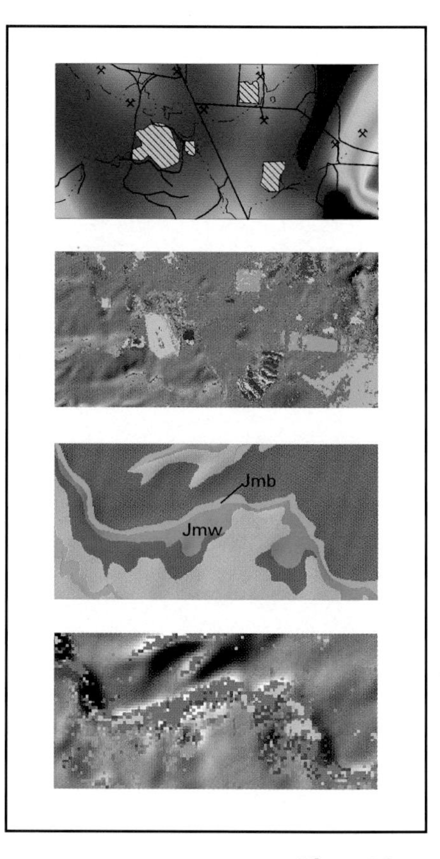

Plate 62.
Geologic map showing uraniferous units, marked by Jmw and Jmb. Yellow marks surface mines and magenta is found where uraniferous units outcrop. Magenta is also found at one of the tailings seen in Plate 61.
(Detecting Contamination From Uranium Mines)

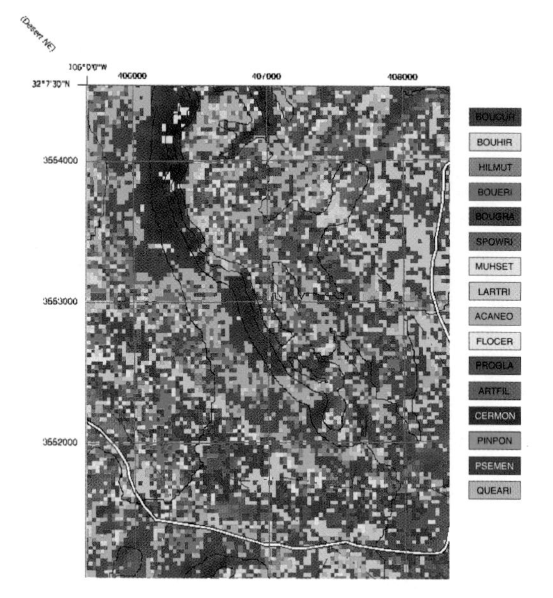

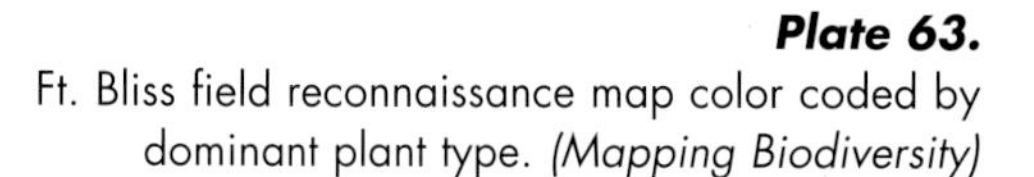

Plate 63.
Ft. Bliss field reconnaissance map color coded by dominant plant type. *(Mapping Biodiversity)*

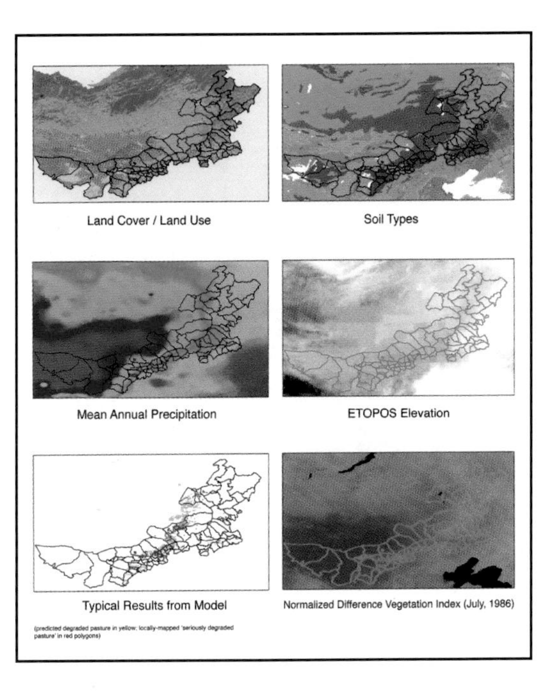

Plate 64.
Mapping pasture degradation: illustration of model inputs and typical results. *(Remote Sensing and GIS for Monitoring Grassland Environments)*

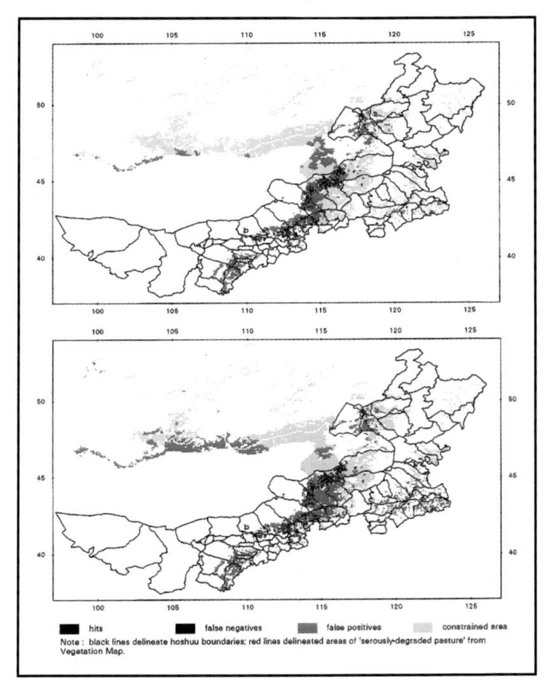

Plate 65.
Predicted degraded steppe in Inner Mongolia, 1986 and 1990. *(Remote Sensing and GIS for Monitoring Grassland Environments)*

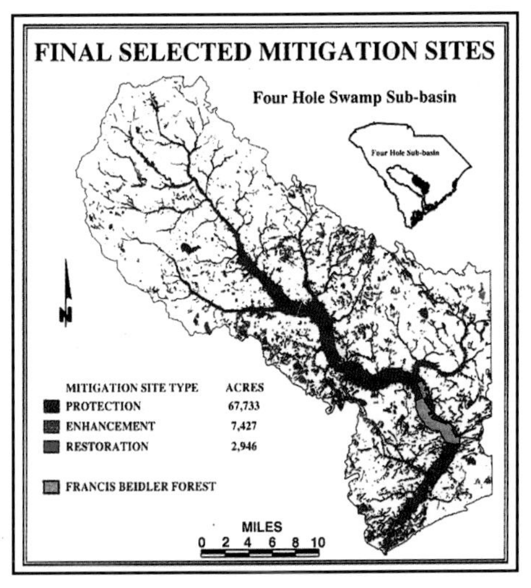

Plate 66.
Composite map illustrating importance of mainstem area. *(A Method for Identifying Wetlands Mitigation Sites)*

Plate 67.
Solitario Dome unsupervised classification. *(Vector and Raster Analysis of the Solitario Dome and Terlingua Uplift, West Texas)*

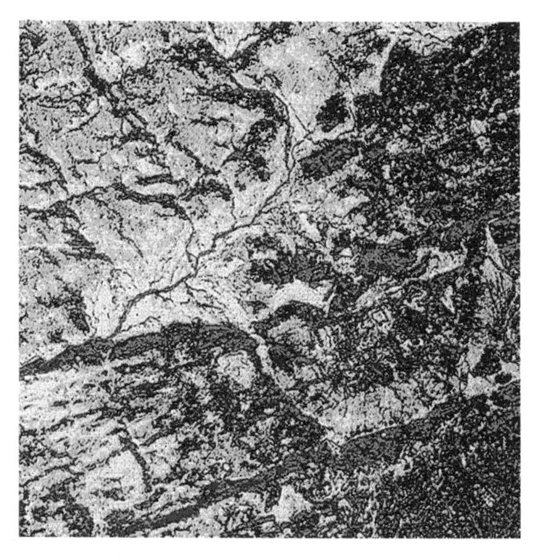

Plate 68.
Terlingua Uplift unsupervised classification. *(Vector and Raster Analysis of the Solitario Dome and Terlingua Uplift, West Texas)*

Plate 69.
Solitario Dome grid. *(Vector and Raster Analysis of the Solitario Dome and Terlingua Uplift, West Texas)*

Plate 70.
Terlingua Uplift grid. *(Vector and Raster Analysis of the Solitario Dome and Terlingua Uplift, West Texas)*

Plate 71.
Distribution of sparse and dense forest in Queensland. *(Developing a Multi-scale Forests Database for Australia)*

Sparse Forest
Dense or Closed Forest

Plate 72.
Detail of an integrated forest (see inset in Plate 71). (Developing a Multi-scale Forests Database for Australia)

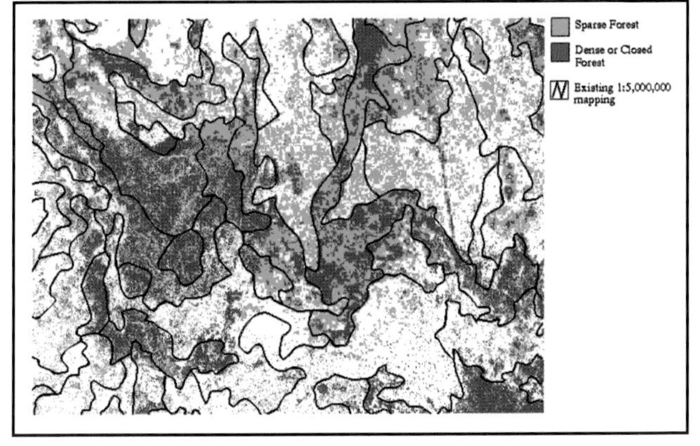

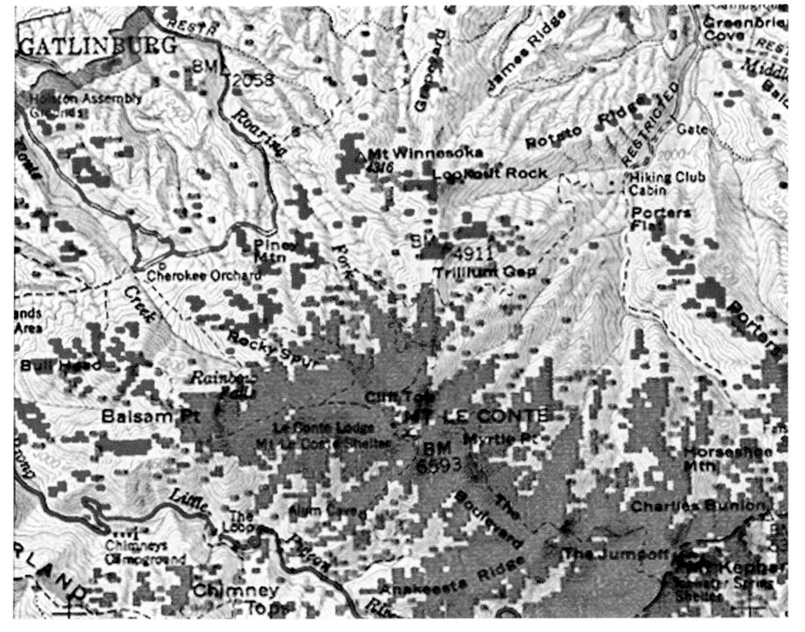

Plate 73.
Needleleaf forests in the Great Smoky Mountains National Park. Red is pine, green is hemlock, and blue is spruce-fir. *(Mapping Needleleaf Forests in Great Smoky Mountains National Park)*

Plate 74.
Post-logging SPOT image overlaid with a slope classification (in degrees) generated from a DEM. Disturbed sites on steep terrain may require special monitoring or rehabilitation. *(Information Technologies to Support the Licensing of Forestry Activities)*

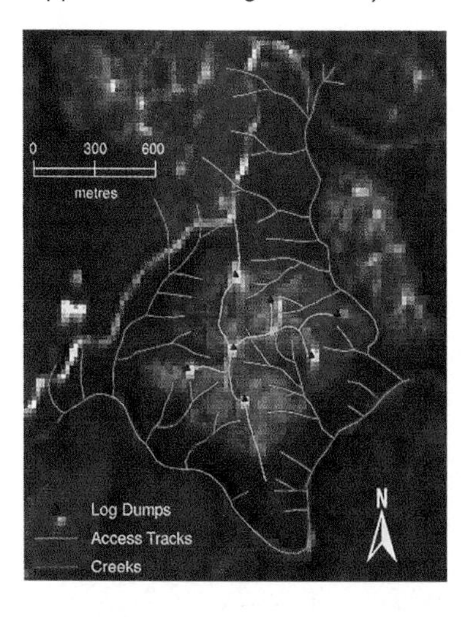

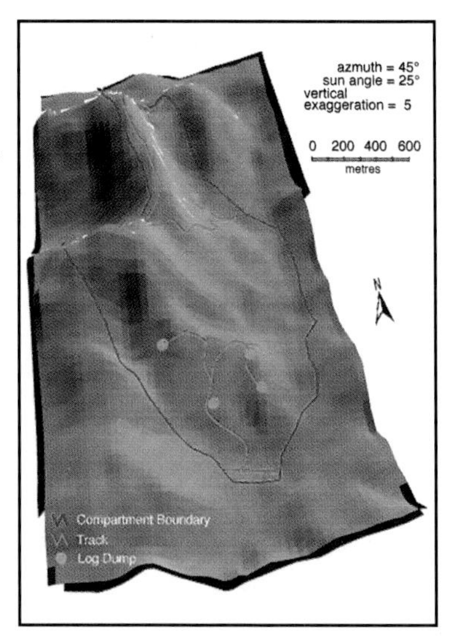

Plate 75.
Terrain view using shaded relief surface model derived from a DEM. Vector layers of proposed harvesting plan features are draped to provide the observer with a catchment perspective view of the proposed siting of activities. *(Information Technologies to Support the Licensing of Forestry Activities)*

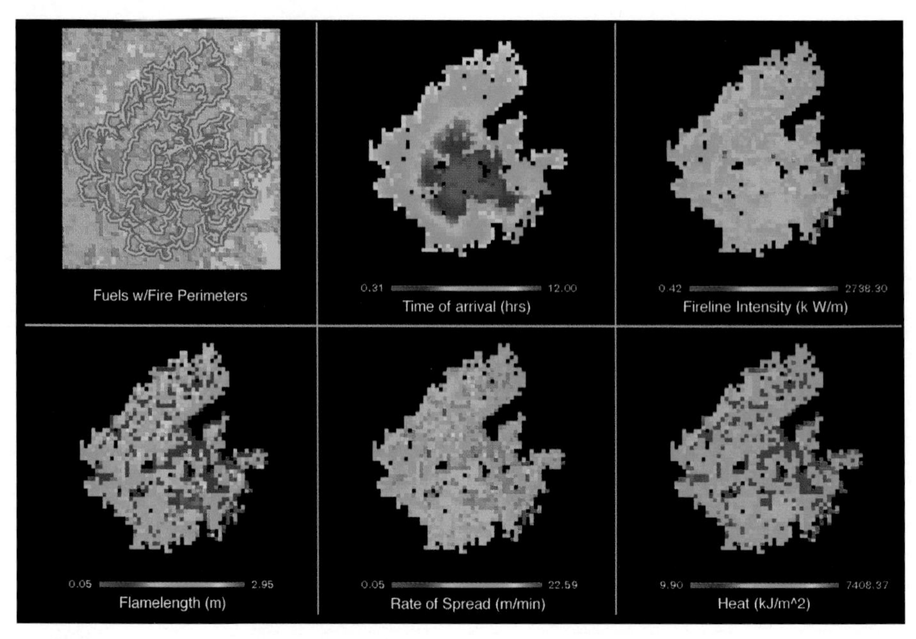

Plate 76.
Simulation of flame length, fireline intensity, time of arrival, heat per unit area, and rate of spread of the fire for every pixel within the burned perimeter.*(Fire! Using GIS to Predict Fire Behavior)*

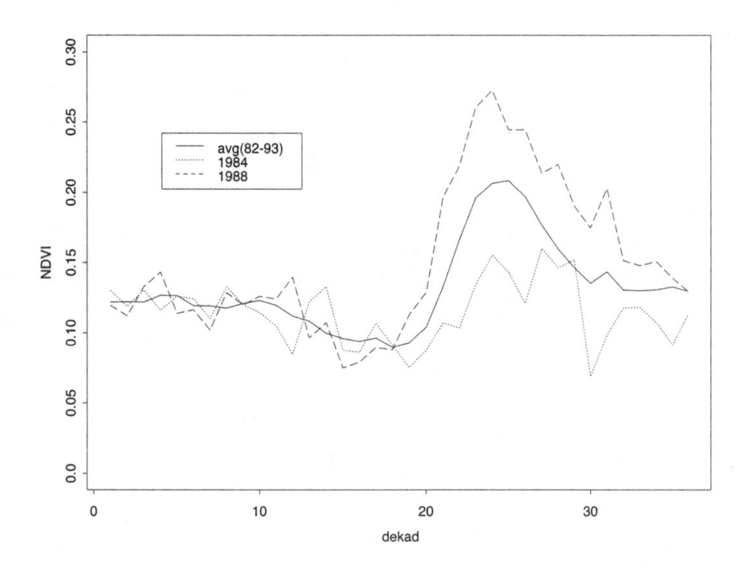

Mean (1982-1993) dekad NDVI for En Nahud, Sudan, with NDVI data for 1988 (good rainfall year) and 1984 (drought year).

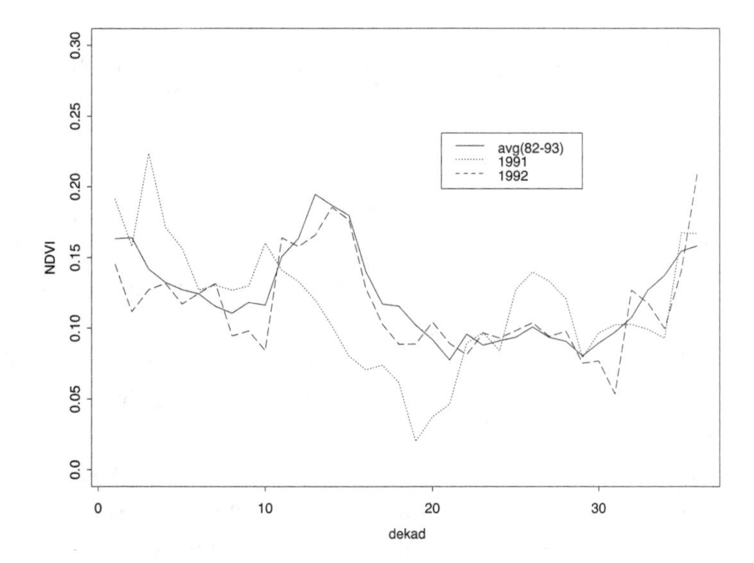

Mean dekad NDVI for Marsabit, Kenya, with the drought season of 1991-92.

The NDVI can be used to indicate deficiencies in rainfall and portray meteorological and/or agricultural drought patterns in both a spatial and temporal extent in a manner that is not possible by using ground observations. Drought will continue to occur, but the application of NDVI as a tool for decision making will allow better integration and more timely planning of methods to promote food security in the greater Horn of Africa.

CropWatch: Monitoring Irrigation Water Use in the Desert

Tom Elder, Department of Water Resources, Phoenix, Arizona

Challenge

For many decades the groundwater table under central and southern Arizona was depleted because enormous quantities of water were pumped and used for agriculture, mining, industry, and a rapidly growing population. In some areas the decline of the groundwater table has actually caused the ground to sink and subsidence fissures to form, while in others water has become more expensive to pump because wells have to be drilled deeper. In many areas the groundwater table is over 1,000 feet deep and any recharge from natural precipitation or stream flow may take decades. The Arizona Department of Water Resources (ADWR) was established in 1980 to provide comprehensive groundwater use regulation and planning with the ultimate goal of halting the depletion of the groundwater table around the Phoenix and Tucson urban areas by 2025 and at least minimizing the depletion in central Arizona where agriculture remains the predominant economic activity.

One of the main challenges to meeting this goal is monitoring over 750,000 acres of irrigated agricultural land throughout a large arid region extending from, and including, the entire Phoenix metropolitan area, continuing south through Casa Grande and the entire Tucson metropolitan area all the way to the international border at Nogales. Irrigated agriculture accounts for 80 percent of all water used, but the Department of Water Resources only has a small field staff to cover the entire area.

Data and Methodology

The CropWatch program was established in 1985 as a pilot, and entered into full production in 1986. The program uses GIS and remote sensing to make crop maps of all irrigated acreage two or three times per year. The primary goal is to enforce the limit on irrigated acreage expansion and to provide crop acreage information for water resource planning purposes. By law, the amount of irrigated land is limited to land that was shown to have been actively irrigated at any time between 1975 and 1980. All irrigated fields over 2.5 acres must have an established right to continue using groundwater for irrigation. The CropWatch Program's three principal products are listed below.

- Seasonal crop maps for monitoring irrigation acreage limits compliance.
- Annual crop history statistics for monitoring water use reporting compliance and estimating irrigation efficiencies.
- Historic crop production and water use statistical trend comparison for long-term water resource planning.

The crop calendar determines when the satellite scenes are ordered. The crops must be close to maturity and have a dense crop canopy to provide the maximum contrast between the crops and surrounding features. Obtaining the satellite scenes too early would show immature crops that are difficult to distinguish from natural vegetation or weeds, especially in the spring. Obtaining satellite scenes too late would not allow enough time to produce the field maps and verify violations in the field before the crops are harvested.

Any source of multispectral satellite imagery can be used. When the program first began, Landsat MSS scenes were used. At present, SPOT XS Level 1B scenes are used because SPOT has more frequent repeat acquisition opportunities to minimize cloud cover, the scenes are delivered within ten days after a successful acquisition, and they show more detail on the ground. Other sources of satellite images that have been evaluated include Landsat TM, IRS-1B LISS1/LISS2, and MOS MESSR. Satellite image vendors are selected through competitive bidding.

The software used for the CropWatch Program includes ARC/INFO and GRID 7.0.3 for vector and raster GIS processing, ERDAS 7.5 for image processing, and ArcView 2.1 and Oracle 7 for the final display and analysis of the crop information. The hardware consists of Sun workstations and various PCs. The CropWatch program is implemented as a series of ARC/INFO menus that call other AML and UNIX shell script programs and ERDAS Audit files in a series of steps. The software used in each step is indicated in brackets below.

1. Once the satellite scenes have been delivered, information about the type of scene in use, including the satellite and sensor, date, path-row designation, and any center and corner coordinates, are entered into a database [ARC/INFO].

2. Each scene is loaded from tape or CD [ERDAS].

3. Each scene is georeferenced to match the projection and coordinate systems used by the GIS [ERDAS]. The scene corner and center points are used for the initial set of ground control points. Visual comparison of the

scene with the field boundaries on the GIS maps and adjustments to the scene origin coordinates in the header provide the final precision registration.

4. Multiple scenes are then combined into a mosaic [ERDAS].

5. A Normalized Difference Vegetation Index (NDVI) image is produced by calculating a ratio of the near infrared and red bands of the combined scenes [ERDAS]. The Arizona climate is one of the keys to producing good results. This hot, arid region receives only seven to 14 inches of rain per annum. Most natural vegetation is very sparse and irrigated crops are large, easily identifiable areas of very dense vegetation. All irrigated field boundaries are digitized and maintained in a GIS database [ARC/INFO]. Information on the ownership and irrigation right status is recorded in a corresponding Oracle database. These boundaries are converted from vector polygons to raster areas based on three irrigation categories (irrigated fields, excluded fields, and urban and desert areas).

6. The entire range of the NDVI image is grouped into four vegetation density categories (none, sparse, medium, dense) and compared with the irrigation categories [GRID]. A crop map is plotted showing the vegetation density and extent as color-coded in shades of green for legitimate crops, shades of red for possible violations, and shades of yellow for dense vegetation in urban or desert areas. The crop maps are distributed to field staff who check for any last-minute field boundary or irrigation right status changes, and past crop maps for similar violations, and schedule field visits to specific locations to measure, photograph, and

document any possible violations. All crop maps are routinely processed and delivered within 24 hours of receiving the satellite scenes.

7. Once all seasonal crop maps for the year are complete, they are assembled into an annual crop inventory linked to annual reported water use. The inventory is used to estimate water use and gross irrigation efficiencies [Oracle and ArcView]. Crop identification is based on multitemporal scene comparison with the local crop calendar instead of multispectral pattern recognition. Three major crops—cotton, wheat/barley, and alfalfa—account for over 90 percent of the total irrigated acreage in this region. All crops can be monitored by acquiring satellite scenes a minimum of twice a year because growing seasons are almost completely separate from each other. Wheat and barley are grown in the spring and harvested by May or early June. Cotton is planted in the spring and harvested in September. Alfalfa is continually grown and cut throughout the entire spring and summer. If crops are observed on the spring satellite scenes but not on the summer scenes, the crop was probably wheat or barley. If crops are observed on the summer scene but not in the spring, the crop was probably cotton. If the crop was observed on both, it was probably alfalfa. Citrus and pecan orchards and vineyards are permanently mapped in the GIS database because these do not move from field to field. Other crops such as corn, melons, sorghum, and winter vegetables are not usually monitored because they are a very small part of total irrigated acreage.

8. The annual crop inventories can be compared to establish trends in water use, irrigation efficiency, and the rate of agricultural land conversion to urban development over time.

Benefits

From the outset, it was understood that using GIS and remote sensing technologies would provide several tangible benefits to a small agency. Benefits are summarized below.

- The time and staff required to check for possible compliance violations on the ground would be greatly reduced.
- Complete coverage of a large region to uniformly enforce irrigation acreage limits would be ensured.
- Satellite scene sets could be used for other projects.
- Crop acreage statistics could be provided that were not obtainable by any other method.

Over 40 maps are produced every season by a single analyst and their use typically uncovers ten or 20 violations. Most of the violations are unintentional and involve very small areas. While some of these cases result in small fines, most are corrected through education. In a very few cases of intentional or repeat violations large fines are imposed and the crop must be plowed under. As the CropWatch Program enters the tenth year of operation, new uses for the information are continually being developed.

Mapping Long-term Change in Irrigated Land

Anthony Morse, William J. Kramber, and David Palmer, Idaho Department of Water Resources, Boise, Idaho

Dewayne McAndrew, U.S. Bureau of Reclamation, Boise, Idaho

Challenge

The population of Boise Valley in southwestern Idaho has grown enormously since 1915, when the U.S. Bureau of Reclamation began delivering stored irrigation water.

Many people currently residing in the valley live on land that was once irrigated, but changes in land use have affected the quality and quantity of local water resources. The Idaho Department of Water Resources (IDWR) and the Bureau of Reclamation are joining forces to use raster and vector data to map the 80-year shift in land and water use patterns covering about 1,000 sq mi (2,590 sq km) of the valley. Results will be used as input to a hydrologic model, and as an integral part of a management plan for the Boise River Basin.

A project to map 80 years of land use change presents many challenges. Cost-effective historical and contemporary data had to be found that could be used in a GIS. The biggest challenge was registering both historical and modern data to a common base. The solution to the registration problem was an innovative combination of raster and vector processing of old maps and new aerial photography.

Data and Methodology

Maps of actual land use, circa 1915, are not available. However, the Bureau of Reclamation was able to provide plat maps drawn on vellum that show the land area originally authorized for irrigation. Each plat covers one square mile at a nominal scale of 1:4,000.

Several options were considered for contemporary data. Such data could have been acquired as color infrared photography and converted into digital orthophotos; acquired directly as digital aircraft data; or generated by fusing Landsat TM and SPOT panchromatic scenes. The project required enough detail to classify present land cover into 24 classes (see table below). Remote sensing specialists at IDWR concluded that scanned aerial photography provided the required detail. The scanned images could be map registered using the CONTROLPOINTS routine in ARC/INFO, and the resulting registered image could then serve as the background on which to interpret land cover and polygon boundaries. The routine also allowed a means of registering the photos to the same base used for the plat maps.

25 Land Use/Land Cover Classes	
Residential-farmstead	Irrigated crops and pasture
Residential-old urban	Irrigated-perennial
Residential-new subdivision	Irrigated-idle
Residential-rural	Land in transition
Commercial and industrial	Feedlot/stockyard
Public	Dairy
Recreation	Abandoned
Transportation	Other agriculture
Sewage treatment	Rangeland
Water	Wetland/riparian
Barren	Canal
Junkyard	Petroleum storage
Unclassified	

Scanned aerial photos and the digitized plat maps were processed using a three-step method. First, ARC/INFO AMLs were used to subdivide the Public Land Survey (PLS) layer of the 1:24,000 DLG into quarter-quarter sections for use as a base. Second, the plat maps were digitized using ARCEDIT by registering each plat to a copy of the subdivided PLS. Nodes formed at the intersections of section lines and subsection lines were used selectively as tics to set up the digitizing. Third, the aerial photos were scanned and registered to the PLS using the ARC/INFO CONTROLPOINTS command. Control points were generated by finding PLS intersections on the photography and matching the intersections to nodes in the subdivided PLS. After using GRIDWARP to warp the scanned photos to fit the PLS, the area within the control points was mosaicked to create images of PLS townships.

The above method registers both the digitized plats and the photos to a common base and limits registration errors. By using the PLS as a rigid framework and registering only the area within the control points, the inevitable registration errors are limited to the confines of the control points and propagation across the whole coverage is avoided.

Results and Conclusions

Plate 46 shows an example of the four data sets generated for each registered section: (1) scanned, registered aerial photography; (2) photography interpreted to show irrigated/non-irrigated land cover; (3) digitized plat map showing land authorized for irrigation; and (4) union of the two interpretations.

The fourth data set contains four classes (including white) that summarize irrigation changes in the valley over the 80-year period. The four categories of change are listed below.

- Class 1: Authorized for irrigation and currently under irrigation (green in Plate 46).
- Class 2: Authorized for irrigation, but not irrigated now (yellow/brown in Plate 46).
- Class 3: Not authorized for irrigation, and not irrigated now (white in Plate 46).
- Class 4. Not authorized for irrigation, but irrigated now (blue in Plate 46).

While section lines were used for economy of processing, they were not always available. The flight lines were not planned with this method in mind, and the scale of the photos was too large to compensate for the offset in flight-lines. When necessary, sections were subdivided into quarter-quarters. Subdivisions were carried out with the use of PLS and ARC/INFO AMLs.

A total of 1,428 frames of 1:12,000-scale color aerial photography were scanned and registered. A 1.5m pixel size was chosen as a trade-off between richness of detail and economy of data. Each township consumes 125 MB of storage space. Once the data for a township were registered and mosaicked, interpretation was initiated.

An ARC/INFO AML was created to streamline the classification process. Using a township of mosaicked photography, the AML was used to draw the boundaries of the land cover/land use polygons on the computer screen. The AML offered a choice of 24 classes from which to choose.

The process of scanning, registering, and mosaicking the aerial photos is time-consuming, requiring about a week for each 36-section township. Using 1:24,000 scale photography would have accelerated processing, but the real efficiencies will come with the new generation of high spatial resolution satellites.

The marriage of raster and vector technologies is powerful. While the chosen approach was not necessarily the best of all possible solutions, it was the best compromise between the need for detail and budget limitations.

METEOROLOGY

Lightning Analysis at Bonneville Power Administration

Robert White, Bonneville Power Administration

Challenge

We have all seen it light up the night sky and heard the thunder that follows soon after. In fact, lightning is one of nature's most ubiquitous phenomena. During a typical summer day, it is not unusual for 25,000 lightning flashes to strike the ground in the Pacific Northwest. While most go unnoticed, others cause devastating wildfires, disrupt airline schedules, and cause power outages. The cost to utilities can be enormous, result of lost revenue, higher insurance deductibles, and repairs to damaged lines. To identify which transmission lines are susceptible to lightning for possible corrective action, the Bonneville Power Administration (BPA) has been studying ways to leverage the strengths of several technologies, including raster and vector GIS systems, satellite imagery, and commercially available lightning flash data.

Bonneville Power Administration (BPA) is the federal power marketing agency for the Pacific Northwest, originally established by Congress to market and transmit the power produced at Bonneville Dam. Today, it sells wholesale power from 29 federal dams to public and private utilities, as well as to large private companies. BPA also constructed and maintains over 15,000 circuit miles of transmission lines, one of the largest and most reliable systems in the United States. BPA's service area includes Oregon, Washington, Idaho, western Montana, and parts of California, Nevada, Utah, Wyoming, and eastern Montana.

Improving the reliability of its transmission system has always been one of BPA's primary objectives, and minimizing the outages caused by lightning is one area getting a lot of attention.

Data and Methodology

To get a better handle on the distribution of lightning flashes occurring in its service area, BPA subscribes to the National Lightning Detection Network (NLDN), a commercial service that maintains a network of lightning strike detectors scattered around the continental United States. Using magnetic direction finding technology and satellite communications, the location of lightning strikes can be viewed as strikes occur on a PC-based system using proprietary display software. The vendor (GeoMet Data Services, Inc.) also makes available historical data in the form of coordinates and attribute information for display or output to a file.

A number of applications were developed to take advantage of the wealth of information supplied by the vendor, including GIS routines to correlate lightning flashes with known outages. The problem is one of time and space. In addition, there was a need to spatially summarize the data to identify historical patterns of flashes, specifically areas of high flash density. This requirement was best addressed by converting the flash coordinates (points in vector format) to a flash density grid (raster format). The resulting grid solved another problem: the sheer volume of data demanded an alternative to displaying and querying the point (vector) data. Display and query of a million records takes time, even on a powerful workstation.

Flash density alone, however, lacks geographic context. The occurrence of lightning is often associated with different land forms that can contribute to unstable air. Therefore, it was important to incorporate satellite imagery (SPOT) and topographic (DEM) information as a backdrop to both the vector and raster maps.

Results and Conclusions

All processing was performed with ARC/INFO software and involved the following steps:

1. Convert the ASCII data file of lightning flashes to a GIS point coverage.
2. Create a vector grid that covered the entire service area with a resolution of one square mile (approximately 400,000 cells).
3. Spatially join the two vector files (point in polygon).
4. Perform a frequency analysis on the resulting point coverage to determine the number of flashes per grid cell (i.e., a count of the number of points with the same unique grid cell number-coverage number in ARC/INFO).
5. Join the frequency table to the point attribute table using the unique grid cell number as the relate item.
6. Use the POINTGRID routine to convert the point data into a raster grid using the frequency (count) as the value for the cell. The resulting flash density grids had values ranging from no hits to 28 hits per square mile.

To make analysis and query easy to use, a desktop mapping application using ArcView was created. For detailed queries, the user has the option to display the raw flash data. For quick analysis, the flash density grids by year are available. All this information can be displayed in conjunction with BPA's transmission system. The real value to the system, however, comes from the inclusion of both the SPOT imagery and topographic information. Being able to view the strikes in relation to the topography and surrounding geography not only adds a sense of realism to the application, but often provides important clues as to the cause of a particularly dense concentration of strikes.

Benefits

The application has already proved to be useful, allowing engineers to isolate lines and even individual towers affected by lightning. This enables the engineers to quickly assess whether preventive measures or structural changes are needed for certain portions of the lines to make them less susceptible to lightning caused outages.

Integrating Raster and Vector GIS with Climate Data for Winter Road Maintenance

Dan Cornford, School of Geography, University of Birmingham

Challenge

Although climate affects the amount of money required for winter road maintenance, it is not currently included in the central funding for such activities. A methodology using both raster and vector GIS is outlined here for the inclusion of climate data into management decisions that incorporates the spatial nature of climate.

Winter road maintenance involves gritting and snow clearance from roads to ensure that they are safe for traffic. In Scotland each regional authority is responsible for keeping respective roads safe, but funding for such activity derives from the central government. Currently, funding for winter road maintenance is distributed based on a weighted count of road length and usage. The objective of using GIS in winter road maintenance is to demonstrate that climate is also an important factor.

The winter climate of Scotland is spatially variable. Some regions, especially higher elevations, experience more severe winters than those near the western coast, which is affected by the North Atlantic Drift. To address this problem, a combined GIS and statistical analysis program is in use. The project also operates over two distinct scales: the regions of Scotland and the internal regions of the Scottish highlands. The methodology of the latter is described, and results from both scale programs are presented.

Data and Methodology

A particular region's winter climate can be summarized by a winter index, which combines the maximum temperature with the number of ground frosts and number of days with snow on the ground. Because this index is a non-linear combination of these climate variables, it must be calculated on a grid cell by grid cell basis. For temperate regions, values typically range between 100 and -200; higher values mean less severe winters.

For the internal region investigation, climate variables were mapped individually on a 500m resolution grid (based on the UK National Grid projection). This procedure required knowledge of several terrain related covariables. These covariables were stored and analyzed in the ARC/INFO GRID module. Several data sets were used, such as a 500m resolution DEM; grids of distance to coast, urban area, and river/stream; and a land use grid derived from (LANDSAT) satellite irradience data developed by the Institute for Terrestrial Ecology. Steps in the analysis are summarized below.

- Create new terrain variables using the GRID functions such as: (1) FOCALMAX, MEAN and MIN to create lower resolution DEMs and for subtracting them from the original DEM to create exposure coverages. This technique could be used, for example, to calculate the height of a location below the maximum elevation within 3 km, which has been found in past studies to be useful for predicting temperature and frost; (2) ASPECT to create an aspect coverage, which was then SINEd to examine solar radiation potential without accounting for shading effects; (3) FLOWACCUMULATION was applied to model the flow of cold air, or katabatic effects. This technique was considered to result in "a little" oversimplification.
- The combination functions were used to create a grid of radiative potential index (the potential for a location to cool) by weighting the satellite-derived land use classes with typical cooling rates under clear sky conditions.
- Some terrain variables such as distance to coast were logged (LN function) to account for non-linear relationships with climate, as well as to create more normally distributed variables.
- Once the above coverages were created, an AML was used to extract the value of the terrain variables for each climate site that measured one of the climate elements.
- Because there were only 50 to 70 sites for measuring each climate variable, a simple linear regression model was constructed based on the terrain variables and location to explain the distribution of the climate

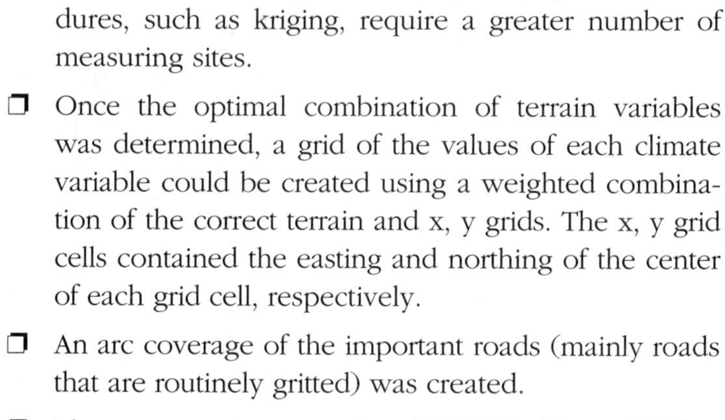

variable. More complex (and explicitly spatial) procedures, such as kriging, require a greater number of measuring sites.

- ❒ Once the optimal combination of terrain variables was determined, a grid of the values of each climate variable could be created using a weighted combination of the correct terrain and x, y grids. The x, y grid cells contained the easting and northing of the center of each grid cell, respectively.
- ❒ An arc coverage of the important roads (mainly roads that are routinely gritted) was created.
- ❒ The arc coverage was then BUFFERed by a 2.25km buffer, and the buffer polygons were then POLYGRIDed to use as a mask. This is important because many regions contain large high-elevation areas (and thus colder winter climates), but roads are located in the valleys (warmer climates on average). These differences are transparent in Plate 47.
- ❒ A polygon coverage of the internal regions of the highland area was converted to a grid.
- ❒ A ZONALSTATISTICS command from GRID then revealed the means (and other statistics if desired) of the region's winter index.
- ❒ The winter index statistics were imported to a statistics package (SPSS for PCs), and analyzed with the expenditure and weighted lane length data.

The results indicate that while weighted lane length is the most important factor in controlling internal regional expenditure, the winter index explains a significant proportion of the variance in the remaining winter road maintenance expenditure differences. To highlight this, the expenditure data are converted from a raw figure in UK pounds sterling to a figure in pounds per kilometer. The conversion removes the effect of weighted lane length and allows interregional comparisons.

The relationship between winter road maintenance expenditure and winter severity. The different symbols show the internal regions.

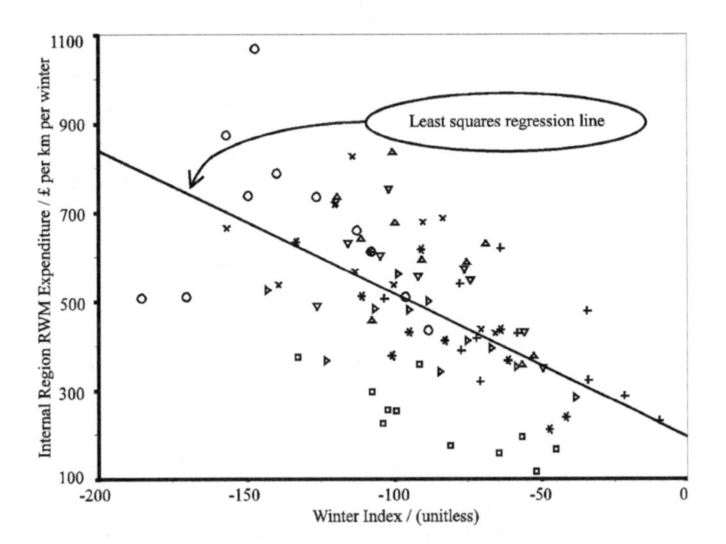

The expenditure per kilometer is then compared to the winter index of each internal region. Note that this procedure compares expenditure per kilometer per winter for several winters and several internal regions. The least squares line represents the best linear fit to the overall impact of winter severity. The correlation coefficient here is 0.59, which is significant (below the 0.01 level), and the slope is -3.22. The result can be interpreted as meaning that for every unit more severe, an extra 3.22 pounds sterling per kilometer per winter must be spent on winter road maintenance. The above figure also reveals that different internal regions respond differently to winter severity. It is also worth noting that the expenditure data are prone to large errors, and several possible outliers are apparent.

A similar study for the administrative regions of Scotland demonstrated very similar results. Therefore, the scale of the region does not affect the influence of winter severity on winter road maintenance spending.

Benefits

The information on the sensitivity of area spending to severe winter weather can be used in two ways. First, by obtaining climate data for the regions, a satisfactory and equitable method for allocating funding for winter road maintenance can be devised. The inclusion of this factor in winter road maintenance expenditure is currently under investigation by the government in Scotland. Second, the information can be used as a management tool because it allows the intercomparison of (internal) regions, and how they respond to severe winters. Regions that respond differently from the majority can be identified and causes can be investigated. This information could then be used to optimize the efficiency of the winter road maintenance services.

Spatial Distribution of Global Atmospheric Water Vapor Storage

Julie Driver, Department of Geography, University of Regina, Canada

Challenge

As the issues of global change and climate variation gain popular understanding, researchers are establishing methods to investigate environmental parameters that are dynamic in time and space. Although satellite imagery is commonly used in climatological research, it was not coupled with the spatial analysis capabilities of a geographic information system until recently. The distribution of water vapor in the global atmosphere is an important attribute that is constantly changing in its x, y, and z coordinates. It is therefore critical to determine relationships between the spatial and temporal distribution of water storage in the atmosphere, and to devise a methodology for modeling these relationships. The GRID module of ARC/INFO was used to address these questions.

Data and Methodology

The International Satellite Cloud Climatology Project (ISCCP) C2 data set was used in this project. Water vapor is represented in the data set by water column amount. Cloud analyses are summarized in daily intervals on a monthly scale over a six-year period (1985-1990). Global coverage is provided by five geostationary satellites (METEO-SAT, INSAT, GMS, GOES-East and GOES-West) and two National Oceanic and Atmospheric Administration (NOAA) polar orbiting satellites, and coverage is presented as the mean water column amount on composite images. The spatial resolution is coarse 280 km x 280 km on an equal area grid.

Fortunately, the data have been preprocessed. Once they have been extracted through the included application software to an ASCII format, a header is added and each ASCII file is converted to a grid in ARC/INFO. The data are continuous rather than discrete, so the floating point option must be used during the conversion. To maintain the integrity of the floating point values, all grid cells are multiplied by a factor of 1000. This is an important step in the protocol to reduce error associated with multiple conversions between floating point and integer, as well as to facilitate faster computation times. The grid cell values are returned to floating point values by division of 1000 at the end of the analysis.

Mean statistics are performed on the daily, monthly and annual temporal scales, a straightforward operation in GRID. The grids are shifted to the latitude/longitude coordinates -180, -90, 180, and 90, and projected to the Cylindrical map projection using the bilinear interpolation option. An attribute table is built for each grid because the command used to shift the grids requires the cell integer values. New grids are created by selecting latitudinal zones. The statistical information obtained can reinforce climatology research on circulation patterns of the atmosphere.

A coastline polygon coverage was overlaid on each grid to aid in visually referencing the variation in mean water column amount by latitude. The coastline coverage can be used to mask out the global land mass, and the mean values over land versus over water can be compared. The maps with overlay were generated in ARC/INFO and exported to ArcView for cartographic fine tuning.

The statistics file within each grid directory gives the range, mean, and standard deviation. This information can be used in a database or spreadsheet to facilitate comparison between the means using statistical tests, such as the Student's T-test.

Results and Conclusions

The spatial patterns can be used to infer temporal patterns because each cell is georeferenced. For instance, the annual mean water column amount for 1989 is seen in the following figure. The cell values range from 0.06 to 5.55cm, and mean value is 2.69cm. These values can be compared to the values for 1985-1988 and 1990 to determine whether there is any variation in annual mean values over time.

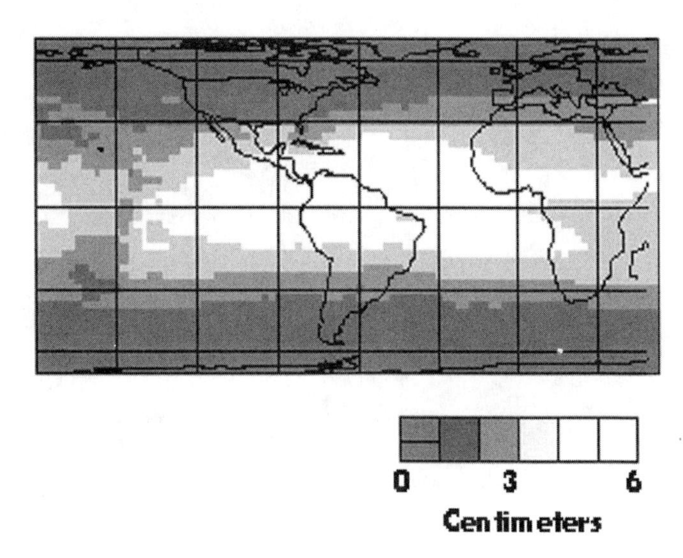

Annual mean water column for 1989.

The next two figures represent the northern hemisphere winter and summer means. The minimums, maximums, and means for January and July 1989 are 0.06cm, 5.82cm, and 2.392cm, and 0.06cm, 5.132cm, and 2.69cm, respectively. The winter mean is less than the annual mean, whereas the summer mean is higher. One might expect the winter and summer means to be the same because the seasons reverse between both hemispheres. However, due to the larger land mass in the northern hemisphere, the difference in the three means is meaningless. These maps may also be compared to monthly means from other years.

Mean water column, northern hemisphere, Winter 1989.

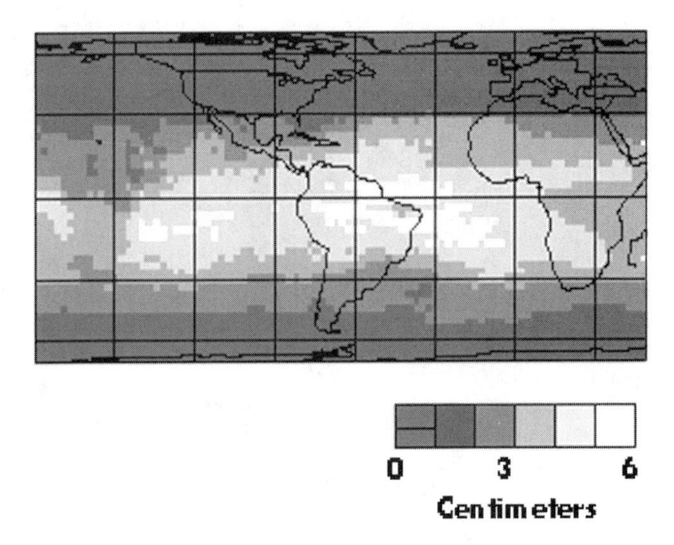

Mean water column, northern hemisphere, Summer 1989.

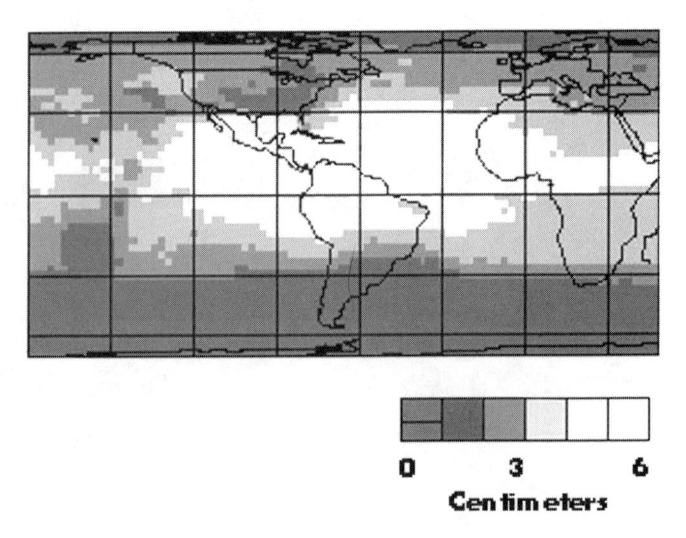

The intercomparison between the seasons and years is more useful with a larger data set. Once data have been collected by satellite over decades and centuries, the results will become more valuable.

Land Use and Planning

Land use and land planning constitute a topic area whose boundaries blend with neighboring chapters. In this chapter, the editors have organized the case studies into subsections on *Landscape Analysis, Land Use/Land Cover,* and *Urban and Regional Planning*. While these are familiar topics to most readers, the ingenuity of individual contributions and their combined breadth show a striking sophistication in how raster and vector data sets are being merged and modeled. Landscape analysis using raster GIS is helping society to address management questions of biodiversity, habitat health, the extent of ecosystems, and a host of related terrain issues. Land use and land cover are attributes that lie at the foundation of GIS solutions, and whereas they have become common stock-in-trade, the case studies here show significant progress in techniques development. Advanced applications no longer use raster data just as a backdrop to vector GIS, but are teasing enough information from pixels that cost savings warrant their routine use.

Perhaps the biggest growth prospects among land use topics are in urban and regional planning. These are the areas about which society routinely collects massive amounts of data by equally impressive numbers of public agencies and authorities. GIS is the only reasonable means for managing these data, let alone for modeling and processing them into meaningful planning scenarios. Success lies in data sharing among many stakeholders, in understanding the properties of data sets (metadata) to assess their utility, and in learning how to merge archival with modern data collection systems to produce quality results. The case studies in this subsection illustrate the art of data sharing and the rewards that can be achieved.

LANDSCAPE ANALYSIS

Landscape Structure as Input to Ecological Planning in Northland, New Zealand

Russell L. Watkins, Department of Geography, University of Auckland, New Zealand

Challenge

With the widespread awareness of the concepts and implications of sustainability and the maintenance of biodiversity, resource managers and regional planners are compelled to consider increasingly larger and more complex databases. These data include both biophysical and socio-economic variables. Managing and analyzing such databases can be a daunting task for typically underfunded and understaffed resource management and planning agencies. The primary need is for accurate and concise biophysical and cultural information for development of effective management policies and programs. Based on this need, a methodology has been developed to provide timely, cost-effective landscape level information on a range of relevant variables.

The Regional Resource Evaluation Project (RREP) is a multiscale, multiyear project focused on analysis of the sustainability of land based production systems (e.g., pastoral agriculture and forestry). One of the objectives of the RREP is to evaluate the spatial analytic capabilities of geographic information system and remote sensing technology in the context of resource management and sustainability. This case study reports on the application of these technologies to the description of landscape structure in a predominantly agricultural ecosystem in the Northland region of New Zealand. The aim is to develop a database and report that can be used to support policy development and implementation for ecological sustainability and diversity. These policies and programs are part of the mission of the Northland Regional Council, and the New Zealand Department of Conservation, as defined by the New Zealand Resource Management Act (RMA). The RMA has shifted resource management responsibilities from the national to the regional and local levels as a means for more explicitly accommodating spatial variations in biophysical characteristics and socioeconomic needs. The legislation emphasizes the need to protect and manage natural ecosystems.

Data and Methodology

In this study, landscape structure was determined by analysis of the spatial heterogeneity of land cover types. The term "spatial heterogeneity" refers to absolute and relative location; area, perimeter (edge), and shape; and proportion of the total landscape area. Analysis of these parameters at the landscape level provides information on distribution, fragmentation, consolidation, and frequency of occurrence of land cover types.

To provide timely information at the landscape scale, a land cover classification derived from satellite imagery was combined with existing map information within a GIS environment. Portions of a Landsat MSS (1978) and a SPOT XS (1993) were classified into eight land cover types. Prior to classification, the SPOT image was resampled from a 20m to a 79m cell size. This is consistent with the MSS image resolution, and facilitated comparison of the images. ERDAS IMAGINE software was used to rectify the raw imagery, and a supervised classification was undertaken. Information for the selection of training sites for the classification was obtained from color infrared aerial photographs, and various thematic maps. Thus, a rapid classification of an approximate area of 3,290 sq km was achieved, with an overall accuracy of approximately 80 percent for the MSS image, and 90 percent for the SPOT image. The different dates allowed for a time series comparison of spatial heterogeneity.

Ecological boundaries (delineating districts of similar biophysical characteristics) provide a logical basis for resource management and ecosystem comparison. Portions of five such districts are incorporated in the study area (Plate 48). The classified MSS and SPOT images were exported in an ARC/INFO GRID format. The grids were then subdivided using the ecological boundaries, initially digitized in a vector format. This enables spatial and temporal comparison both within and between ecological districts.

The ecological district grids were converted to ASCII files for subsequent analysis using the FRAGSTATS statistical software. FRAGSTATS is a spatial pattern analysis program for quantifying landscape structure, written by Kevin McGarigal and Barbara J. Marks, and available from Oregon State University. It consists of a series of landscape metrics which describe landscape structure and composition. For a given landscape mosaic, FRAGSTATS computes a range of statistics at the patch, class, and landscape levels, providing a landscape description over a hierarchy of scales.

Results and Conclusions

Review of the landscape statistics in this study indicates several trends. At the landscape scale, there is a decrease in the number of land cover patches (polygons), an increase in the mean patch size, and a decrease in the variance of patch size for the 1978-1993 time period. In addition, there is a decrease in land cover diversity. The inference is that the landscape is becoming more homogenized, and variations in ecological districts are apparent. Departures from these trends were detected in two of the five ecological districts regarding variance in patch size and land cover diversity. Ongoing evaluation will consider the class and patch level statistics to determine which land cover types are changing, the type of changes that are occurring, the spatial distribution of these changes, and the type of variation occurring within the ecological districts.

Structural changes in the landscape are important considerations in the preservation of biological diversity and ecological integrity. Possessing a low level of biodiversity relative to overall size, New Zealand can ill afford to lose additional species. Adding to the conservation imperative are the high levels of endemism in species and the impacts of introduced predators on these species. There is also the issue of genetic resources to maintain biological diversity and enable reestablishment of plant and animal populations impacted by structural changes in habitat. These impacts include microclimate modification, invasion by weeds and introduced species, and trampling by grazing animals. Another consideration is impact on soil and water quality occurring as a result of the removal or modification of native vegetation.

Benefits

The methodology used for this study provides a rapid, timely, and low-cost means to map large areas at the landscape scale. It allows for comparison of structural changes in land cover over time, commonly referred to as *change detection*. As an input to the ecological planning process, the combination of raster and vector data within the GIS environment provides information for development of alternative management scenarios. The projection of impacts and changes occurring as a result of management policies and programs facilitates evaluation of their efficacy and appropriateness. The composition and display capabilities of GIS facilitates the decision making process and the effective consideration of the stakeholder concerns that inform this process. In addition, the range of data manipulation capabilities facilitates the sharing and exchange of relevant landscape information among agencies and interest groups.

Acknowledgments

Thanks to Dr. Chris Cocklin for his helpful editorial assistance, and to Mark Patterson and David Skillcorn for their continuous and dedicated efforts in data processing and manipulation. The RREP is supported financially by grants from the Foundation for Research Science and Technology, the University of Auckland, Lottery Science, and the New Zealand Vice Chancellors' Committee.

Mapping the Wilderness Continuum Using Raster GIS

Steve Carver, School of Geography, University of Leeds, United Kingdom

Challenge

Is there any wilderness left in Britain? The answer is almost certainly "No," at least not in comparison to places such as Greenland and Alaska. Wilderness as an entity is very difficult to define, and making statements about it can be very tricky. Most definitions stress the natural state of the environment, the absence of human habitation, and the absence of other human influences and impacts. Clearly, few such areas exist in Britain today, and where they do exist, they take the form of small and isolated pockets where a natural ecosystem has remained largely unaltered by human activity. By adopting a more flexible definition of wilderness, it is argued that it should be possible to identify a continuum from the most altered and accessible to the most natural and remote. With this in mind, this study applied raster GIS techniques to perform a remote survey level reconnaissance of the wilderness continuum for Britain. This survey utilized both raster and vector data sets within an integrated GIS and multiple criteria evaluation (MCE) approach.

Data and Methodology

The idea of wilderness is largely one of personal experience and perception and not one of measurable standards. The wilderness continuum concept states that true, pristine wilderness is one extreme on the environmental modification spectrum. At the opposite end of this spectrum is the indoor and totally urbanized environment of the city center shopping mall or office block where the person is entirely isolated from all aspects of the natural world. At all stages in between it is possible to identify various environments with varying levels of human modification and naturalness, such as a city park, suburbia, farmland, forest, and mountain. Identifying the point on the spectrum at which wilderness begins in absolute terms is a difficult task when trying to define wilderness. The key issue is reflected in Roderick Nash's statement that "one man's wilderness may be another's roadside picnic ground" in that different people have different perceptions as to what wilderness is. This not only refers to that point along the continuum at which a person considers that wilderness begins, but also to the relative importance a person may place on particular factors affecting the wildness of the landscape.

A flexible GIS/MCE approach to defining the wilderness continuum is therefore much more appropriate than other more prescriptive methods commonly used in GIS-based analyses, since it is not restricted by the necessity to specify rigid thresholds or criteria. Studies by the National Wilderness Inventory in Australia have used GIS to successfully map wilderness areas on the basis of four factors: remoteness from settlement, remoteness from mechanized access, apparent naturalness (lack of construction and other human impacts), and biophysical naturalness. These are mapped according to specified threshold criteria and overlaid to define the boundaries of wilderness areas meeting minimum levels of remoteness and naturalness.

This study adapted the Australian approach and used similar factors, but within a GIS/MCE framework to identify the wilderness continuum in Britain. Whereas Australia retains large tracts of wild and primitive landscapes around which lines can be drawn for management and preservation, Britain cannot possibly adopt or afford such a luxury. Applying the Australian criteria in Britain would result in a blank map. There are areas of the country that do, however, retain certain wilderness qualities, such as lack of obvious human structures, remoteness and inaccessibility. These qualities can be identified using the continuum concept and highlighted for appropriate management. Raster based GIS/MCE techniques are best suited for this purpose because they are fully able to cope with the continuous nature of the data and the mapping problem. Their main advantage over discrete overlay mapping techniques is that they allow user preferences to be attached to individual factor maps. In this manner, trade-offs between factors can be performed in a single operation across the entire study area.

Several existing digital data sets were used to create the four factor maps describing remoteness from population, remoteness from access, apparent naturalness, and biophysical naturalness. A 1km grid based on the Ordnance Survey's National Grid was used for all data sets, and all data manipulations were carried out using ESRI's ARC/INFO GRID module with custom MCE macros.

Remoteness from population is based on the 1991 UK Census of Population. Rather than base this factor simply on the distance from nearest settlement weighted by settlement size, remoteness from population is defined on a linear weighted distance decay model for all populated cells within a 25km radius of the target cell. This radius was chosen because it represents the distance an individual can reasonably walk in a day over rough terrain. The radius is therefore more representative of true population accessibility in that it is not influenced by the size of the nearest settlement but rather more by the total population of settlements within 25 km of a given site and the distance of those settlements from that site.

Remoteness from access is based on distance from mechanized access, namely road and rail transport. A similar weighted linear distance decay model to the above was used to calculate inaccessibility on the basis that minor roads and railroad stations are weighted lower than highways and main line stations, respectively.

Apparent naturalness is defined here as the unweighted linear distance from highly visible human-made features such as dams, power lines, masts, railway lines, roads, and so forth. A true national picture of apparent naturalness will always be difficult to obtain due to the impossibility of the task of spatially referencing all evidence of human activity at its various levels from discarded beer cans and footprints through to HEP developments and six-lane highways. There is always a need to rationalize data requirements for GIS based analyses and the current application was no exception.

Biophysical naturalness is defined as the naturalness of the vegetation cover. This was derived from land cover data available on the UK's Countryside Information System (CIS). The Landsat TM-derived land cover data provided by the CIS gave total hectares per square kilometer of particular land cover types found within each 1km grid square in Britain. These figures were used to derive a map indicating the likelihood of finding natural ecosystems based on the weighted sum of areas of different land cover types in each grid cell. Again, there is some doubt as to the appropriateness of these data, but they are considered suitable for the national level survey described here.

The above data sets were standardized and combined using user-specified factor weights and a simple weighted linear summation model using a suite of ARC/INFO macros developed by the author. By applying different factor weights, different continuum maps were produced reflecting different experiential values concerning people's ideas of wilderness. The wildest areas of the country can be highlighted simply by reclassifying the resulting continuum maps to identify areas with the highest "wildness" values. A step-by-step description of this procedure follows:

1. Select data sets describing relevant factor maps.
2. Standardize factor maps using a common scale (i.e., 0-1.0).
3. Specify factor weights according to user perceptions.
4. Combine using weighted linear summation algorithm.
5. Save results.
6. Repeat stages 1 through 5 using different factor weights reflect different user perceptions.
7. Compare results.

Results and Conclusions

Analysis results are shown in Plates 49 and 50 using two sets of hypothetical weighting schemes. Plate 49 shows a continuum map based on factor weights that stress remoteness from population and access, while Plate 50 shows a continuum map based on factor weights that stress apparent and biophysical naturalness. The difference between the two maps is obvious and serves to illustrate just how different people's perceptions affect the mapped outcome. The key point is that both maps are essentially correct, at least in the eyes of their originators, and both could equally be used as a basis for further work on wilderness and associated landscape conservation policy in Britain. From a wider perspective, such an approach is easily transferable to other study areas using the same or modified factor definitions.

The final individual mapped surfaces presented in this paper can be regarded as one representation of the environmental modification spectrum for Britain which emphasizes certain wilderness qualities. While there may be no true wilderness areas represented on the surfaces, they do provide an indication of the wildness of the landscape on the basis of the chosen factors and weights. Although the proposed GIS/MCE approach to mapping the wilderness continuum for Britain and to identifying areas possessing high wildness values is not without problems, it does provide a framework for wildland identification and management. One interesting point arising from this study is that the bulk of Britain's wildest areas, particularly in the northwest highlands of Scotland, do not have protected area status and as such are potentially at risk from development. However, like many areas identified in the northwest highlands, the wildest are under private estate management. As long as this situation persists and land management practices employed are responsible and sympathetic to the environment, formal environmental conservation policy will not be necessary.

Defining Biophysical Land Units (BLU)

Crista S. Carroll, U.S. Department of Interior, Bureau of Land Management

Challenge

The natural resources and ecosystems of public lands are subject to escalating competition and conflict over uses. There is heightened concern for conserving and preserving lands that historically have been considered only for disposal. In order to manage these lands, it is crucial to understand and predict natural processes, the interplay of cultural uses and impacts, and the ever-changing spectrum of dynamics within ecosystems. Ecological change is continuous and inevitable. Quantifying, monitoring, predicting, and subsequently protecting natural change, or directing desired change, must be scientifically evaluated to help resolve conflict over ecosystem management and use. The Albuquerque District of the Bureau of Land Management (BLM) continues to use and develop scientific models using a GIS framework.

Data and Methodology

Biophysical land units (BLU) have been developed using both raster and vector data, and are the quantifiable spatial representations of the location, extent, and dynamics of multiple ecological components. These components are the pieces of an ecosystem or ecotype. The pieces may include soils, geology, terrain features such as a limited elevation range or degrees of aspect or slope, vegetation/land cover, surface water, or whatever is pertinent to the geographic location. For example, in New Mexico, some of the pieces may be conifer stands on steep slopes of volcanic cinders, or open shrub and forbs on low slope, or highly erodible soils subject to violent storm runoff.

Alternately, a coastal ecosystem may include pieces such as geology, vegetation, water depth in an estuary, or tidal flow dynamics. In simple terms, BLUs are a graphic representation of ecological responses in a single map layer.

Historically, land managers have described the existing environment of regions or administrative units like wilderness areas or range allotments by extrapolating from field surveys that covered only small proportions of the ecological components in question. Conversely, BLUs are delineated through GIS Boolean logic from a matrix or cross-referenced ecological attributes without regard to administrative or ownership boundaries or personal bias. This means that GIS sets no limitations on the number of ecological layers that can comprise the BLU matrix, and can be exploited to represent hierarchical ecological structure. The matrix of GIS-defined components draws the BLU boundary lines. Therefore, human perspectives do not predetermine the spatial delineation of BLUs. Any ecotype component combination not contained in the matrix drops out of the model. This results in a blank space on the map that is revisited by single component analysis and field verification of the site. In this way, previously unknown anomalies, ecotypes, ecological responses, or disturbances (hot spots) may be identified.

Satellite remote sensing data are currently used in the BLU model for vegetation/land cover and surface geology/soils. Satellite data provide total spatial coverage of a large area as a snapshot in time. However, vegetation/land cover is only the superficial expression of a system. Using multitemporal snapshots, changes in system dynamics or responses can be detected and analyzed using raster based Boolean operations.

The initial iteration of core BLUs was based on a matrix combination of vegetation/land cover, soils, surface hydrology, and terrain characteristics. This is flexible enough to provide common ground to a wide spectrum of resource specialists. Incrementally more detailed BLUs can be tailored for specific questions or conflict analyses by adding layers of biophysical or cultural information and/or site-specific data. Data collected and geocoded to a particular site might include observed assemblages of flora and fauna, rain gauge or other climatic data, or a particular localized use or management practice.

Results and Conclusions

A simple, practical example of BLU monitoring is change detection referenced to potential plant communities. A potential plant community is the biotic community that an undisturbed site is capable of supporting based on the site's physical characteristics. The GIS plots in the following illustration show outlines of potential plant communities derived from soils and terrain data. The hatched areas depict a comparable BLU. The left plot displays the BLU in June 1984, and the right plot displays the same BLU in June 1988. There has been a change in the location and size of the BLU indicating a change in ecological response or condition. A larger part of the potential area has been occupied in 1988 as a result of reduced vehicular travel, good vegetation growth in a wet year, and subsequent reduced grazing pressure. With BLUs in a GIS, one can intersect data layers and analyze acres of actual vegetation condition, potential, and what may be a desired condition. For example, if there are 1,000 acres (50 percent of potential in 1984) and 1,500 acres (75 percent of potential in 1988), a 25 percent improvement occurred. The same BLU monitoring concept can be used in relationship to potential, existing, and desired habitats, range sites, or whatever program is of interest. In addition, all these programs can be compared to each other for conflict analysis. The causes of change can then be analyzed by overlaying specific natural and cultural layers.

The delineated area shows potential distribution of a vegetation type. Left panel shows actual distribution in 1984, and the right panel, actual distribution in 1988 based on identical BLU matrix components.

Benefits

- ❒ Ecosystem management requires understanding of dynamic energy exchanges and processes.
- ❒ A GIS never tires of repeating the processes at different points in time, or trying different scenarios.
- ❒ Flexibility in scale or resolution.
- ❒ Many mapping models or strategies are two-dimensional, but BLUs are not only three-dimensional, they incorporate time. BLUs provide a methodology to link past and present data sets with future predictions.
- ❒ BLUs can be used to track and evaluate the amount, direction, and rate of responses.

- BLUs have been shown to document trends from a patchy to more homogeneous landscape as a measure of biodiversity.

The Definition of Topographic Divisions for Italy

Fausto Guzzetti and Paola Reichenbach, Consiglio Nacionale delle Ricerche, Institute for Hydrogeological Protection (CNR-IRPI), Perugia, Italy

Challenge

The availability of digital elevation models (DEMs) has provided geomorphologists with new tools for quantitative analysis of topography. Numerical classification of landforms has been attempted by computer manipulation of DEMs either by automating previously proposed manual classifications or by devising new ones. The automatic characterization of terrain to recognize divisions that maximize internal homogeneity and external heterogeneity is difficult. The large variety of observed topographies must be expressed well enough in quantitative or analytical terms so that different landscapes, formed by contrasting processes, can be distinguished and individually described and interpreted.

An approach is described for partitioning Italy into eight topographic provinces and 30 sections. Topographic provinces are first order divisions with distinct or unique geomorphological characteristics that distinguish them from neighboring areas. Boundaries between provinces were easily traced and correspond to major morphological and geological features or coastlines. Sections are the minor topographic divisions within provinces. Their boundaries are less distinctive and generally more open to interpretation. Starting from a nationwide DEM, topographic divisions were defined step-wise in a semi-quantitative approach. This approach combined an unsupervised, three-class cluster analysis of four derivatives of altitude, visual interpretation of morphometric maps, and comparative inspection of small-scale geological and structural maps at various scales.

Data and Methodology

In the late 1980s, an archive of mean height values for Italy was compiled by the Italian Geological Survey. The archive was generated by estimating mean elevation values using both manual and machine methods, at a scale of 1:25,000. The manual data were transferred from topographic maps using a square grid template and assigning each point an altitude value to the nearest meter by averaging contour lines and spot height within each square grid. Machine gathered data were obtained by computer interpolation of digitized contours.

In 1991 a joint Italian National Research Council and U.S. Geological Survey project assembled a DEM for Italy at a ground resolution of 230m by mosaicking all 280 files of the mean elevation archive. A test was performed to assess the quality of the elevation data. It was discovered that quality varied according to terrain type and the technique used to capture the information. The frequency distribution of elevation was inspected for minor modes corresponding to contour interval multiples. Data gathered by manual inspection were shown to be more precise than those obtained using the automated approach. As the graph demonstrates, this was particularly evident in gentle terrain where topographic gradient is low and the interpolating algorithms are less efficient. In a few places (<1% of total) missing or mislabeled data and incorrectly coded pixels were identified and corrected.

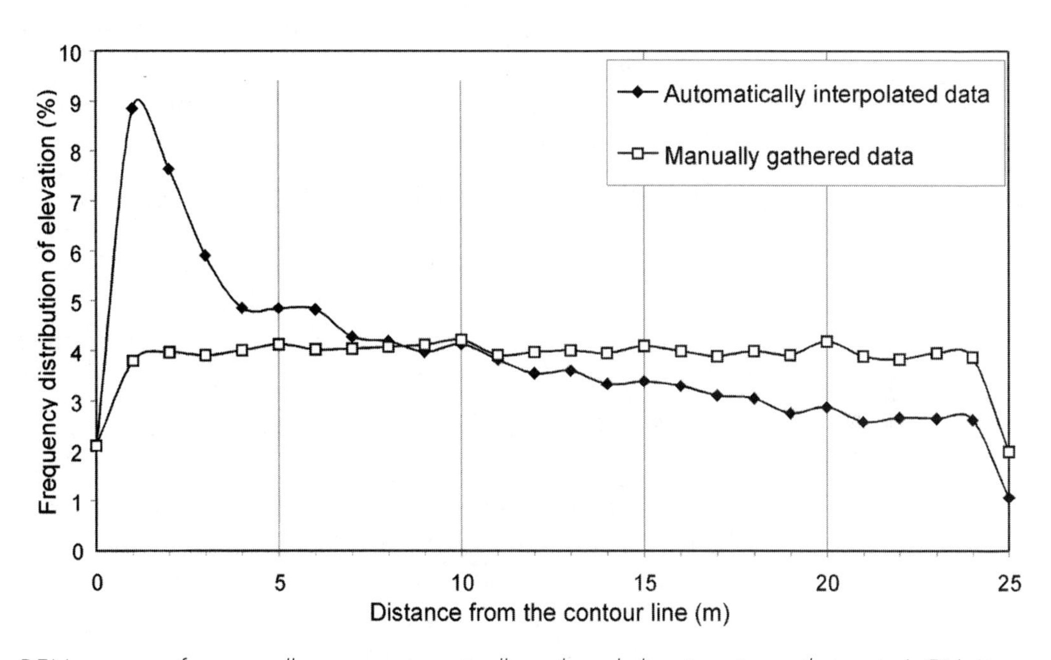

DEM accuracy for manually versus automatically gathered elevations in gentle terrain (±5°). Lines show vertical distance of each elevation from its nearest contour.

A wide range of morphometric criteria can be extracted from a DEM. Four main criteria extracted from 11 variables were used to divide Italy into topographic provinces and sections. Colored digital maps were prepared for each of the variables to facilitate evaluation of spatial patterns, and these maps were used to subdivide the country. Maps of all variables played at least a qualitative role in the classification. The 11 variables are described below.

- *Minimum and maximum elevation* were mapped by assigning the lowest and highest height value within a predefined neighborhood to the central pixel of a moving window of variable size and shape.
- *Local relief* is the difference between the highest and lowest elevation within a given window. It has long been used to quantify terrain contrast and surface roughness. The arbitrariness of the window size

represents a problem. A window too small to include at least one major ridge and valley is unlikely to represent the local topographic wavelength, and local relief becomes a measure of gradient.

- *Mean elevation* generalizes raw altitude by averaging the DEM within a sampling window moved through the elevation grid. Although not differing significantly from the elevation, the resulting map has less detail, which makes elevation more comparable with other generalized variables for regional analysis.
- *Elevation-relief ratio* is a measure of the elevation skewness, that is, a measure of the degree of dissection of a landscape, or the extent to which topography has been opened up by erosion.
- *Slope* at a point is the angle between the surface tangent and the horizontal. Terrain slope, which controls the gravitational force available for geomorphic work, is possibly the single most descriptive measure of mesoscale topography.
- *Aspect* (or exposure) is the direction that the slope faces.
- *Slope reversal*, also known as slope direction change, is a measure of terrain intricacy. It is defined by the inflection in profile slope from uphill to downhill (a ridge top) and vice versa (a valley bottom).
- *Frequency* of slope reversal, the inverse of ridge-to-valley spacing, is a function of the drainage density and represents the only measure of topographic planform available in the study.
- *Curvature* is the rate of change of slope over a distance and describes one aspect of the terrain roughness.

- *Shaded relief* is based on the first derivatives of altitude comprising aspect and gradient. The computer generated image resembles a cloud-free black-and-white aerial photograph and is a detailed synoptic representation of topography. Each pixel is a theoretical reflected light intensity computed from a mathematical relation between ground slope and direction, sun position, and location of the observer.

Shaded relief image of Italy. Sun azimuth is 270°, sun elevation is 30°, no vertical exaggeration. (This image is a reduced version of a map prepared by Reichenbach P., W. Acevedo, R. K. Mark, and R. J. Pike, 1992, Landforms of Italy, 1:1,200,000 scale, CNR-IRPI, GNDCI Pub. No. 581, Rome, Italy).

Results and Discussion

Formal topographic divisions, physiographic units, or types of land surface forms have not been identified in Italy. The topographic categories shown in the figure depicting Italy's topographic division were derived in a semi-quantitative step-wise manner. The process involved multivariate statistical classification of morphometric maps; visual interpretation of the shaded relief image and of other digital morphometric maps; study of descriptive statistics; and refinements of division boundaries by inspection of small-scale geological maps and reports.

Topographic division of Italy. Patterns represent different topographic provinces.

In the first step, the topographic types were ascertained by numerical taxonomy adapted from remote sensing classification of multispectral images. Cluster analysis grouped pixels according to their similarity at the same location on input data sets. An unsupervised cluster analysis was used of four maps considered as input variables. Two, four, and ten class statistical groupings were tried using four to six input variables. The three types resulting from the elevation, curvature, frequency of slope reversal, and elevation relief ratio were judged to be the "best" result. These four variables were the least correlated of the 11, and their spatial pattern and level of detail make the most sense for the country as a whole. The resulting topographic classes are the three most common morphological types: highland or mountain terrain, upland or hilly terrain, and lowland or plains.

In the second step, boundaries of topographic divisions were drawn manually by visual inspection of morphometric maps. The map portraying results of the cluster analysis and morphometric maps were used to sketch first-order provisional boundaries where differences in topographic pattern were distinct and consistent across all digital maps. In the next step, areas were outlined that comprised a systematic mixture of two topographic types as well as divisions that were particularly distinctive on individual maps.

In step three, the descriptive statistics were analyzed on the four input variables as well as other morphometric variables computed for each topographic division. The analysis revealed some unduly large standard deviations that indicated excess dispersion in the data, that is, inconsistencies in the position of the topographic boundaries which were locally modified. These statistics also revealed the need to add entirely new subdivisions to further reduce excess dispersion in the terrain measures. Lastly, topographic boundaries were compared for control with important features on small-scale geological and structural maps and were adjusted locally to match regional or structural lineaments.

Benefits

Based on the distributions of morphometric parameters and particularly the dispersion of elevation and computed slope, the 30 topographic divisions can be grouped into the four main classes or terrain types: plains, hills, low mountains, and high mountains. The two extremes, plains and high mountains, show distinct morphometric attributes representing low and gentle versus high and steep terrain. Between these two extremes, low mountains and hills constitute two separate groups.

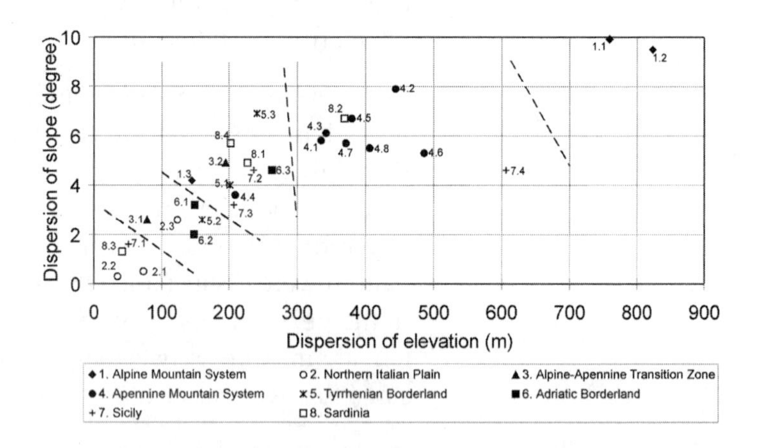

Graph of the dispersion of elevation versus dispersion of computed slope values. Topographic units are grouped into four main terrain types: plains, hills, low mountains, and high mountains. (Modified from Guzzetti F. And P. Reichenbach, 1994, "Towards a definition of topographic division for Italy," Geomorphology 11: 57-74.)

The proposed division reflects existing physical, geological, and structural differences in the Italian landscape. Topographic boundaries can be refined using alternative morphometric criteria by improving the statistical analysis or possibly by using a more detailed DEM that would permit the description of topographic meso- and microforms.

Morphometric parameters have interest in themselves, but are more widely useful in context with other thematic variables like soil type, vegetation type, or geology. The description of topography has many purposes. Morphometric characterization is used to derive trafficability maps for military operations and to automate the creation of hill-shaded images. Map applications include assessment of geomorphic hazard and risk, analysis of regional morphology and structure, evaluation of energy resources, hydrological modeling, and regional planning.

Regional Forestry and Biodiversity in Tierra del Fuego, Chile

George R. Carlson, Robert J. Henry, and James A. Pugh, Breedlove, Dennis & Associates, Inc.

Challenge

The project contained several inherent problems that immediately ruled out using conventional technologies and approaches. The remoteness of the project site, the size of the land area to be analyzed (approximately 450,000 ha or 1 million acres), the absence of accurate, up-to-date base maps, the time schedule and seasonal window, and the breadth of technical disciplines to be addressed drove the decision process and project planning to developing project-specific applications that merge satellite imagery with GIS and GPS technologies. By combining the analytical power of pictures of the area (raster data) with the spatial data management and mapping power of ARC/INFO (vector data) and the positional accuracies afforded by GPS, a successful effort resulted. The objectives of the project were as follows: (1) provide the first accurate vegetation and land use map of Tierra del Fuego, Chile; (2) develop the GIS database for the first biodiversity inventory of the island with the largest scientific expedition in the region since Charles Darwin; and (3) assist in completing the first forestry project environmental impact statement (EIS) in the country's history.

The next generation of work will involve developing monitoring programs, and long-term wildlife and modeling studies. Tierra del Fuego, lying at the southern end of South America, consists of one large island, Isla Grande de Tierra del Fuego, which is shared by Chile and Argentina, and many smaller islands belonging to Chile. This archipelago (approximately 76,000 sq km, or 29,330 sq mi) is separated from the mainland of South America by the Strait of Magellan. The climate is highly changeable and can be quite severe. The availability of cloud-free imagery was a problem, along with locating different seasonal stages of the vegetation. The location, near latitude 55° south, allows many opportunities for satellite coverage, but most of them are cloudy. In addition, this southern extreme produces triangulation problems for GPS satellite configurations and does not always allow acceptable positional readings.

Breedlove, Dennis and Associates, Inc. is an environmental consulting firm specializing in the use of advanced technologies to solve environmental and natural resource management problems. The firm was hired by a U.S. company to provide a map of forest resources on its holdings in Tierra del Fuego.

Data and Methodology

Because the end user customers were PC-based and would be using data in either ARC/INFO (for spatial analyses), ArcCAD (for map production), or ArcView (for management decision making), ERDAS IMAGINE 8.2 was used for all raster processing. This software has built-in compatibility with ARC/INFO and a wide variety of image processing and map-making capabilities. ARC/INFO was a natural choice because of its wide use and PC support. Trimble Navigation's Pathfinder System was used for all GPS data collection. A Trimble Pathfinder community base station with a 500km omnidirectional range was established to provide the base files required for differentially correcting the ground data and improving positional accuracies.

The first step was to collect ground truth and control points on the property for classifying and rectifying satellite imagery. Because of the large areas to be covered, the number of data fields to be collected, and the unfamiliarity of many of the foresters and scientists with GPS equipment, the data loggers were programmed with a unique data dictionary and equipped with bar code scanners. Use of bar codes simplified data collection and provided a high degree of confidence in a consistent data collection process. Several dBASE programs were written to provide an intuitive menu system, allowing foresters and scientists to focus their attention on data collection rather than technical concerns. Data points were collected by traversing the site on foot, ground transportation (trucks and horseback), and helicopter surveys. In total, over 8,000 data points were collected. The data were sent to Orlando, Florida, where, after a control sample was extracted, the points were converted to an ARC/INFO point coverage. This coverage was overlaid on the satellite data to aid classification.

GPS technologies were also used extensively to map the existing infrastructure on the island. With GPS mounted on top of all-terrain vehicles, even the most primitive trail could be recorded, converted to a vector format, and overlaid on the raster image. Existing bridges, water crossings, and road conditions were recorded to better assist forestry operations. They were digitized in ArcCAD and converted to line coverages. These coverages were used to build a lattice and digital terrain model in ARC/INFO. Slope and aspect values were calculated and used to help classify the satellite images. These values also provided important data to the foresters developing a harvest plan.

Several satellite scenes were used to classify vegetation on the island. The first was a Landsat TM. This was a late winter/early spring 1986 image, in which snow covered much of the low lying plant communities. The leaves on the deciduous trees, including the principal commercial species, were off. Another commercial species, an evergreen, was clearly distinguishable. Most of the lakes were frozen, resulting in very high reflections that swamped the spectral signatures of the surrounding communities. The low angle of the sun, innate in any winter image at this latitude, caused shadows on the southwestern side of the mountains and southern coast which also hampered classification efforts. A second image was acquired for February 1986 to facilitate a more accurate classification. This image provided a summer/leaf-on view of the vegetation communities and better resolution of the wetland communities. This scene contained a higher degree of cloud cover, however, especially in a major river valley on the western side of the island. The scene was also shifted more to the west than the first image and did not cover a large critical area in the southeastern part of the property. An October 1994 SPOT image was acquired for this area. Because the SPOT image has finer spatial resolution and a later date than the Landsat image, it provided better coverage for stands in that area.

The western half of the property was classified from a 12-band composite of the winter and summer scenes. Initially, using an unsupervised classification, the scene was divided into 283 classes. A supervised classification was then performed using GPS ground truth points, slope, aspect and elevation data. A thematic classification containing 15 classes was produced (see Plate 51). The eastern area of the property used only the winter image and was classified into the same 15 classes through the methods used on the western image, and the SPOT and summer images were employed wherever possible. The two images were then merged and edited for discrepancies. Using the control ground truth points extracted earlier, an accuracy assessment was completed which indicated accuracy of greater than 80 percent. The resulting classification became the base vegetation and land use map for all subsequent work and analyses. The image went through a further intensive, area-specific detailed classification in several high interest areas. This work provided the basis for constructing 1:20,000 scale harvest/inventory maps.

The recommendations and constraints developed for the project by an independent scientific commission contained several restrictions regarding the locations of timber harvest units. Commercial forest areas had to be identified which fell within a 15m or 30m buffer along the major rivers and their tributaries, and within 10m of wetland communities. Several other constraints were identified, such as isolated areas of forest less than 10 ha, located within wetland areas, and areas within specified slope and elevation categories. Buffering the wetland areas using ARC/INFO 6.0 was not possible due to the number of arcs in each polygon. ERDAS IMAGINE was used to create a 30m buffer around all wetland polygons. This raster coverage, as well as the vegetation cover, was subset by parcel and vectorized into ARC coverages.

An ArcCAD AUTOLISP routine was developed to identify the isolated forest areas and overlay the buffer, slope, and elevation layers. From this analysis, a composite coverage containing fields for each type of constraint was produced. The FREQUENCY command was employed to sum the vegetation data by constraint. This process was automated and applied to each of several parcels. The resulting databases were imported into Microsoft Excel 5.0 and placed into pivot tables. This allowed for greater flexibility in producing tables showing each vegetation type by each constraint, either by individual parcels or property-wide. The tables became critical in facilitating the production of harvest plans and the EIS.

Results and Discussion

A series of maps was produced in IMAGINE combining the classified satellite image and various vector layers such as property boundaries, roads (both GPS and digitized), contours, rivers, watersheds, buffer zones, study areas, and wildlife data. Converting the vegetation/land use map into a composite grayscale image allowed hydrologic and elevation features to become prominent. Using this as a backdrop, scientists were able to observe river systems and mountain ranges. They were then able to "heads-up digitize" the hydrologic basins across the island. Vector maps of meteorological conditions, archaeological sites, site-specific study areas, and GPS data points were also produced. ENCAD's NOVAJET III was used to plot all maps and images.

Benefits

When faced with similar project constraints, project managers, GIS analysts, or project scientists should not accept the "advertised capabilities" of any raster or vector tool. Each project is unique, and some can present what seem to be insurmountable obstacles. At this point, trying to develop specific applications for the technologies at hand can be extremely cost effective and, perhaps, the only way to solve the problem. In the case history outlined above, the project could not have been completed in one field season if raster and vector tools had not been employed. This translates to considerable time and dollar savings, especially for a company trying to meet the laws and regulations (and their imposing time constraints) of a foreign country. On a day-by-day basis, everyone was presented with situations that precluded using standard, off-the-shelf applications and left the scientists/analysts to unleash their imaginations to develop a unique application, algorithm, or procedure. The technologies allowed the project team to map a remote corner of the world, providing an international team of scientists the means for developing the first biodiversity database of the island. They completed the EIS on schedule.

LAND USE/LAND COVER

Enhancement, Identification, and Quantification of Land Cover Change

Douglas Stow and Dong Mei Chen, Department of Geography, San Diego State University
Robert Parrott, Research Division, San Diego Association of Governments

Challenge

With the pervasive implementation of geographic information systems (GISs) by public and private sector entities, an important and expanding use of raster image data is to update GIS databases and analyze temporally dynamic landscapes. An efficient means for updating and analyzing dynamic components of these geographically referenced databases is provided by multitemporal remote sensing. Multitemporal remote sensing data are most commonly acquired as digital images by sensors on aircraft or satellites at more than one instance in time. These image data are useful for monitoring land surface changes at most spatial scales (from patch to global) and many temporal scales (from days to decades). Such scales of remote sensing fall within the domain of high- to medium-resolution airborne and polar orbiting satellite systems.

The processing and analysis of multitemporal image data for purposes of mapping and quantifying land surface changes is often referred to as "change detection." In the strictest sense, change detection refers to a fairly simplistic process of highlighting or enhancing land surface features which have appeared to change. It was coined by defense reconnaissance interpreters whose job was to detect changes in enemy military personnel and weaponry. The term is used more broadly to include "identification" and "quantification" of the type of land use and land cover change for civilian purposes.

If remote sensing is to play a significant role in monitoring land surface dynamics and updating GIS databases, then some signal pertaining to changes in the reflectance or emittance of the surface must at least be detectable and preferably, identifiable or quantifiable. In addition to changes in captured energy associated with changes in land surface materials, other environmental variables influence changes in remotely sensed energy. Spatial and temporal variability in atmospheric properties also cause variations in the energy received by the sensor. Artifacts and uncertainties may result from environmental effects on radiation transfer and the effects of the sensor and its mode of sampling. Even when the same instrument is used for multitemporal acquisitions, the earth-platform-sensor geometry relationship and the sensor calibration are different for each data acquisition. Without properly correcting for or reducing these environmental and sensor noise effects, the resultant temporal variability of image data values may falsely detect or mask actual land surface changes.

Data and Methodology

Transforming multitemporal image data into useful information pertaining to land use/land cover change consists of a sequence of several image processing tasks. These can generally be grouped into geometric, radiometric, and change detection processing procedures. Geometric processing entails correction and alignment of multitemporal images such as to allow for precise image-to-ground and image-to-image comparisons. The most critical of these operations for multitemporal monitoring is spatial registration or image-to-image matching. Normal spatial registration and georeferencing accuracy are generally equivalent to 0.5 picture element or pixel. The most important requirement of radiometric correction is to ensure that changes in image brightness values for corresponding pixels of a multitemporal image sequence are predominantly caused by actual changes in spectral reflectance or emittance of the surface.

Once multitemporal images are geometrically and radiometrically aligned, another type of procedure is usually applied to convert them into digital maps which enhance relative degrees of changes or portray transitions of land cover categories. The analyst generally has four choices for such procedures: (1) multitemporal overlay, (2) multitemporal differencing, (3) post-classification comparison, or (4) multidate classification of transition sequences.

Multitemporal overlay is a type of change detection enhancement. It is usually achieved by displaying the same, single band or monochromatic image acquired on two or three different dates in different color planes (RGB) of an image display. For example, red band images from registered SPOT multispectral (XS) images taken five years apart are displayed simultaneously, with the early image date in the red color plane and the latter date in both the blue and green planes of an ERDAS image processing system (Plate 52). For a three-date overlay, each date is displayed in one of the RGB planes. Land surface changes resulting in substantial surface reflectance or emittance, and therefore, image brightness change between image dates, show up as hues on the color monitor or subsequent hardcopy image. For the two-date SPOT XS example shown in Plate 52, changes from lower reflectance to higher reflectance (vegetated to bare soil) are portrayed as cyan hues and higher-to-lower reflectance changes (bare soil to urban residential) as reddish hues. More extreme changes in image brightness often correspond to changes in land use, but may correspond to other surficial changes such as differences in cropping practices, condition of vegetation, or presence/absence of standing water or snow. Portions of the image where land surfaces and their reflectance or emittance are essentially unchanged, show up as gray tones with an absence of a predominant hue.

Procedures for change detection.

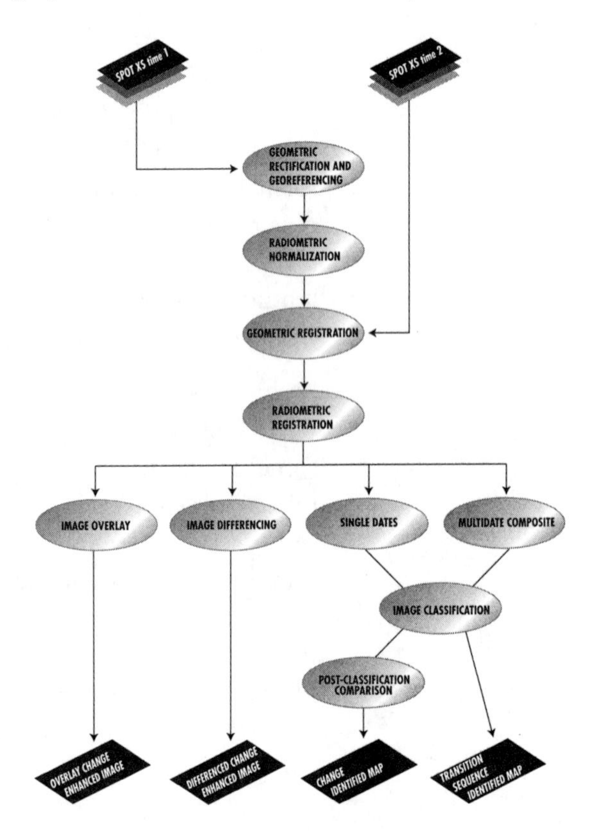

Another type of change enhancement technique is multi-temporal image differencing. Image brightness values from one date are subtracted from the values of the corresponding pixels (in a spatial sense) of a second date. If the first date is subtracted from the second, large positive differences in brightness values correspond to increases in surface reflectance over time (vegetated to bare soil) and large negative differences are associated with a temporal reduction in surface reflectance (bare soil to urban residential). Differences may be displayed as change detection images in monochrome, such that dark pixels are negative changes, bright pixels are positive changes and mid-gray pixels represent little to no change (Plate 52). The single band change detection image can also be color enhanced by density slicing. An advantage of differencing is that it can be applied to multiple image bands and differences for up to three bands can be displayed at one time. In addition, difference thresholds may be determined to generate simple maps depicting change/no change or positive change/negative change/no change.

The post-classification comparison approach goes beyond detection or enhancement of change and is an attempt at identifying types of land cover change. It simply involves the comparison of two land cover type maps independently generated from computer-assisted classification of two co-registered images captured on different dates. Categorical comparisons are made for each pixel to determine if change has apparently occurred and if so, to identify the old and new land cover categories. Clearly, the degree of success of post-classification comparison depends on the reliability of the maps generated from image classifications. Many analysts incorporate contextual and spatial proximity rules and other extent spatial data from GIS databases to reduce the number of false changes identified from post-classification comparison.

An example of categorical comparisons based on unsupervised classification of a multidate Landsat-TM is shown in Plate 52.

A less frequently utilized alternative to the post-classification comparison approach is multidate classification of transition sequences. The multidate classification approach entails the simultaneous classification of both image dates which have been merged into a single multiband-multidate data set (termed a stack layer in IMAGINE). Standard supervised and unsupervised approaches used to guide image classification routines based on single date images can be applied to multidate stacks, with the major difference being that the patterns are land cover transition sequences (or non-change state). Supervised methods rely on the analyst designating specific pixels known to represent particular categorical transitions. The unsupervised approach is based on statistical clustering of spectral-temporal signatures followed by interactive labeling of transition sequences by the analyst (Plate 52).

The specific utilization of change detection products depends on the objectives of the image analyst, GIS manager, or end user. One objective may be to assist in updating land cover or other layers of a GIS database. For updating a vector GIS layer in ARC/INFO format, raster change enhancements, change/no change maps, or change identification maps may be incorporated into "on-screen digitizing" procedures to augment visual interpretation of georeferenced image data on a color display monitor. This can be achieved with the Image Integrator in ARC/INFO or with the Vector Module in ERDAS, which is compatible with ARC/INFO formats. A more automated approach would be to vectorize the raster version of these maps and then overlay the revised vectors and category labels to a dated vector layer (usually with a great deal of interactive editing required to fix artifacts generated by automated classification and vectorization). Raster GIS layers may be updated directly by replacing "change" pixels with the newly identified land cover category.

The other principal objective for generating change detection products is to generate tabular, graphical, or map data on the amount and location of land surface changes. Examples of tabular statistics would be areal extents of land cover types at two different points in time or changes in extents during a specific time interval. These statistics may be reported by unit such as census tracks or watersheds. In addition, they can be depicted in graphical bar or pie charts rather than tabular form. Maps of change/no change or land cover change sequences are the direct products of several of the change detection procedures described above.

Updating Vector Land Use Inventories Using Multi-date Satellite Imagery

Robert Parrott and Sue Carnevale, San Diego Association of Governments (SANDAG)
Douglas Stow, Department of Geography, San Diego State University

Challenge

Between 1980 and 1990, population in the San Diego region increased by more than 636,000. Nearly 226,000 new housing units were constructed, and more than 406,000 jobs were added. By the year 2000, an additional 506,000 people, 109,000 housing units, and 54,000 jobs are projected. All of this growth has significant impacts on land use patterns. To assist in conducting its regional planning responsibilities, the San Diego Association of Governments (SANDAG) utilizes ESRI's GIS software to compile generalized digital vector land use inventories for the over 4,200 sq mi (10,878 sq km) comprising the San Diego region.

SANDAG compiled inventories of generalized land use in 1971, 1975, 1980, 1986, and 1990. Efforts are currently underway to update the 1990 inventory to represent land use conditions in the summer of 1995. In addition to tracking trends and changes in existing land use over time, these inventories are used in conjunction with other databases for a wide variety of projects. Examples include land use inventories as inputs to the Regional Growth Forecasting models; calibrating trip rates by land use type for transportation forecasting models; and using inventories in combination with biological databases to study potential wildlife habitat preservation areas.

Before 1990, land use inventories were prepared "from scratch," mostly by interpreting aerial photographs. More than 90 photos at a scale of 1:24,000 (or 1" = 2,000') were first interpreted and delineated on the photos. The land use polygons were then transferred to stable base maps and digitized. To ensure accuracy of the land use classification, secondary source materials were used such as USGS topographic maps, Thomas Brothers maps, telephone books, Criss Cross directories, reference lists, other databases, and field checks. Due to the cost of aerial photos and the time required for photo interpretation, transferring, and digitizing, inventories prior to 1990 covered only the western portion of the region.

Data and Methodology

1990 Inventory

Many of the same techniques and sources used to create prior inventories were used to produce the 1990 Generalized Land Use Inventory. However, to avoid starting from scratch, the 1986 vector land use database was updated where change occurred by using satellite image change detection. Satellite imagery and change detection procedures eliminated the need to "photo interpret" the entire region. Moreover, since the satellite imagery provided similar resolution to the aerial photos and was less expensive for a full regional coverage, SANDAG was able to expand the 1990 inventory to cover the entire 4,200 sq mi of the region.

In 1990, a grant from NASA's Earth Observation Commercial Application Program (EOCAP) enabled the San Diego State University (SDSU) Geography Department to evaluate the feasibility of using satellite imagery to assist in updating a vector land use database. Cooperative efforts among SANDAG, SDSU, ERDAS, and ESRI allowed SANDAG to use multidate satellite scenes, image change detection procedures, and the ERDAS-ARC/INFO Live Link software to update the vector land use database.

An ERDAS Databundle package was purchased to do this work. The Databundle included a SUN workstation with ERDAS, ARC/INFO, and Live Link software, and co-registered 1990 geocoded data with terrain-corrected satellite imagery tiled to 15-minute quads covering the San Diego region. The imagery consisted of SPOT Panchromatic and LANDSAT multispectral Thematic Mapper (TM) scenes. The SPOT panchromatic image shown in the figure below depicts a 15-minute quad, and illustrates the size of a work area (working window).

USGS 7-1/2 minute quads representing working windows for the San Diego region.

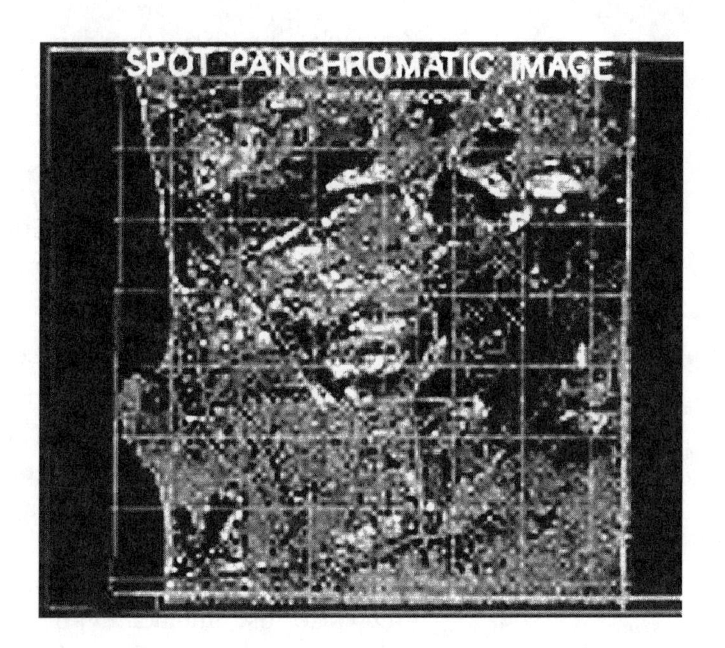

The satellite image change detection process used in the 1990 update involved SPOT panchromatic imagery from 1986 and 1990 (see next figures). The update was conducted in two phases. In phase one, the areas of change were detected, and in phase two the land use was classified.

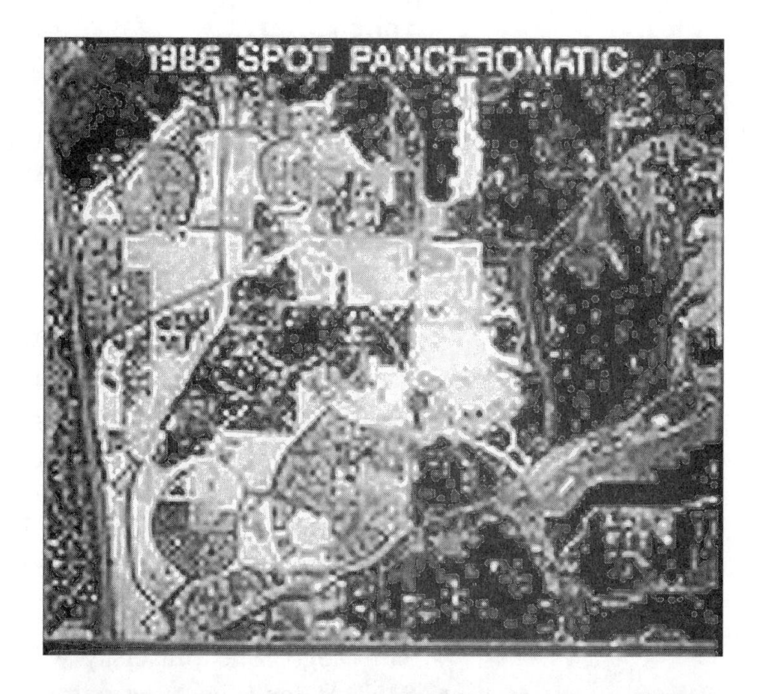

1986 SPOT panchromatic image of one working window in northwest San Diego county.

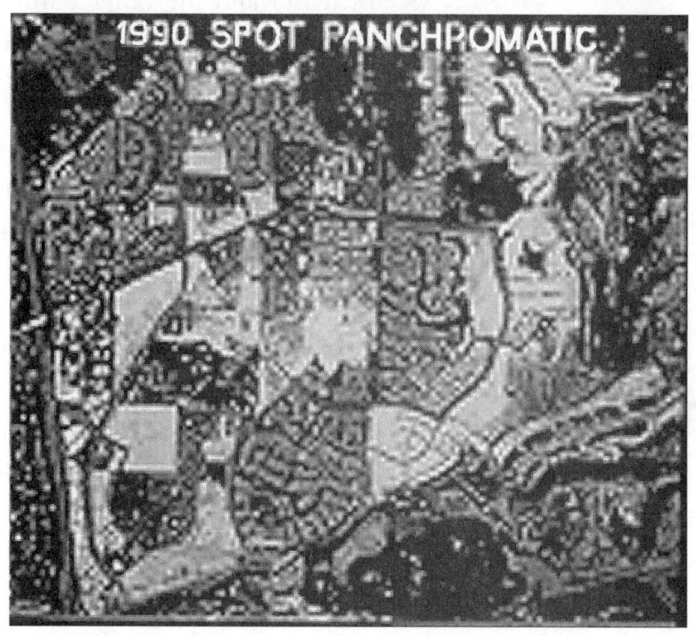

1990 mate to 1986 working window.

1995 Inventory

Based on the success of the 1990 update project, SANDAG is again using SPOT satellite imagery, ERDAS image processing software, and ESRI's ARC/INFO GIS software to complete the 1995 update of the generalized land use inventory, as well as to update the region's vegetation database.

A license for both the 1995 SPOT panchromatic and multispectral digital imagery, terrain corrected and geocoded, was obtained from SPOT Image Inc. by a partnership of five public agencies. The imagery was acquired for the entire San Diego region and the Tijuana River Valley watershed (two-thirds of the 1,700 sq mi watershed is in Mexico) during the summer of 1995 to correspond to the 1990 imagery of the San Diego region.

San Diego State University will process the imagery for SANDAG and the other partners using ERDAS IMAGINE software. In addition to creating a composite image for displaying changed areas, polygons of changed areas are being created from the imagery. Since the composite image with its pixel-to-pixel representation of change tends to overrepresent areas with true land use change, the changed area polygons will be produced with a process to reduce "noise" and fragmented small areas of change.

Two image processing phases outlined below are underway for the 1995 vector land use inventory update. In the first phase, preprocessing of the imagery is conducted and the 1990-1995 change detection composite is produced. In the second phase, the changed areas are being extracted and vectorized.

Preprocessing of the imagery

❒ Tiling of the 1990 and 1995 SPOT panchromatic imagery into 15-minute quads.

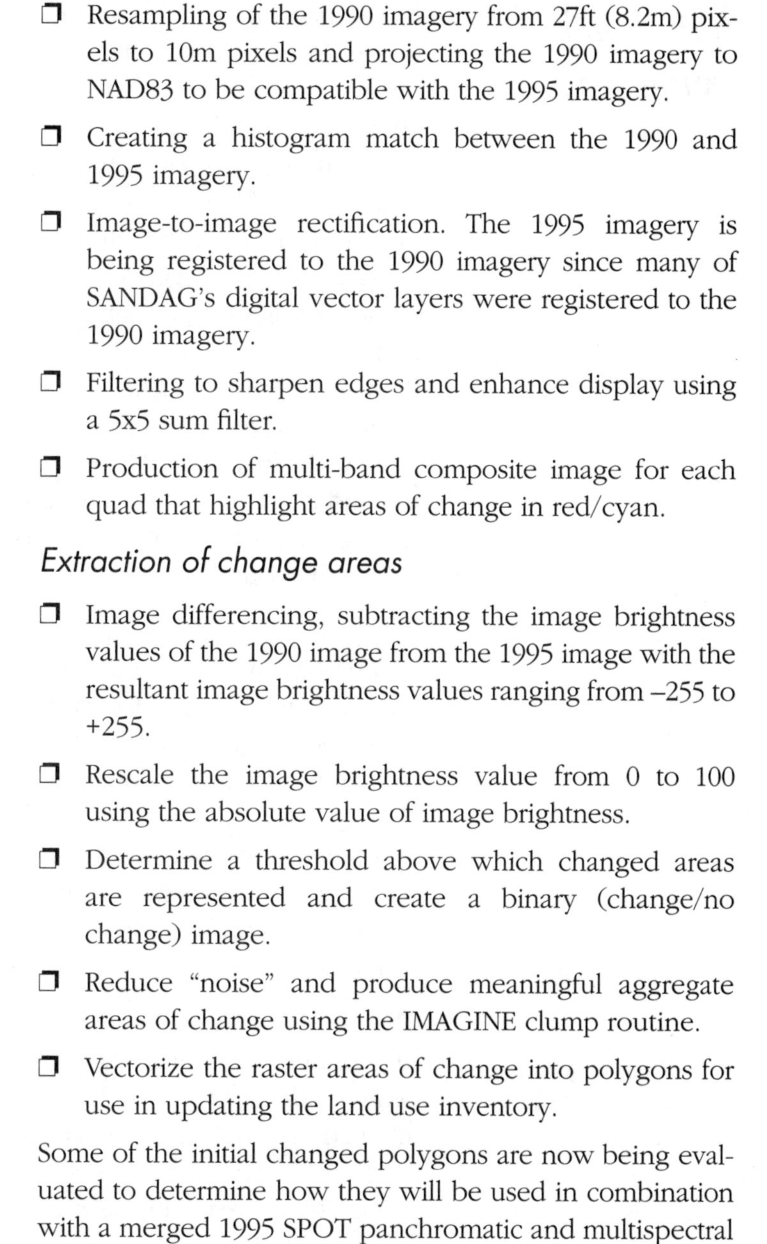

- Resampling of the 1990 imagery from 27ft (8.2m) pixels to 10m pixels and projecting the 1990 imagery to NAD83 to be compatible with the 1995 imagery.
- Creating a histogram match between the 1990 and 1995 imagery.
- Image-to-image rectification. The 1995 imagery is being registered to the 1990 imagery since many of SANDAG's digital vector layers were registered to the 1990 imagery.
- Filtering to sharpen edges and enhance display using a 5x5 sum filter.
- Production of multi-band composite image for each quad that highlight areas of change in red/cyan.

Extraction of change areas

- Image differencing, subtracting the image brightness values of the 1990 image from the 1995 image with the resultant image brightness values ranging from –255 to +255.
- Rescale the image brightness value from 0 to 100 using the absolute value of image brightness.
- Determine a threshold above which changed areas are represented and create a binary (change/no change) image.
- Reduce "noise" and produce meaningful aggregate areas of change using the IMAGINE clump routine.
- Vectorize the raster areas of change into polygons for use in updating the land use inventory.

Some of the initial changed polygons are now being evaluated to determine how they will be used in combination with a merged 1995 SPOT panchromatic and multispectral image product to update the 1990 land use inventory to 1995 conditions. The 1995 land use inventory update project is scheduled for completion by the summer of 1996.

Results and Conclusions

In the first phase, a composite of the two panchromatic images from the different years was displayed on the image display monitor. Areas where physical characteristics have changed are highlighted in color. When viewed in composite mode, areas of land use change were shown in red and cyan. Red highlights areas that were cleared for development in 1986 and are urbanized in 1990. Cyan indicates areas that were vegetated in 1986 and are cleared and graded for urban development in 1990. Areas of no change appear in grayish tones. The 1986 land use vectors were then registered to the composite display of the satellite imagery and were displayed in the graphics plane using the Live Link. The analyst could usually identify new land use and digitize its boundaries directly off the video display through a process called "heads-up" digitizing. Plate 53 shows the composite image display for the 1986 land use vectors in blue and the updated 1990 vectors in yellow.

To assist analysts in digitizing and focusing their attention on areas most likely to change, ARCPLOT AMLs were developed. When displaying the imagery, areas urbanized in 1986 were masked out and land uses that were likely to change, such as low-density residential, were highlighted. Unshaded areas had the highest likelihood for development (e.g., vacant and agricultural lands). In addition, lands under construction in 1986 were flagged for classification with a vector overlay on the imagery.

The needs of the transportation modeling process have increased the complexity of land use inventories over the years. Advances in technique and technology have allowed the original 16 categories of land use cataloged in 1971 to be expanded to over 75 categories in 1990 and 1995. This increase in the number of categories leads to the second phase of land use classification. By combining panchromatic and multispectral imagery of same-date (1990) satellite images, the computer display resembles a high resolution color infrared photo. This technique often provided enough detail to allow specific identification. If not, the secondary sources mentioned above were used.

Benefits

Six advantages have been identified for using multidate satellite image change detection over traditional aerial photo procedures:

- Satellite change detection allows SANDAG to update the earlier developed vector layer, and to concentrate on areas that changed. This will improve the comparison to future inventories.
- Imagery provides an up-to-date base map for a fast growing area like San Diego where the USGS quad maps are, for the most part, 20 years old.
- The procedure eliminates the need to delineate land use polygons on the aerials and then transfer the polygons to mylar. This new process gives SANDAG the capability for "heads-up" digitizing directly on the screen.
- The procedure allows the land use database to be expanded to the entire region.
  - The procedure permits other applications for digital satellite imagery, such as registration and maintenance of a street-centerline, geocode database, and the development of a vegetation layer.
  - Future updates can be accomplished in a matter of months compared to traditional methods that typically take a year or more.

The U.S.-Mexico Border GIS: The Tijuana River Watershed Project

Richard D. Wright, John F. O'Leary, and Douglas A. Stow, Department of Geography, San Diego State University

Challenge

The Tijuana River Watershed, covering a 1,725 sq mi (4,468 sq km) area, two-thirds of which is in Mexico, lies astride the California-Baja California section of the U.S.-Mexico border (Plate 54). The watershed is a diverse geographical area with a wide range of topography, climates, biological resources, land uses and sociopolitical institutions. It is also located within the San Diego-Tijuana region which has a population of over 4 million that is expected to increase to more than 5 million by the end of the century. Abutting the border on the U.S. side is the Tijuana River estuary. This is the largest functioning wetland in southern California, providing critical habitat for birds on the Pacific flyway.

The watershed and contiguous area are a quagmire of differing agendas, cultures, economic classes, and political systems. Mexico is struggling to improve its economy with tremendous growth occurring in the region due to increasing economic development along the border. The Tijuana-San Diego region expects to experience the most significant growth resulting from the North American Free Trade Agreement (NAFTA). However, the region's urban infrastructure is inadequate to support such growth. Although the two countries share many watersheds, the international boundary has reinforced an approach in which both countries address social and environmental problems as national responsibilities, rather than international obligations. The United States can afford to place environmental protection as a high priority on its political agenda. Mexico, however, is overwhelmed by current and projected demands on its limited infrastructure and resources.

With startup funding from the National Oceanic and Atmospheric Administration (NOAA) and subsequent support from the United States Geological Survey (USGS) and the Department of Defense (DoD), San Diego State University (SDSU) and El Colegio de la Frontera Norte (COLEF) in Tijuana are jointly preparing a GIS for the Tijuana River Watershed (TRW). The purpose of the GIS is to address a set of environmental concerns related to the watershed shared by Mexico and the United States. Cooperation between SDSU and COLEF has been formalized in an agreement that provides for data sharing, coordinated data development and scientific research, and joint use of the data for education and research.

The project consists of five complementary components: GIS database development, community outreach, education, scientific research, and watershed management. The purpose of the GIS component is to develop a spatially referenced socio-environmental database to facilitate scientific research, education, and decision making within the context of shared watershed management. The database is approximately 75 percent complete and includes hypsography, hydrography, geology, geomorphology, soils, land use, climate, vegetation, and demographic characteristics. Most of the coverages are designed at scales and resolutions equal to map scales of 1:24,000 to 1:50,000.

Efforts are underway to develop a strong community outreach component so that the GIS development will not occur in a sociopolitical vacuum. In addition to holding user needs assessment workshops, the project team envisions a coordinator responsible for interacting with the public, government, and private sectors.

Education for a wide range of students and the lay public is another essential component. Educational initiatives planned for the project include a video of the project and the watershed, a digital flyover through the watershed, one or more interactive multi-media educational modules using ArcView, and a small monochromatic atlas of the watershed.

Numerous research initiatives have been identified as an important project component to receive attention in the near term. Some of the more critical studies include environmental and toxics risk assessment, water and air pollution analysis, multiple species habitat modeling, evaluation of service and infrastructure needs, and land use analysis.

Effective resource management policies require a binational, participatory decision making process that incorporates the temporal changes expected to occur throughout the watershed as a result of economic development. The GIS under preparation can be used to facilitate a dialog among watershed stakeholders about the possible future impacts of human activities and resource distribution in the watershed. A series of workshops focused on using the GIS is envisioned to raise stakeholder awareness of complex, interrelated border issues. These workshops would use the spatial relationships and visualization to emphasize problems as they relate to long-term, sustainable, binational management practices. In conjunction with this component, there will be a feasibility study to establish a binational nonprofit organization aimed at promoting improved management practices in the watershed.

Data and Methodology

A comprehensive strategy was adopted for mapping vegetation and land cover types in the Baja California portion of the watershed. The strategy paralleled one used by San Diego in its Multiple Species Conservation Planning (MSCP) effort to map 594,000 acres (240,392 ha) in southwestern San Diego County. Mapping south of the international border utilized MSCP decision mapping rules, some of which were modified to correct deficiencies or to better accommodate interregional differences.

Color aerial photos of the watershed at 1:50,000 were acquired by NOAA in August 1994 and enlarged to 1:12,500. Minimum mapping units (MMUs) were chosen to be consistent with MSCP results and to aid mappers in delineating features on the enlargement. An MMU of one acre was selected for wetland and riparian vegetation types due to their relative scarcity in the overall region. All other vegetation and land cover types were mapped at an MMU of five acres.

Field and laboratory mapping occurred from August 1995 through March 1996 beginning in the western extremity of the watershed and systematically progressing eastward. While most of the mapping was accomplished by photo interpretation in the laboratory, a modest amount of field mapping and a substantial amount of ground truthing occurred throughout the mapping effort. Mapping initially began in the field and progressed slowly in order for mappers to gain increasing familiarity and confidence with the mapping procedures. Individual units (stands) of vegetation and land cover types were identified and delineated as polygons on clear acetate overlays placed upon aerial photos. An acetate template imprinted with various geometric forms for both MMUs was used to help determine whether a specific vegetation or land cover unit met the MMU thresholds. Care was taken to avoid mapping the outer four-inch perimeters of aerial photos in order to reduce distortion and possible mismappings of polygons due to poorer color quality that occurs near the edge of an aerial photograph.

The automation process for creating a digital vegetation layer for the TRW GIS database was performed using an "on-screen digitizing" or "heads-up digitizing" approach and was based on SPOT image data. This was achieved using ERDAS IMAGINE software to generate a vector GIS layer in ARC/INFO format (see the following illustration).

Vegetation and land cover, Tijuana River watershed.

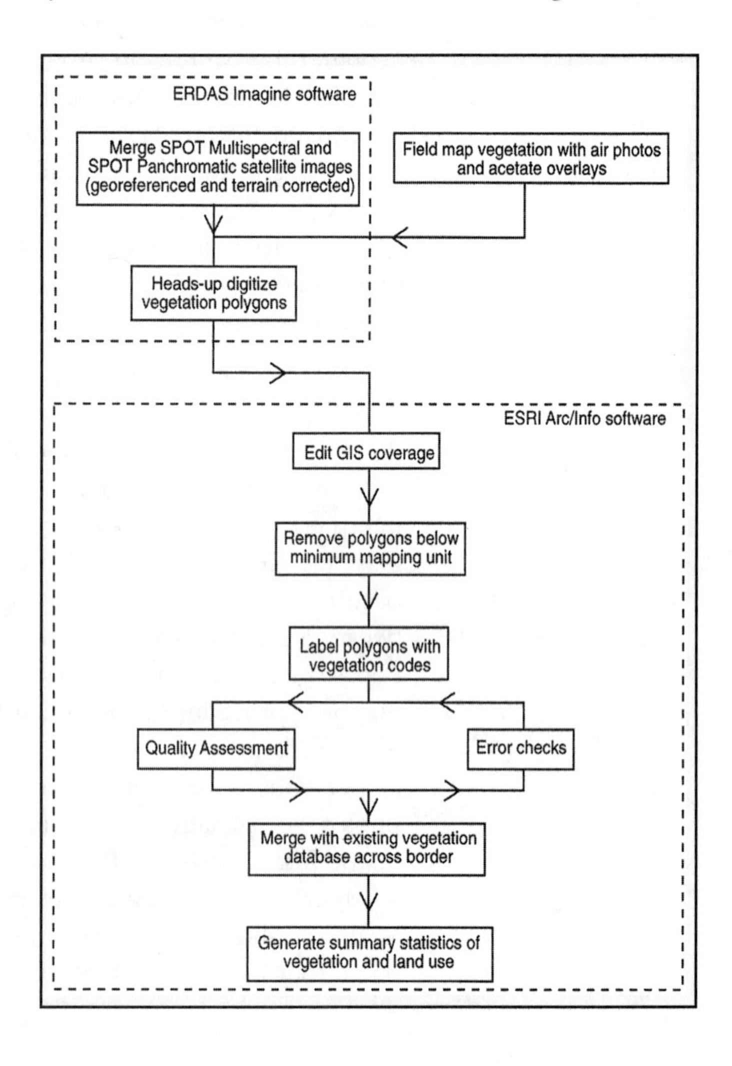

SPOT panchromatic (Pan) and multispectral (XS) data covering the TRW were acquired during summer 1995. These data were purchased from SPOT Image in orthorectified form, meaning that the image data had been rectified to remove sensor and earth distortions, corrected for terrain displacement (using digital elevation model data), and georeferenced to the UTM coordinates. An image fusion procedure in IMAGINE based on an intensity-hue-saturation (IHS) transformation was applied to merge the SPOT Pan and XS data. This technique was effective in integrating and exploiting the high spatial resolution of Pan data with the multispectral color discrimination advantages of XS data.

The resultant merged and georeferenced SPOT image data enabled vegetation polygons to be interactively digitized and georeferenced on the image display monitor via on-screen digitizing. This was performed using the ERDAS Vector Module (and may also be achieved using the ARC/INFO Image Integrator). Vegetation polygons on the air photo transparencies provided the basis for interactively digitizing boundaries and encoding vegetation category labels. Color contrast, edge, and texture enhancement routines in IMAGINE facilitated boundary delineation. Polygons were digitized in the graphic overlay plane of the color monitor by using a conventional mouse for location control. An AML program was developed and used to efficiently add the vegetation category codes and any qualifier descriptions (as INFO attributes) to label points for each polygon.

Results and Conclusions

The final step in the automation process was to harmonize the Mexican portion of the database with existing digital versions of vegetation maps north of, and contiguous to, the international border to create a comprehensive map of the entire watershed. The integrated data were then processed using topological building and editing functions. Quality assurance was based on careful analysis of tabular and map products. Plate 55 shows the pre-integration vegetation polygons and Plate 56 shows the post-integration polygons.

This project illustrates the integration of field mapping methods and aerial photograph interpretation with spatial technologies to prepare one of the layers in a binational GIS. Vegetation is an important theme because it is an essential ingredient of all components of the project—GIS database development, community outreach, education, scientific research, and watershed management. In support of community outreach and education, the vegetation data along with other data will be packaged with a customized version of ArcView to help the lay public and students explore the geographic characteristics of the watershed. It will also be used for various types of scientific research including natural habitat preserve design and water runoff modeling. Finally, this data layer will be employed for binational management of the watershed's natural resources.

Development of a Land Use/Land Cover Map

Jerry L. Whistler, Stephen L. Egbert, Edward A. Martinko, David Baumgartner, and Re-Yang Lee, Kansas Applied Remote Sensing Program, University of Kansas, Lawrence, Kansas

Mark E. Jakubauskas, Department of Geography, University of Oklahoma, Norman, Oklahoma

Challenge

The Kansas Applied Remote Sensing (KARS) Program at the University of Kansas recently completed a three-year project to develop a digital map of land use/land cover for the State of Kansas. During the development phase of the

Kansas State GIS Initiative, numerous state and federal agencies indicated that land use and land cover data were critical to their management and planning missions. Existing land cover information was out of date. It consisted of U.S. Department of Agriculture LUDA data compiled from aerial photography taken in the late 1970s and a 1974 land use/land cover map produced by the State of Kansas from manual interpretation of Landsat MSS images. The 1974 map was compiled and produced at a scale of 1:1,000,000, depicting 12 land use classes using a modified Anderson Level II classification scheme. Despite the scale limitations, the age of the data, and the fact that it was never converted to a digital form, requests for the 1974 Kansas map are still received.

The update was undertaken in 1991. A technical advisory committee was convened to evaluate several options for mapping the state. Factors taken into account included intended application, data form, minimum mapping unit (MMU), data acquisition and personnel costs, classification costs and methods, and total project costs. The KARS Program was contracted to develop the map within a 2.5 year period. Digital classification of TM imagery would result in the quickest production of the land cover database, but the number of classes depicted would be fewer and more general relative to the other options. This was judged to be an acceptable trade-off by the technical advisory committee.

The following land use/land cover classification scheme is based on a modified USGS classification by Anderson et al. (1976).

Urban	Rural
Residential	Cropland
Commercial/industrial	Grassland
Open land (parks and golf courses)	Woodland
Woodland	Water
Water	Other (includes sandbars, quarries, dams, and major highways)

Urban grassland, water, and woodland were maintained as separate entities from their corresponding rural classes to enable users to recode urban areas to a single entity if needed. An MMU appropriate for producing 1:50,000 scale maps was specified at approximately 1 ha (2.5 acres). The MMU for water was set at 0.1 ha or 0.25 acres (about 1 TM pixel). Developed areas were mapped as urban if they exceeded 16 ha (40 acres). An overall map accuracy of 85 percent or greater was required.

Data and Methodology

ERDAS software was used for all image and raster GIS processing, and ARC/INFO was used for vector GIS processing and projection. For optimal distinction of agricultural land use and forest cover, TM data between June 1 and September 30 were used for the project. Sixteen scenes were required to completely cover Kansas. County-sized subscenes were extracted from the raster data, and spliced from two or more scenes where necessary to create a complete county file.

Data were processed county by county for two reasons. First, this approach served to stratify the data into areas small enough that a degree of homogeneity within the spectral classes could be maintained. Second, counties were natural units by which to tile the data for release to end users. Data were not geometrically rectified before processing, as it was found that rectification before classification "smeared" the spectral values for small stock ponds such that they were no longer classified as water. This was unacceptable because stock ponds represent an important water resource in many Kansas counties.

A layered, unsupervised classification was produced. For rural classes, the data were initially clustered into 100 preliminary spectral classes. For urban classes, the data were initially clustered into 50 preliminary spectral classes. Since separate urban and rural clustering and recoding operations for urban and rural classes were performed, it was necessary to create masks defining the boundaries of urban areas. This was accomplished by screen digitizing the urban outlines directly from the imagery displayed on the screen. The two recoded GIS files, one containing five rural classes and one containing five urban classes, were then merged in a GIS overlay operation to create an initial map of land use/land cover.

Early in the project it became apparent that certain landscape features were clearly visible in the imagery but were consistently misclassified (e.g., sandbars in rivers were frequently confused with plowed cropland). Since these features were desirable in the final land cover map, four-lane highways, major dams, sandbars, and large quarries and gravel pits were screen digitized and then merged with the rural and urban GIS maps.

At this stage in the process, the land cover maps were characterized by "speckling" produced by individual pixels or groups of pixels smaller than the MMU thresholds of the project. Isolated pixels and pixels jutting anomalously from homogeneous areas were removed using software written at KARS. Single water pixels were not removed as they frequently represented important small stock ponds.

This initial generalization greatly reduced the number of polygons in the map. This reduction of polygons was important in the raster-to-vector conversion process because the number of polygons in a county map frequently exceeded the ARC/INFO Version 5.0 limit of 10,000 polygons in a coverage. After raster generalization, the county maps were converted from ERDAS raster format to ARC/INFO coverages and further generalized. The ARC/INFO ELIMINATE function was used to remove any remaining polygons below the specified MMU threshold. The urban water and water classes were not generalized.

Image to map rectification was performed on the vector version of the classified data on a county by county basis. Fifteen to 20 GCPs were digitized from USGS topographic quads. Corresponding image column-row positions were transformed to state plane coordinates (measured in feet) using the ARC/INFO TRANSFORM command. Counties were edge matched to each other by "borrowing" a narrow strip, nine pixels wide, from an adjacent, previously completed county. Land cover information from the adjacent county was allowed to overwrite information in the current county. Minor editing of arcs and polygons was necessary after the update to improve the cartographic appearance of the splice.

Because Kansas is split into north and south state plane zones, some counties required temporary projection of the coverage to the Universal Transverse Mercator projection for edge matching. After edge matching, the coverages were projected back to the appropriate state plane zone. Contract specifications required the coverages to be delivered in decimal geographic coordinates, requiring double precision real numbers to maintain spatial precision. The geographic coverage was then clipped with the appropriate double precision county boundary provided by the Kansas Geological Survey.

Accuracy was calculated by comparing classified data with manually interpreted 1:58,000 color-infrared 1985 National High Altitude Photography and 1:20,000 black-and-white 1986 State of Kansas reappraisal photography. A systematic sampling procedure for each county was adopted to ensure that at least five percent of the total county area was covered by sample sites (county area x 0.05). Individual sample sites within a county covered 4 sq mi (2x2-mile or 10.4x10.4-sq km sections). The number of sites per county ranged between three and 18.

Land use/land cover classes were manually interpreted from the photography directly onto mylar, digitized using ARC/INFO, and converted from vector to raster format. The TM-derived land use/land cover map was then cross tabulated with the photointerpreted sites, generating statistics for errors of omission and commission, accuracy by land cover class, overall accuracy, and a Kappa statistic. These statistics, along with information such as image source, image date, image path/row, image scene-id, and date of coverage completion, were provided as data set documentation (metadata) for the county and embedded in the LOG file of the ARC/INFO coverage.

Results and Discussion

A seamless digital land use/land cover database meeting or exceeding all map specifications for the 105 counties in Kansas was created. An eight-county area in northeast Kansas is shown in Plate 57. The actual total cost for the project came to $313,908, or $3.82/sq mi ($1.47/sq km). The spatial accuracy of the data is well under 30m. Random comparison of polygon boundaries to a limited number of digital orthophotographs (original photo scale, 1:12,000) has shown spatial accuracy in the range of +33 feet. Overall accuracy (co-occurrence) of the classification for the entire state was 91 percent, with a Kappa value of 0.81.

The land use/land cover data set has been transferred to the Kansas Data Access and Support Center (DASC), a central library and clearinghouse for distribution of the state GIS database. From November 1991 through May 1994, 60 requests for the data set had been made from federal, state, and county government agencies, universities, and private industry. The Kansas GIS Technical Advisory Committee Land Cover Work Group has recommended that this data set be reviewed every five years, with updates occurring every ten years or less. Future updates may use multitemporal data in order to provide the greater detail users have requested.

NOTE: *The county level ARC/INFO coverages created by the Kansas Land Cover Mapping Project are available via anonymous ftp. Contact DASC at (913) 864-3965, or at the WWW site,* http://gisdasc.kgs.ukans.edu.

A Structural Vegetation Database for the Murray Darling Basin

Kim Ritman, Bureau of Resource Sciences, ACT Australia

Challenge

The Murray Darling Basin is a catchment area exceeding 1 million sq km and comprising 14 percent of Australia's land mass. This catchment forms an agriculturally important basin responsible for nearly 50 percent of Australia's gross agricultural productivity. It also comprises a diverse range of land covers, from alpine forests and bogs to desert shrubs, extensive wheat and sheep farms, and many types of forest and woodlands.

Location map of the Murray Darling Basin. The box indicates the area covered by the binary image.

A project was initiated by the Murray Darling Basin Commission to assemble a digital database of vegetation information over the entire basin. A significant outcome of this project is the structural vegetation database derived by interpreting TM satellite images acquired between 1989 and 1991 and supported by field data and aerial photograph interpretation.

Structural vegetation is nested within a broad land cover classification, with the remnant vegetation characterized by type, density, and growth form. This database is extremely versatile and can be interrogated for a number of specific themes. This study focuses on forest extent as one of those themes. The binary image shows a section of the forest extent theme over the Australian Capital Territory.

Forest extent (shown in black) over the Australian Capital Territory.

Forest extent was mapped in the database at a grid cell resolution of 25m. The forest cover was filtered to a minimum patch size of 0.25 ha (four grid cells) using a simple routine developed in ARC/INFO GRID. The entire data set covers 1.06 million sq km. The structural vegetation database is held in GRID format and contains metadata items at both the tile and grid cell level. Prior to delivery as final data, each tile of the database went through a set of quality assurance tests written in GRID AML. In addition, an attribute accuracy assessment method was devised specifically for the data. A specifications manual describes the database and includes database design and a data dictionary.

Data and Methodology

The specified format for the database is ARC/INFO GRID. Most of the image processing was performed using ERDAS IMAGINE, and then transferred to GRID. For several reasons, no attempt was made to convert to polygons. Because the source data (satellite images) are in raster format, GRID was immediately suitable. Maintaining the data in as primary a nature as possible is important. Outputs from the database can then be in any number of formats, comprise different attribute themes, or be generalized to suit specific uses.

The raster environment provides the capacity to import data from other software systems and to structure the themes of attributes efficiently without the generalization procedures usually associated with raster-to-polygon conversion. The INFO relational table makes the GRID data model almost transparent with ARC/INFO vector databases; thus, polygonized vegetation data were easily merged.

The process for interpreting structural vegetation attributes began with digital classification of the forest extent. This part of the database forms the spatial template for the identification and attribution of vegetation type, growth form, and density. The digital classification of forest extent is achieved by producing an unsupervised classification of satellite spectral signatures. Manual class assignment follows digital classification and is aided by hierarchical clustering ordinated on NDVI, field work, and aerial photographs.

Structural vegetation attributes were assigned to the database from both existing and new sources. Forest extent, the template within which these attributes are placed, is the only completely new data set added to the database. The integration of the various data sources is undertaken in the GRID environment according to a number of rules governing logical consistency.

The process of assembling the full structural database includes 12 steps. There are a number of variations possible on the basic method. The "woody" vegetation data are a preliminary step in identifying forest extent.

Flow diagram of methods to obtain structural vegetation data. The ideal method uses all 12 steps.

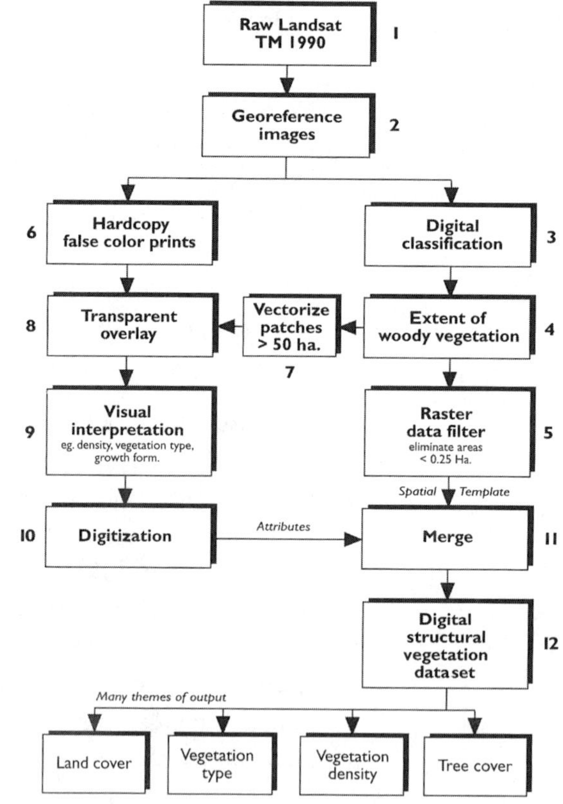

Method Variations

A - no step 7, all others unchanged

B - substitute on-screen display at step 6, conduct visual interpretation at 9, and eliminate step 10.

C - substitute existing data (eg. API derived) for steps 6, 7, 8 & 9.

D - digital classification for density at step 3, visual interpretation of other attributes at step 9.

E - complete digital classification at step 3, therefore no steps 6, 7, 8, 9, 10 & 11.

The forest extent data set was filtered to remove patches less than 0.25 ha. This was necessary to eliminate erroneous classifications of smaller areas caused by edge effects and spectral mixing. The data retained a grid cell size of 25m following filtering. The filter used was designed to remove groups of cells according to the following rules:

- Remove patches less than 0.25 ha (four grid cells).
- Retain linear features as much as possible.
- Assume connectivity for patches satisfied through orthogonal and diagonal connection.
- Retain boundaries in original position.
- Replace patches under threshold with dominant focal class.
- Do not alter classification of patches greater than or equal to 0.25 ha.

An automated process was written in ARC/INFO AML to process entire tiles of the data. Other filters (e.g., modal and majority filters) available in GRID were tested and found unable to satisfy the list of rules above. Because the devised filter only operates on binary grid cell data and experiences processing overloads with extremely large arrays of data, processing was restricted to overlapping tiles.

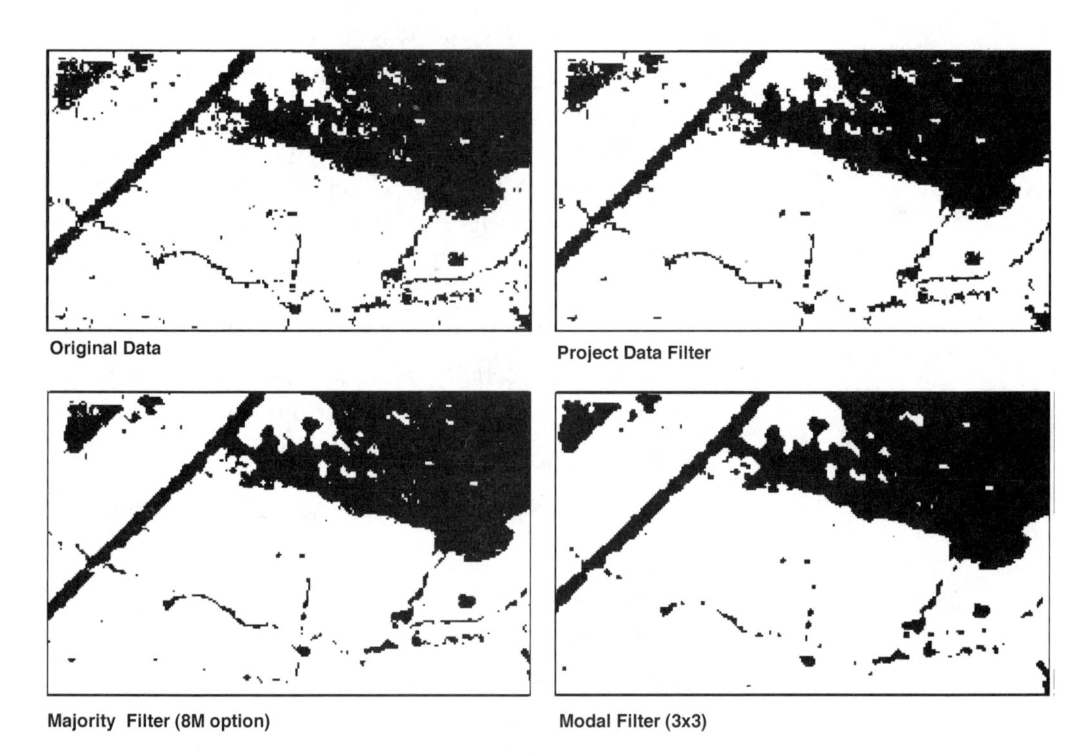

Comparison of three data filters with orginal binary image of forest extent (black). The project data filter causes minimal change to boundaries while still retaining linear features.

Metadata items are essential components of the electronic database. Metadata are information about the data without which the user is poorly informed and consequently risks misusing the data. Metadata referred to here must not be confused with directories of metadata, although the contents have considerable overlap. GRID was found to be amenable to incorporation of metadata; however, some improvements are still necessary.

The key to processing metadata is to provide the user with essential information but not overload the database with information that can be sourced elsewhere. In addition, metadata should be embedded pragmatically in the database—a capability that is slow in development. When the data set is copied, the metadata should follow, even if the data are converted into another format.

Two groups of metadata have been identified from exhaustive emerging national and international standards. One group of metadata relates to the data set at the tile level (e.g., documentation, custodian details, date of survey). Seventeen such general metadata items were identified for this study. Metadata information was included as an ARC/INFO file (*filename.MET*), but this facility could be improved.

The second group of metadata includes information referring to the fundamental spatial unit, in this case the grid cell. These more specific metadata deal with information that can vary across a tile (e.g., data source, satellite date, path, and row), and can be added as extra items in the INFO tables. Only six metadata items were identified as necessary.

The attribute accuracy assessment method for the forest extent data set is based on a multistage systematic sample design. The design has the practical advantage of simplifying the acquisition and handling of the aerial photographs used to acquire the pseudo-ground truth values for the accuracy assessment. Most of the data preparation and presentation for assessment is achieved using ARC/INFO GRID.

In the first sampling stage, a square grid of points was placed over the entire Murray Darling Basin. The grid was oriented to incorporate the linear trends in land cover types peculiar to the basin. These grid points formed the primary sample. At each primary sample point, a secondary grid of points was created. The secondary grid was scaled to sample relatively local pseudo-ground truth values by covering the equivalent of only a few aerial photographs.

The raster GIS was used to produce the primary and secondary grid of points, overlay the grids on the forest extent data set (reference data), interrogate the reference value (forest or non-forest), print a transparent overlay containing the secondary sample points at photo scale, record the pseudo truth and reference values at each secondary sample point, and display the spatial variability of attribute accuracy (Kappa statistics) for the basin.

Results and Discussion

The structural vegetation database for the Murray Darling Basin was produced from satellite imagery and managed as digital raster data. ARC/INFO GRID was used to undertake a number of key steps including data source integration, data filtering, quality assurance testing, embedding metadata, and assessing attribute accuracy. The data were maintained in raster format, which enhanced versatility and established the same as a source of primary data from which many themes of land cover and vegetation could be extracted.

Benefits

The cost of producing the structural vegetation database for such a large area is a little over $3.10 (Australian dollars)/sq km. This includes the cost of imagery and takes into account the in kind as well as hard cash contributions from the 16 participant state and federal government agencies.

Comparison with two similar projects indicate that three to four dollars Australian/sq km is a typical cost range. The other projects involved mapping land cover in Great Britain and Georgia, USA, both involving a land area much smaller than Murray Darling Basin.

URBAN AND REGIONAL PLANNING

Measuring and Modeling Urban Growth in Kathmandu

Barry Haack, David Craven, and Susan McDonald Jampoler, Department of Geography, George Mason University

Challenge

A significant and recent problem in the Kathmandu Valley in Nepal is the loss of very productive agricultural lands due to uncontrolled urban growth. This problem is compounded by the lack of information available to responsible government organizations needing to know where, and at what rate, this growth is occurring.

An automated raster-based GIS for the Kathmandu Valley was constructed with cooperating Nepali scientists to evaluate the location and extent of urban growth. The resulting GIS provides planners with information on where growth has occurred and how different planning controls might affect future growth.

The absence of current information on land cover in the valley created a concomitant need to use spaceborne data from the SPOT Image satellite to identify the current urban growth areas in Kathmandu as input to the GIS.

Kathmandu Valley

Kathmandu Valley is in the central region of Nepal encompassing an area of approximately 30 by 35km. The floor of the valley is at an average elevation of 1,300m, and the sides of the valley are very steeply sloping to elevations of over 3,000m. The valley is home for the largest and most rapidly growing population center in the country. The population in 1981 was 739,000; it is currently estimated to be over 1,000,000.

Kathmandu Valley is also one of the most productive agricultural regions of Nepal, providing 97 percent of the food grain needs for the valley in 1981. Most of the valley floor, and much of the terraced valley walls, are intensively cultivated. The temperate climate and availability of water for irrigation allows for almost continuous crop production. Major crops include rice, wheat, maize, potatoes, mustard, and a number of other seeds used for oil production. A large variety of vegetables is grown throughout the year, providing fresh produce to the local population.

The attractions of the national capital for business, commercial, and social interests, and its proximity to some of the most agriculturally productive land in the nation, are inducements to migrants from other areas in Nepal and neighboring countries. Much of this population growth has occurred without effective planning causing serious problems, including environmental pollution, rising unemployment, inadequate infrastructure facilities, and conflicting land use demands. The loss of prime agricultural land to residential and commercial uses has impaired the ability to feed the current valley population, a problem which previously did not exist.

A multilayer GIS which includes current land cover conditions and updated by use of spaceborne raster imagery, can document change, determine suitability or capability, and allow spatial modeling. However, while a GIS can assist decision makers in developing land use regulations and in monitoring compliance, it cannot directly enforce these regulations.

Data and Methodology

The Kathmandu Valley GIS was constructed on PC-based hardware and ERDAS software available at both George Mason University in the United States and the Nepal National Remote Sensing Centre. This compatibility allowed for easy exchange of data layers and analysis results.

Significant data layers in the GIS are listed in the following table. Included are layers showing the extent of Kathmandu's built-up area for seven dates between 1954 and 1989 from a variety of sources. Each data layer is raster based using a 70m cell size and all layers are registered to the Universal Transverse Mercator geographic coordinate system. Significant data were obtained from the Canadian-funded Land Resources Mapping Program (LRMP) maps published in 1979. These maps were produced at a scale of 1:50,000 using Survey of India maps of Nepal as a base and 1978-79 aerial photographs with field verification to obtain land use/land cover, land forms, and capability information. Plate 58 is an example of the LRMP land use layer.

Selected Kathmandu Valley GIS Data Layers

Land use	Rivers
Land capability	Urban areas (1954-1989)
Land forms	Geomorphology
Soils	Geology
Soils texture	Slope failures
Ground water	Administrative units
Slope	Population density

Road Network

There were very limited amounts of spatial data for the Kathmandu Valley, and obtaining those data for the GIS was often awkward. It was difficult to identify existing maps and in many cases, only single copies of thematic maps existed and no accurate method of reproduction was available. Additional problems encountered included varied scales, lack of consistency and definitions for mapping parameters such as land cover types, and absence of a consistent geographic coordinate system.

Three dates of multispectral SPOT imagery were obtained for Kathmandu: March 22, 1987; October 19, 1988; and July 22, 1989. In developing maps of the urban area from these data, four issues were considered: (1) roofing materials are changing from traditional thatch and clay to concrete and metal; (2) new building developments are seemingly random in location and highly variable in density; (3) there is spectral confusion between urban features and highly reflective sands and gravels along the rivers; and, most importantly, (4) the cropping patterns are complex and create spectral confusion between urban features and bare agricultural soils.

By focusing attention on the July image, a strategy could be developed for minimizing the urban/bare soil confusion. During July, there are no dry fallow fields which are spectrally very similar to the roofs of structures. The urban areas are light-toned on this image. There are fields of maize which are very easily identified together with the forested areas as red colors. The majority of the other fields are in the early stages of rice cultivation. The paddy fields are flooded or fully saturated so that they are spectrally very dark and distinct from the new urban areas.

Minor spectral confusion remained between the urban areas and highly reflective sands and gravels in the river beds. This confusion was eliminated by using the GIS to mask the rivers, and a buffer along the rivers, from consideration. To assess the accuracy of urban mapping using the SPOT imagery, 500 sample points were selected and field checked. Based on ground observations for these points, the visual interpretation of the SPOT data had an accuracy of 94 percent.

The final raster image map for Kathmandu was converted to a GIS file for comparison with other GIS layers. This process involved digitizing the map into vectors, converting the vectors into a cell- or raster-based GIS layer, and then geometrically rectifying that layer to spatially register it with the other data layers. After adding the urban map of Kathmandu based on the July SPOT interpretation into the Kathmandu Valley GIS, a variety of spatial analyses could be accomplished.

Results and Conclusions

Initially, a simple growth map for the urban area was constructed. This was created by overlaying the seven individual year urban maps available in the GIS. The trend in the urban expansion is, as expected, one of continuous growth. Between 1954 and 1989, the urban areas expanded by over 3,500 ha or over 300 percent.

As a second level of analysis, the urban growth was cross indexed with other GIS layers such as existing land use, soil type, slope, and proximity to roads or rivers to show where the growth has occurred. For example, a comparison was made between urban expansion and land systems. Most of Kathmandu is found on one of two land systems, either alluvial plains or elevated dissected and non-dissected ancient lake and river terraces. The 1954 map of the urban area indicates that 18 percent was in the valleys and 82 percent on terraces. Between 1954 and 1989, the percent of urban expansion in the valleys has increased and on the terraces, decreased, so that by 1989, more urban growth was on the alluvial floodplains than the uplands. Not only does this valley expansion decrease the agricultural productivity of the valley, structures on the floodplains are very susceptible to damage or destruction in flooding and are potential sources of groundwater pollution.

A third level of GIS analysis, and one which truly demonstrates the power of spatial data integration, is model building for the prediction of future growth, or for conducting suitability or capability analyses. A model on where future urban growth should be encouraged was created using the Kathmandu Valley GIS. The model consists of several basic steps involving various GIS layers. First, it was assumed that new growth should occur in proximity to existing urban areas and major roads. Areas within 980m of existing urban areas and 560m of major roads were located independently from the current SPOT-based urban GIS layer and the roads layer. These areas were then combined in a new GIS layer of potential growth.

Next, it was assumed that there were certain conditions which should exclude urban expansion. These conditions included the following: (1) slopes over 30 percent, (2) areas where the depth to groundwater was under 2m, (3) protected forests and plantations, (4) existing rivers and streams, and (5) soils which were less than 50cm to bedrock. These areas were identified from the five relevant GIS layers and combined to a new GIS layer of areas not suitable for urban development. These areas were then excluded from the urban and road proximity map of potential growth.

This layer was then overlaid with the existing land use map to identify which lands may be consumed by future growth (see Plate 59). The 7,818 ha (19,310 acres) of current land use consumed by future growth according to this model include 5,534 ha of existing tar agriculture, 789 ha of valley floor agriculture, 788 ha of terrace agriculture, and only 707 ha of lands not currently classified as agriculture. This model suggests that a relatively small amount of valley floor agriculture will be consumed because areas where the depth to groundwater was under 2m were excluded from growth. This model differs from the reality of where growth is occurring because much of the growth is on the floodplains. The ability, however, to alter the parameters of a GIS model makes it a very effective planning tool.

Benefits

This study illustrates how SPOT Image data can be used to map urban growth in developing economies. Use of any satellite data, however, is dictated by the crop calendar and cloud conditions, some of which can be overcome by using other spatial data in a GIS context.

This study has also illustrated how spatial data from a variety of sources can be combined to examine the location and rate of urban growth, to identify what type of lands are being consumed by the growth, and to develop models of predicted urban expansion and land suitability studies.

Merged Raster and Vector Data for Regional Environmental and Land Use Planning: Tamaulipas, Mexico

Mindy Roberts, Keith E. Miller, and Philip R. Chernin, Camp Dresser & McKee Inc.

Challenge

Mexico's Secretaría de Desarrollo Social (SEDESOL), Instituto Nacional de Ecología, hired CDM International Inc. (CDM) to conduct an ecological regionalization study near San Fernando in the northeast part of Tamaulipas state. The goal of the study was to facilitate regional planning based on the rational use of natural resources and promote sustainable economic development. Specific tasks included delineating natural resources, productive economic activities, and areas of conflict between resources and activities. The regional study required spatial data processing over a large area for which no digital data were readily available.

CDM drew from several data sources, both vector and raster, to describe the 1.2 million ha (3 million acre) study area. Located on the Gulf Coast south of Texas, the San Fernando region relies on agriculture and ranching, as well as fishing from ports located within Laguna Madre as primary economic activities.

The project team used GIS for three functions during the study: managing a large quantity of spatial information from sources that varied in scale, age, projection, and format; interpreting land cover and morphologic regions of the landscape; and analyzing relationships between physiologic, ecological, and anthropomorphic conditions.

CDM selected ARC/INFO as the primary GIS due to its ability to manage and integrate both vector and raster data through the software's GRID module. The team also used IDRISI software to classify the satellite imagery into land cover types because the software algorithms could analyze spectral signatures and develop land use/land cover classifications based on these signatures.

Location of study area.

Data and Methodology

In accordance with information provided by SEDESOL, all analyses were based on physiographically homogeneous units. Delineation of these units required the integration of both vector and raster data because neither could provide all delineation criteria (geomorphology, hydrology, vegetation, soils, and land use). Following delineation, the units were grouped by attributes based on the vector and raster data to create maps of resources and activities.

Sources of vector data included topography, soils, and groundwater maps from the Instituto Nacional de Estadística, Geografía e Informática. Because digitizing or scanning all soils, topography, and groundwater maps was not feasible, the project team preprocessed the data to distinguish attributes that would be used in the diagnostic analyses. For example, only major soil categories and low-, moderate-, and high-salinity areas were digitized. This provided a significant time savings without losing the salient features of the data.

Because no recent land cover or land use maps were available, the project team turned to satellite imagery. Initially, CDM evaluated data from both EOSAT/Landsat and SPOT Image to determine the best source of satellite imagery data. The team found that SPOT had the most recent contemporary coverage without substantial cloud cover for the entire study area. CDM purchased six SPOT multispectral scenes from March 1987 that covered most of the study area.

The team purchased the raw imagery files from SPOT because of project budget limitations, but later in the project, after several false starts at georeferencing the imagery files, the team recognized that spending the additional money to purchase "GIS-ready" data would have been time- and cost-effective. CDM succeeded at georeferencing the images by selecting recognizable physical features from both the satellite image and the topographic maps, and "transforming" the image grid based on these common points. Prominent features along the coastline were very useful for obtaining common points. Intersections of dirt roads and inland water bodies in the interior of the study area, however, proved more problematic because they appeared to have "moved" between the time the topographic map was developed and the date of the satellite imagery.

Due to limited ground truthing resources, the project team needed an image processing expert with experience in Mexican landscapes and vegetation. Dr. Jeffrey Jones, the creator of the IDRISI software package, and a recognized expert in land cover classification, was retained to classify the San Fernando imagery. The large size of the study area and associated size of the electronic data files (a three-band raster file using the SPOT 20m cell size, or approximately 96 MB) prompted a decrease in resolution from the available 20m to 40m pixels. CDM determined that doubling the cell size, and thereby reducing analytical time by four, would not significantly affect the results of the analysis because of the regional nature of this study. To create a grid for the entire study area, the team normalized the individual images to account for differences in sensor variations between the individual scenes.

Initially, an unsupervised classification, without ground truthing data, was created to identify the degree of plant material coverage. This classification was drawn from the team members' previous experience. While in some cases, an unsupervised classification may provide satisfactory results, a supervised classification is necessary when spectrally similar elements must be differentiated. The project team needed to conduct a supervised classification to differentiate cultivated pastures from coastal grasslands, and natural vegetation from areas subjected to cattle grazing. Therefore, the team selected 66 diagnostic training sites based on field observations and the unsupervised classification.

Because cropping and cattle ranching are important and competing economic activities in the San Fernando region, it was important to distinguish areas occupied by each of these land uses. Field observations verified that few non-agricultural areas existed without vegetative cover, even in the dry season. Consequently, agriculture was distinguished from other land uses on the basis of soil and residual vegetation characteristics. Dirt roads and small towns were the exception, and were distinguished based on geometry using available topographic maps and raster spectral differences. Although variations in tillage practices and fallow fields should produce different spectral signatures, these could not be distinguished with the 1987 SPOT data because no simultaneous (1987) ground truthing information was available. The team theorized that because tillage practices have changed since 1987, the 1993 ground truthing information could not be used. Consequently, the team decided that agriculture could not be subdivided in the image classification.

Delineating the extent of cattle grazing was difficult because most cattle in the region roam freely over a wide area rather than being confined to feed lots or other intensive facilities. Cattle grazing generally occurs in areas covered by unimproved vegetation, and the team was challenged with how to differentiate signatures for undisturbed and disturbed (cattle-grazed) vegetation. Using the data collected during the ground truthing, the team was able to develop a spectral signature for "vegetated areas with grazing" as differentiated from undisturbed vegetated areas. This was possible only because in this landscape, grazing cattle substantially reduce the underbrush, leaving low ground cover and mesquite trees. Determining this difference in spectral signatures made it possible to avoid sending an army of field crews throughout the region (nearly the size of Connecticut) to map individual cattle grazing areas.

The team used several different classification strategies for analyzing the data collected at the 66 training sites. An initial grouping of all sites by their reported uses resulted in a set of signatures with broad standard deviations despite the use of two different and powerful classification methods: minimum distance and maximum likelihood. Reviewing the notes and photos taken by the field crews at each site revealed that much of the land cover in the region was not homogeneous, and that land cover within a single cover "class" might have numerous different physical forms. For example, some pastures were made up of sown grasses, others a combination of grass and shrubs, and still others were composed of shrubs and small trees. The solution was to subdivide the 66 training sites into 30 subgroups based on differentiation within the land cover classes.

The final classification (see Plate 60) used a combination of methods. Land use was classified primarily using the minimum distance classifier applied to the 30 signatures. For certain categories, such as urban areas, the digital image classification methods were complemented by a texture analysis, and for a small number of "problematic" areas, the classified data were complemented with vector data.

Once the final land use/land cover classification was performed, the team applied ARC/INFO's GRIDPOLY function to the resulting data grid and then merged it with the vector data. The two data sets were combined not only to map land cover and land use, but also to analyze the interplay between resources and competing economic pressures.

Results and Conclusions

Vector and raster based coverages were merged to produce multiple analyses. For example, the vector based coverage defining soils appropriate for agriculture was merged with the raster based agricultural land use coverage to develop a coverage identifying inadequate land use. Polygons in which agriculture occurred, but the soils were not appropriate, were termed inadequate land use. This coverage emphasizes the strength of merged data for diagnosing conflicts between productive activities and limited natural resources. The following list describes coverages created in the GIS. Once the vector, raster, and merged coverages were complete, four maps were created.

- The base map presented the regionalization (natural units) upon which the remaining maps were based.
- The natural resources map included several "natural" vegetation categories, soils appropriate for agriculture, and groundwater resources.
- The productive activities map included agriculture, cattle grazing, and fishing ports.
- The diagnostic map presented inadequate land use, large vegetated areas threatened by development, and other phenomena.

Each of the four maps consisted of 18 sheets at 1:50,000 scale to convey the necessary information. The team created a final coverage and a map to describe the overall management policies that should be implemented throughout the region. Environmental management units were defined by grouping natural units with different attributes (land cover, resources) for which similar management strategies should be implemented. Each unit was assigned one of three policies—development, conservation, or restoration—to describe SEDESOL's goals.

Benefits

CDM could not have achieved the project objectives without the use of merged vector and raster data. Combined vector and raster data enabled the efficient use of project resources to map a large study area, take advantage of the full set of data (older, less changeable soils versus newer, more dynamic land cover and use data), and meet the objectives of the study by distinguishing areas of cattle grazing and cropland, delineating natural resources based on land cover, and mapping areas with unsustainable practices. SPOT images saved project personnel weeks of mapping land cover and land use in the field, and due to the study area's size and remote location were less expensive than project-specific orthophotography.

Coverages used to evaluate natural resources, productive economic activities, and conflicts between the two drew from both vector and raster data sets.

Vector data sets	Raster data sets	Merged coverages
Regionalization (natural units)	Agriculture	Inadequate land use (soils appropriate for agriculture)
Water (perennial, intermittent)	Pasture	Cattle grazing (disturbed spiny forest)
Streams	Low brush	Large vegetation areas (large contiguous areas of Matorral, pasture, and halophytic vegetation)
Canals	Spiny forest (dense mesquite with underbrush)	
Towns	Disturbed spiny forest (mesquite without significant underbrush)	
Municipalities	Halophytic vegetation (coastal grasslands)	
Roads and highways	Mud flats	
Airstrips and microwave towers	Sand	
Study limit	Water	
Latitude/Longitude	Human settlements	
Aquifers/groundwater sources		
Wells (freshwater/tolerable/saline)		
Soils appropriate for agriculture		

Mapping Urban Growth in Metropolitan Beirut

Yousef Nizam, Faculty of Engineering and Architecture, American University of Beirut

Challenge

Following two decades of civil war that virtually destroyed Beirut, the city is beginning to rebuild and is undergoing a period of rapid urban growth. Urban and city planners have very little map and demographic information to draw upon in devising growth management policies that will regulate urban renewal and expansion. In the absence of traditional mapping and census data, the urban engineers are using multitemporal SPOT satellite imagery and GIS analysis in ARC/INFO to map the city and track changes in population density patterns between 1984 and 1993.

In the 1970s and 1980s, Beirut was devastated by civil war and occupation by neighboring countries. The city, virtually destroyed in 1982, is now experiencing a period of growth and reconstruction. The population has risen from 700,000 in 1980 to an estimated 1.1 million in 1990. Rebuilding basic infrastructure while attempting to manage growth patterns has proved a difficult task.

As part of a collaborative research project between the Department of Architecture at the American University of Beirut (AUB) and the Computer Resource Laboratory (CRL) at Massachusetts Institute of Technology, a foundation for growth management policies is being developed with SPOT imagery and GIS.

The decision to use SPOT imagery was primarily an economic one. The cost and time required to collect map and growth information by traditional survey techniques have made systematic updating of urban databases beyond the reach of many planning authorities. The GIS portion of the project would not have been nearly as effective without the satellite image data.

Data and Methodology

AUB/CRL participants divided the research project into three related development areas: infrastructure, transportation, and urban planning. The research has aimed at demonstrating how satellite imagery and GIS technology can provide decision makers with the information needed to understand relationships among distributions of land value, population growth, and construction constraints.

Using the imagery and GIS, the researchers examined three specific issues:

- *Estimation of land value.* This is a key factor that must be considered in predicting, directing, and controlling growth patterns. In the absence of the data required to calibrate traditional land value models, the GIS interpreted and quantified some of the factors that influence land values in Beirut.
- *Identification of land use change.* Understanding the type and rate of urban growth in Beirut required an analysis of changes in land use patterns, especially in vacant and agricultural areas recently transformed into dense urban and residential areas.
- *Calculation of a growth saturation index.* This index gives planners an estimate of the number of years required for a certain municipal region to reach the point where no further urban development can be sustained.

The saturation index may be one of the most important applications of the satellite imagery and GIS because it helps planners determine how limited budgets should be allocated for reconstruction efforts. For instance, additional funding can be directed at areas experiencing high growth rates while funding may be reduced for another area as it approaches the saturation limit. The timely information provided by satellite imagery allows the planners to keep ahead of growth rates.

The project area included the geographic boundaries of metropolitan Beirut, an area of roughly 500 sq km (193 sq mi). For land use comparison purposes, a 1984 paper map of Beirut was used in the study. Municipal boundaries, transportation networks, and topographic map data were also available for inclusion in the mapping exercise. A three-dimensional TIN (triangulated irregular network) map was also used to identify slopes and non-developable land and to conduct a land visibility study.

Multitemporal change detection requires acquisition and processing of satellite imagery collected over the same area at different times. The images can be compared by computer or by eye to locate and quantify changes in urban land use and population. In addition, this information allows planners to predict the need for new infrastructure to accommodate growth.

For this purpose, AUB/CRL researchers purchased SPOT multispectral satellite images acquired over Beirut in May 1988 and January 1993. Each scene has a 20m spatial resolution and covers 60 x 60 sq km, most of the Beirut urban area.

IDRISI PC-based software was used for processing the satellite images. IDRISI is a grid-based system developed by the Graduate School of Geography at Clark University.

After rectifying and registering the image data, AUB/CRL researchers applied supervised classification procedures to the imagery in IDRISI. The first step in the process was to determine the number of classes that were desired from the imagery. For the Beirut project, four major classes were considered: high density housing, medium density housing, sea water, and open space.

Training sites for each spectral class were identified and matched with appropriate pixels using the on-screen feature of the IDRISI COLOR module. The resulting training sites were then saved as a vector file. Once the training sites were determined in a particular band, the corresponding categories were characterized across all bands to create a spectral response pattern for each class. This was carried out in the IDRISI MAKESIG module.

The AUB/CRL researchers used the maximum likelihood classification method and the signatures for each class to classify the full image by determining the most likely class for individual pixels.

There was some dissatisfaction with the accuracy of the image classification, which was traced back to poor choice of training sites. Training sites are usually identified with the aid of registered aerial photos or intensive GPS surveying, but these tools were unavailable for this project. Future projects will invest more time in accurately locating training sites.

The images were reclassified to eliminate the water classes and improve the accuracy of the open space, high housing density, and medium housing density classes. The reclassification allowed open space areas to be subclassified as roadways, concrete, gardens, and trees.

Results and Discussion

Change detection is a simple process with automated image processing software such as IDRISI. The classified 1988 SPOT image was subtracted from the 1993 image. The resulting image displays all areas that have changed from one image to the next. This procedure identified several areas in Beirut where change was occurring most noticeably.

The change image showed that a significant amount of open space land in the middle of the city and along the northern coastline had changed to high density housing. A moderate amount of medium density housing had changed to high density in the eastern suburbs. And not surprisingly, at the very fringe areas of the metropolitan region, extensive tracts of open land had converted to medium density housing.

GIS analyses of change detection information is considered a critical part of the study because GIS puts the change data into a meaningful perspective. By importing the SPOT change detection images into GIS, the Beirut researchers could better understand the economic, political, and topographic factors contributing to particular land use changes.

The images were imported into ARC/INFO and registered using the IMAGE INTEGRATOR module. This created a base map upon which other vector layers could be overlaid in ARCPLOT. For example, political boundary layers were overlaid on the satellite image so that change could be attributed to a given neighborhood. For political reasons, annotating these boundaries was extremely important for the correct allocation of funds.

Researchers were able to place values on property by tracking growth trends in the neighborhood. Areas where population growth was fastest were the most appealing and were considered to have rising values.

In addition, the existing 1984 land use map was overlaid on the change image for updating. Combined with the change detection information from the satellite imagery, the planners could extend their study of growth to 1984. The older land use map in vector format was able to be updated on screen using the satellite images.

The satellite images were draped on a 3D terrain model built using the TIN module in ARC/INFO. By viewing this, the planners gained insight into how topography constrained growth in some areas. They also could determine which remaining open landscapes would be suitable for future construction.

This information along with the housing density data allowed the researchers to make judgments about saturation points. The 3D imagery clearly showed where urban development had reached the confines of the local topography related to water bodies or difficult terrain.

Benefits

The SPOT satellite imagery provided the AUB/CRL researchers with map information that would have been too expensive to acquire through field surveys or other mapping processes. The imagery was successfully integrated as vector data in ARC/INFO for analysis with other types of data.

The AUB plans to expand this project to include all of Lebanon. Four panchromatic SPOT scenes will be purchased for inclusion in the GIS. The current objective is to expand the scope of the GIS by inputting additional layers of correlative data so the construction budgeting and planning process can be refined.

Environment and Mineral Exploration

GIS is exploding into the architectural and engineering (A&E) communities as they develop efficient ways to visualize environmental impacts of industrial and public works. In this chapter, and to some extent in Chapter 9, case studies are presented that illustrate the growing number of ways raster and vector data are used to approach environmental and resource exploration issues. Some of the case studies are site-specific, while others are regional in scope. Some describe rather involved analysis protocols signifying the complexity of application models, while others are straightforward approaches to solving more general questions. Some are merely exploratory. Developed nations are ever more sensitized to the environmental impacts of economic and transportation activities, and as regulatory, licensing, clean-up, and other requirements evolve, the technologies for managing data and visualizing alternative scenarios expand. "Engineering geography" or "spatial engineering" are perhaps other ways to describe these emerging needs, whether they seek change detection or the search for hazardous materials.

Detecting Contamination from Uranium Mines

Paul R.H. Neville, Earth Data Analysis Center, University of New Mexico

Challenge

This study evaluates the utility of remote sensing and GIS for detecting possible contamination from uranium mines and milling operations. The study area is part of the Grants Uranium Belt known as the Ambrosia Lake District. For decades, the Grants Uranium Belt ranked as the nation's major producer of uranium ore, and the Ambrosia Lake District was one of the key producing districts with 35 mines and four refining mills. In the 1980s, with increasing competition from abroad and a waning nuclear industry in the United States, the uranium market collapsed. This region was hit hard by the collapse; at present, there is no active mining with the exception of a couple in-situ leaching operations. In 1988, the uranium processing mills began reclamation of over 700 acres (283 ha) of mill tailings.

Most of the mine shafts are sunk to depths below the Ambrosia Lake District to the Jurassic age Morrison Formation. The only surface mines are found in the older Todlito limestone outcrops and on the southern edge of the district, where uraniferous sands of the Morrison outcropped on the cliffs of some mesas. Two mills were in the Ambrosia Lake District and two to the south near the town of Milan. All used acid leaching methods to process Todlito ore except for the one nearest Milan, which used an alkaline leach method.

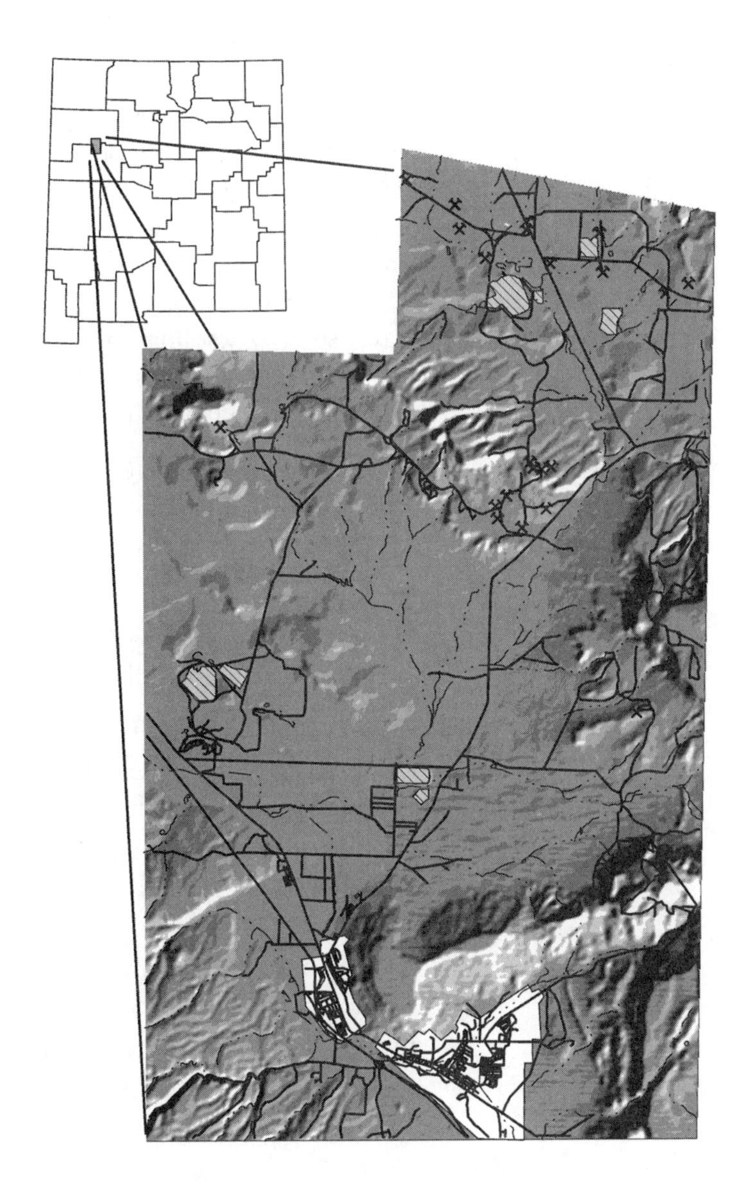

Shaded relief map of the study area and inset showing its location in New Mexico. The towns of Grants and Milan (to the northwest of Grants) are found at the bottom of the map. Mills are shown as polygons with diagonal lines, roads as solid lines, drainages as dashed lines, and mines as crossed picks and shovels.

Data and Methodology

A Landsat Thematic Mapper (TM) image acquired in 1988 was atmospherically corrected and rectified to a UTM grid. Certain surface features were expected to be associated with possible uranium contamination. The tailings have a high albedo response because the host rock consisted of gray to white sands or limestone. There was a possibility that pyrite associated with the ore would have oxidized causing contaminated material to become iron stained. Clays, such as kaolinite and montmorillonite, have been found with the ore. These mining and milling activities were also believed to have a deleterious effect on vegetation.

The above features were individually enhanced using principal component analysis (PCA) on specific bands to create the following images:

- TM bands 1, 2, 3, 4, 5, and 7: The first component with strong positive loadings from all channels is interpreted as the albedo image.
- TM bands 1, 3, 4, and 5: The fourth component, after reversing the image, had the loadings required to enhance iron staining.
- TM bands 1, 4, 5, and 7: The fourth component, after reversing the image, had the loadings required to enhance clays.
- TM bands 3, 4, 5, and 7: The second component had the loadings required to enhance vegetation.

All component images were combined into a single file along with the thermal band. Despite its poorer spatial resolution when compared to other TM bands, TM band 6 is generally helpful in discriminating items that should be warmer, such as iron-stained soils, from items that should be cooler, such as the tailings, which have water sprinkled on them regularly to keep them stabilized.

These images were displayed both separately and in false color combinations to ensure that proper features were enhanced. The combinations were compared to what was known about the area from field visits and by using the data in the GIS database. The ISODATA clustering method was used on the image to gather 54 signatures. These signatures were used to classify the image into 54 separate classes using the maximum likelihood classifier.

Numerous coverages were created in ARC/INFO to produce a GIS database. The coverages included road and stream networks, town boundaries, mill tailings, and tailing pond polygons based on 1979 photo-revised 1:24,000 scale maps, point locations of mines, and geology and soil information digitized at 1:24,000 scale. Available 7.5′ Digital Elevation Models (DEMs) were imported from the U.S. Geological Survey. A missing DEM was created by digitizing the contours into ARC/INFO and using the IDW module to create a grid. The grid was then imported into ERDAS IMAGINE where it was mosaicked with the other DEMs. A grid and raster image representing the overall geologic structure was made using the above technique on contours of the base of the younger Dakota Formation. Finally, points from an earlier environmental study that sampled the soil near the Ambrosia Lake mills were digitized and attributed with readings in μ/l for each of the six different radionuclides.

In addition, air photos were scanned and turned into images, rectified to the Landsat TM scene, and digitally merged with the TM bands. The TM/air photo merged images were used to provide better detail; the default display for the merged photos was 1:9,000 scale compared to the 1:90,000 scale TM scene. These merged photos were viewed alongside the classified image and helped interpretation of the classes.

Results and Discussion

The ARC/INFO hydrologic modeling software was used to model potential flow direction for both the alluvial aquifer and the lithologically confined aquifers. The DEM mosaic was used as an approximation of the alluvial aquifer. The lithologic structure surface was used as an approximation for the lithologically confined aquifers. These flow directions, as well as the surface stream network, were displayed with the potential contamination classes. From these overlays, potential contamination pathways were drawn.

Classes were interpreted using the following methods:

- ❐ Photo interpretation based on field knowledge of the area.
- ❐ Comparison of each class response with each of the feature components.
- ❐ Overlaying the ARC/INFO coverages, especially those relating geologic structure, soil, mines, and mills.
- ❐ The points from the environmental study were overlaid onto the classification.

The relationship between the point data and the classification was hard to discern at first because of the wide data range for the different radionuclides. Using the ARC/INFO KRIGING routine, a surface model was created for each of the radionuclides. The surfaces for each of the radionuclides proved to be similar and showed the highest radionuclide concentration to be highly correlated to a particular class (see Plate 61). This class has an anomalous high clay content and the second highest vegetation response. Field investigation using GPS equipment revealed that this site has a high montmorillonite content and is covered with rushes and other hydrophilic plants. Apparently, the source of the water and radiation derived from constant mine water drainage over many years. This same class is also found in the main drainage below the two major mills in the Ambrosia Lake District.

Five classes, all with high albedo readings, represented mine and mill tailings. One particular class was found at the alkaline leach mill. Other classes represented the differences between dry and hydrated material. Several mines not found on the original maps were located by the classification. One class with both high albedo and high clay response was found in the tailings and at outcrops of the uraniferous units within the Morrison Formation (see Plate 62). TM band 6 helped to differentiate between tailing pond water and other bodies of water in the area. Classes with high vegetation responses were found to be important for two reasons: (1) a high vegetation response in this semi-arid environment was indicative of tailings water leakage or mine water drainage; and (2) these classes showed the proximity of agricultural fields to other contamination indicators. None of the iron-stained classes directly implicated uranium contamination despite successfully enhancing the iron-stained lithologies found in the area. An anomalous iron-stained area over one of the mills marked where reclamation was initiated by using iron-stained soil as a cover.

TM data might be considered too coarse for a study of this kind. However, because of spectral qualities, especially in the mid-infrared bands, these data helped identify major areas of concern. More expensive and elaborate surface and sub-surface testing could then be used to better quantify the threat at these identified sites. In addition, the PCA-enhanced images facilitate interpretation of features on the surface that cannot be readily seen, and the images can be used to test assumptions about where contamination might be located and how it might be represented.

ARC/INFO data also provided several advantages to this study. The conversion of all hardcopy data to digital form provided a quick way to collate the image results either by overlay or by side-by-side viewing at a scale of choice.

This process would have been difficult to accomplish using hardcopy data alone. In the case of the soil map, the original data were in 1:24,000 map format for one county and in air photo format for another; the counties' soil names were not the same. Using ARC/INFO, all mapped data could be joined and similar soil types from the different counties could be combined as the same soil type by using the image as an underlay.

The surface modeling capabilities of ARC/INFO's KRIGING and IDW were helpful in several ways. The IDW was able to produce a surface DEM where one was not available. The IDW was also used to create a surface for geologic structure that in its original form as contour lines on several 1:24,000 scale geologic maps was difficult to visualize. KRIGING took point data and created a surface not apparent in the original data. This surface showed a pattern of radionuclide distribution and pointed out one area of high radiation correlated to a particular class.

The hydrologic modeling capabilities of ARC/INFO allowed flow directions for both the alluvial and lithologically controlled aquifers to be created. The final comparison of the stream networks and the aquifer flow directions with the classified images illustrated the location of surface problems and their relationships to surface and groundwater flow.

Acknowledgments

The author would like to thank Teri Bennett and Tom Budge, specifically, and the staff at the Earth Data Analysis Center, in general, for their help on this project.

Mapping Biodiversity

Teri Brotman Bennett, Earth Data Analysis Center, University of New Mexico
Esteban Muldavin, New Mexico Natural Heritage Program, University of New Mexico

Challenge

The U.S. Department of Defense (DOD) is caretaker for 25 million acres (10,117,500 ha) of public land, including a broad spectrum of landscapes not represented by other public lands and many quickly disappearing ecotypes. Over the past decade, congressional mandates and regulations have been adopted to establish DOD goals for long-term land management. The Biodiversity Initiative is a cooperative project between the DOD and The Nature Conservancy to develop a natural resource management strategy for military installations. In compliance with these directives, Ft. Bliss is building a GIS of spatially nested maps.

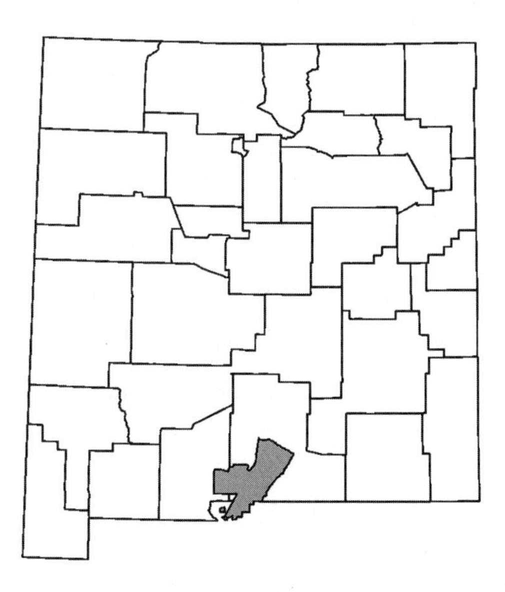

Location of Ft. Bliss shown in gray with New Mexico counties and state boundaries.

Maps for a baseline vegetation study were produced jointly by the New Mexico Natural Heritage Program (NMNHP) and Earth Data Analysis Center (EDAC) for Ft. Bliss in compliance with the long-term adaptive strategies of the Biodiversity Initiative. Mapping levels began with dominant species and progressed to co-dominants. One of the maps shows plant community compositions. Ecological associations that include physical landscape characteristics were mapped to create the mapping units.

Data and Methodology

The mapping process required multidisciplinary experience with several tools such as global positioning systems (GPS), GIS, airborne videography (ABV), and remote sensing (RS). The procedures outlined below are chronological and present the general methods and technologies used to map and create the required database.

Ft. Bliss comprises approximately 1.1 million acres ranging from arid to mesic temperature regimes influenced by elevational changes within the Tularosa Basin. Most of Ft. Bliss lies within a basin bottom and adjacent alluvial plains bounded on the west by the Organ Mountains and on the east by Otero Mesa. Elevation ranges from 1,200m to 2,600m (4,000ft to 8,500ft). Most of the land cover is desert shrub dominated by honey mesquite and creosote bush, interspersed with grama grasses. The mountain region covers are complex and diverse grasses, forbs, and shrubs. Higher elevations are dominated by piñon and juniper woodlands. Threatened and endangered species have been identified within Ft. Bliss borders.

The large land area and the diverse geology, surface soil compositions, and sparse plant canopies dictated the need for multispectral data. An initial classification of plant cover was accomplished using TM imagery processed with ERDAS IMAGINE software. Field reconnaissance maps were color coded by dominant plant type at 1:24,000 scale (see Plate 63). Vector data sets developed using ARC/INFO were overlaid on the classification using the dynamic link capabilities of ERDAS IMAGINE. The vector data sets utilized were roads, hydrology, and elevation contours. A random, stratified sampling technique was used for the field surveys. Data collected for each of the 465 survey sites, such as geomorphic surface features, soil texture, plant identification, and percent cover, were recorded along with GPS coordinates.

Field data and the differentially corrected GPS locations for each plot were entered into The Nature Conservancy's Biological Conservation Database (BCD). The BCD was imported into ARC/INFO to construct a framework that would spatially represent the database, provide quality control tracking of image processing results, and guide statistical clustering techniques for processing satellite imagery.

Species composition and occurrence rate, soil texture, slope, elevation, aspect, and geomorphic position were BCD items added to the ARC/INFO point data for each survey site. A supervised classification was generated from the concurrent display of the ARC/INFO database attributes and vectors and the statistical "signatures" for each of the plots. The ARC/INFO attribute table for each survey site included descriptions about the plant communities, such as ". . . a blue grama grass community on a rocky, north facing slope which continues in a band across the slope . . ."

All attribute information was used to control shape, size, and direction of the "seed," which in turn was used to classify the satellite image. The statistical distance used in gathering the seed was added to the ARC/INFO attribute file as an additional item. Classifications in IMAGINE were performed and the attribute information from ARC/INFO was copied to the ERDAS raster attribute files. Various combinations of plant associations were grouped to provide mapped distributions of ecological characteristics. Additional mapping units were developed by grouping other characteristics of plant distributions within the context of geomorphic and soil properties.

During the field reconnaissance season, the Southwest Regional Office of the U.S. Forest Service flew over 600 mi of ABV above the base. The videography was digitally encoded with a GPS displaying latitude/longitude references. The flight line map for these flights was generated with IMAGINE and ARC/INFO as a seamless integration through dynamic link. Using the dynamic link, a tabular printout was developed that listed occurrences where the flight lines intercepted field plots. The list generated survey site, plant dominance and co-dominance, latitude/longitude, and flight line. The videography interpreters could then train themselves to identify unique communities. A library of still videography was archived, and plant communities were identified in the videography. Geographic positions of plant communities were recorded and later created as an ARC/INFO point coverage used to validate the accuracy of the maps.

Results and Discussion

Past resource management by the DOD has often reacted to changing social, political, economic, and departmental factors. However, scientific research has often been detached from the socioeconomic impact and military missions of DOD installations as a result of relying largely on static views of the environment obtained by identifying "climax" communities and isolating factors such as multiple-use withdrawals, military missions, or endangered species. The DOD needed more adaptive, integrated resource management tools. In light of recent studies, departmental initiatives, and progressive resource management goals within the military, a broader array of ecosystem approaches is under consideration. Maps developed in this study provide diverse perspectives on the distributions of ecological elements and plant communities within Ft. Bliss.

The dynamic link capabilities between ERDAS and ARC/INFO were integral in the supervised classification procedures. The size, shape, and direction of the "seeds" used to gather statistics could be quality controlled. Field information was interactively accessible in this process, and supplemental information such as statistical distance could be added. The flexibility of the dynamic link for accessing multiple combinations within the ARC/INFO database allowed for quick post-processing of survey site to flight line occurrences for videography training.

The GIS will be used for tracking and monitoring changes through time in response to management programs and policy objectives. Within this context, multidisciplinary expertise and integrated geographic technologies contributed many reliable tools for characterizing and modeling these physical landscapes, as well as adopting appropriate management strategies. As a result, a more inclusive, proactive approach to conservation management of public lands should be possible.

Acknowledgments

We are grateful to the field ecologists of the New Mexico Natural Heritage Program—Sarah Wood, Steven Yanoff, Sanam Radjy, and Norman Douglas—for their assistance. We also thank the Ft. Bliss Directorate of Environment, especially Kevin Von Finger, for sharing knowledge of the base and granting access to library resources. At the Earth Data Analysis Center, we especially thank Paul Neville, Tom Budge, Laura Gleasner, and Jeanette Albany for their help in this project.

Remote Sensing and GIS for Monitoring Grassland Environments

Mark Chopping, Geography Department, University of Nottingham, England

Challenge

Seriously degraded grassland accounts for an estimated one-third of the over 400 million ha (988 million acres) of grassland in the People's Republic of China (PRC). The most affected areas are those of the colder, drier, northern grasslands. In the mid-1980s, the survival of grasslands bordering the desert regions of north and northeast PRC was recognized as precarious, even without human interference.

Inner Mongolia Autonomous Region (IMAR) defines the northern border of the PRC with Mongolia and Siberia and provides some of the most dramatic examples of desertification. It is estimated that, of the 791,530 sq km (305,530 sq mi) of natural grassland in the region, 5 percent is completely desertified; 21 percent is seriously damaged; 14 percent is degraded arable land; 49 percent is intact steppe; and 10 percent is intact arable land. The challenge has been to develop a method for regional monitoring of the grassland of IMAR using remote sensing and GIS techniques. The area is so vast that ground- and air-based surveys cannot provide the same long-term and synoptic views afforded by space remote sensing in conjunction with other environmental variables.

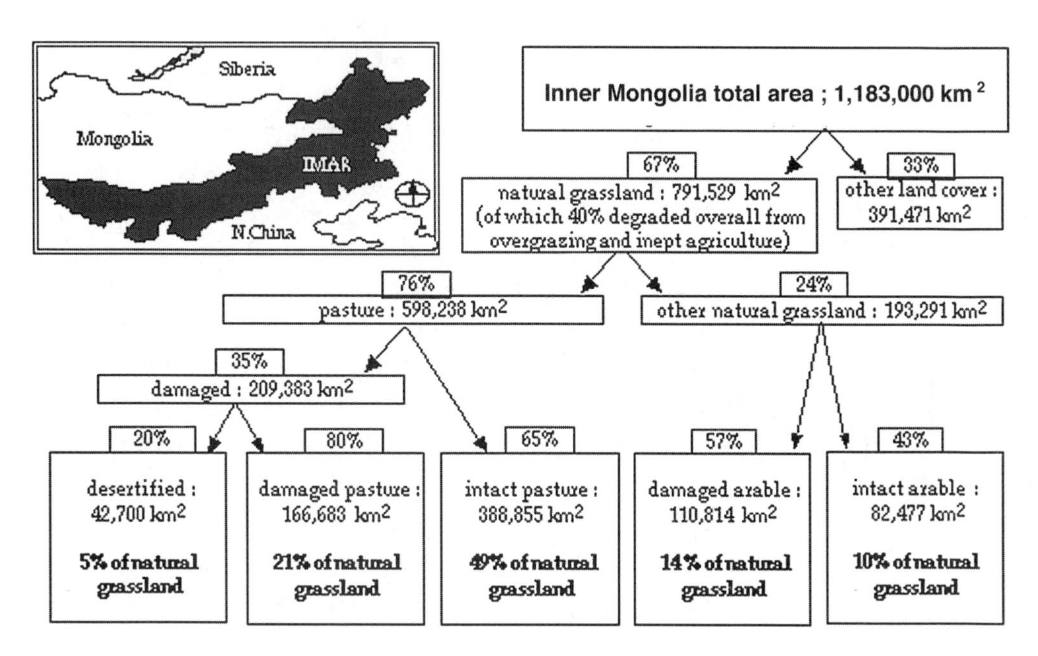

1990 inventory of natural grasslands, Inner Mongolia.

In spite of the limited resources available, results are corroborated by both local assessments of serious and continuing grassland degradation in Xilingol, Ih Juu, Chifeng, and Jirem *aimags* (departments) and the Global Assessment of Human-Induced Soil Degradation database (ISRIC, 1990). Achieving such results is the outcome of the ERDAS IMAGINE software's flexibility and the generosity of the NOAA/NASA Pathfinder Land AVHRR project. The IMAGINE Graphical Model Editor and Spatial Modeler Language (SML) were flexible and time-saving tools, enabling combination of data in raster and vector structures.

Data and Methodology

The aim of this study was to determine whether there was a consistent trend in grassland degradation in IMAR over the 1986-1990 period. The attribute to be tested was the normalized difference vegetation index (NDVI). This is derived from the red and near infrared (NIR) channels of the AVHRR sensor using the following formula:

$$NDVI = \frac{NIR - Red}{NIR + Red}$$

NDVI has been shown to be highly correlated with green vegetation (leaf area index, photosynthetically active radiation, and plant cover). Plant communities in degraded grasslands are simpler in structure and diversity than natural grasslands, having low vegetation cover and height. A lower NDVI might therefore be expected from degraded areas in relation to non-degraded areas in similar eco-climatic zones.

Sixty ten-day maximum value AVHRR-NDVI composites for the 11th to 21st of each month from January 1986 to December 1990 were acquired from the NASA/NOAA Pathfinder archive. These data were formatted as global images mapped onto the Goode Interrupted Homolosine projection. The following data preparation steps were required to extract and format the IMAR data from the global scenes.

1. The 5,004-column by 2,168-row, 8-bit, global NDVI images were imported into ERDAS IMAGINE 8.1 as generic binary files.

2. A series of SML scripts was used to extract subsets representing Goode regions 2 and 4 from the monthly global NDVI images and to stack them into two files for each year.

3. The ERDAS IMAGINE Image Info and Projections Editor dialogs were used to register the stacked image sets to the Mollweide and Sinusoidal projections with 30° central meridians for Goode Regions 2 and 4, respectively.

4. A further subset of the stacked, registered images was extracted to include only the extremities of IMAR.

5. The IMAR images were reprojected (using Raster → Resample) to geodetic coordinates with a nominal cell resolution of 4km. A union overlay was used to combine the Goode Region 2 and 4 image sets into a single set for each year.

6. A median filter (IMAGINE Focal Analysis dialog) was used to remove spike "noise" in the images.

7. A series of IMAGINE Spatial Modeler Language scripts was used to rescale the NDVI to its normal range (-1 to +1), from the nominal 8-bit range of 3 to 253 used for archive purposes.

Using SML scripts to subset, stack, and rescale images was advantageous in compiling a multiyear analysis of monthly files. Processing for a 12-month image set was prototyped using the IMAGINE Graphical Model Editor from which an SML text script was generated. The original SML script was then copied and edited for the remaining 12-month sets simply by changing the input and output file names. The addition of a 30° central meridian Mollweide projection to the set of external projections supplied with IMAGINE 8.1 was accomplished through text file editing in the *$IMAGINEHOME/etc/expro* directory.

Locating degraded steppe using NDVI requires determining the deviation of this indicator from a "normal" or expected state. However, identifying this expected state at any given place or time is difficult because rainfall is highly variable. Therefore, satellite measurements had to be tied to a relatively invariable measure. This "constant" was retrieved from a local map compiled from ground surveys and high resolution remote sensing data. At a scale of 1:1,500,000, the vegetation map of IMAR shows light, medium, and seriously degraded pasture (SDP) across the region ca. 1986. SDP may be considered as a more or less constant NDVI reference point for two reasons. First, it requires many years for recuperation. Second, the relative magnitudes of NDVI can be compared to see if the NDVI for degraded steppe is consistently below that of the non-degraded steppe. To avoid setting too high an NDVI threshold, only polygons defining SDP were digitized.

With IMAGINE, this task was straightforward. Digitizing was done manually using the Arc Digitizing System (ADS) on a VAX/VMS system. The SDP coverage was then checked and edited in ARCEDIT, reprojected to geodetic coordinates, and then exported as a single interchange file for transfer to a UNIX target machine. Since ARC/INFO vector routines are embedded in IMAGINE 8.1, no special routines were needed to import the coverage. IMAGINE's Vector to Raster routine was used to rasterize the SDP coverage for use in SMLs.

The vectorized SDP, IMAR counties, rivers, road/rail, and urban area coverages were displayed over the land use pseudo-color image. A number of larger SDP polygons representing specific steppe types were selected and added to separate AOIs (areas of interest). AOIs of similar dimensions to these polygons were created over areas of non-degraded pasture in the same vicinity, taking care to select areas of the same land use/land cover and away from the influence of rivers, road/rail routes, and urban areas.

IMAGINE's Pixel-to-Table function was used to extract the AOI NDVI data from all 12 mid-monthly layers in the five annual NDVI image sets. The data were transferred to a spreadsheet, and the mean NDVI for each AOI in each month was calculated. When plotted on a series of graphs, the mean NDVI of the digitized SDP from the vegetation map clearly shows consistently lower NDVI values than the mean for neighboring non-degraded areas.

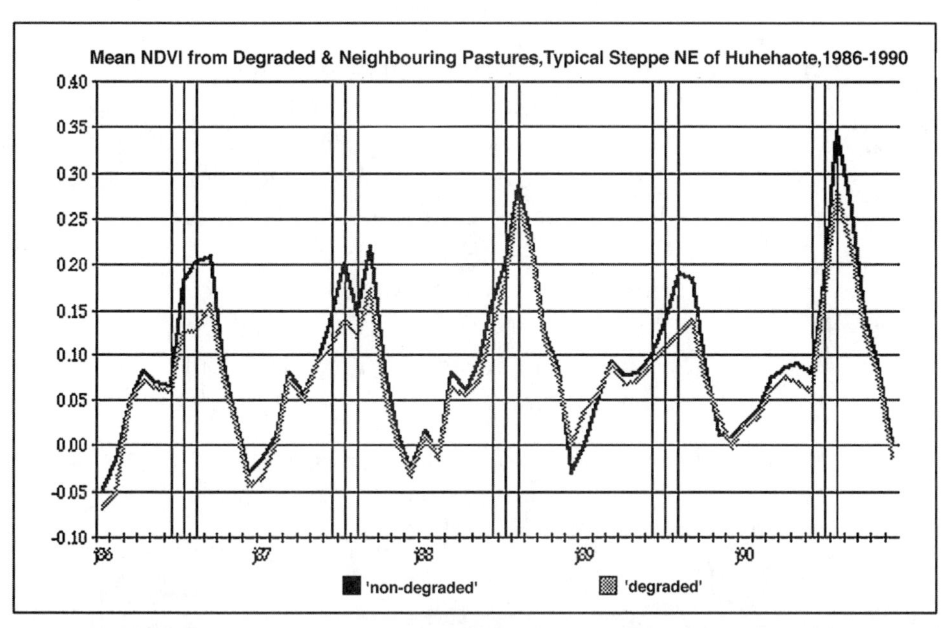

Source : NDVI means calculated from 10-day Maximum Value Composite NOAA/NASA Pathfinder Land datasets for the 11th-21st of each month, 1986 - 1990. Data for 'degraded' areas derived from a selected area of 'serious degraded pasture' digitized from Series Resource Maps of IMAR (Vegetation map); data for 'non-degraded' areas derived from neighbouring area of similar dimensions. Vertical lines show July-August.

Mean NDVI for degraded and neighboring pastures, 1986–1990, showing consistently lower NDVIs for degraded sites.

Other factors influencing grassland vegetation status are elevation, soil type, land use/land cover, and average annual precipitation. These were taken into account by constructing two sets of rules based on the chief characteristics of two major grassland zones: the typical semi-arid steppe and the lowland mixed-usage steppe. Digital maps of elevation (NOAA's ETOPO5), soil type (FAO Soils, 1973), land use/land cover (manually digitized from series resource map at 1:1,500,000 and rasterized), and average annual precipitation (interpolated from long-term meteorological station point data using IDRISI's INTERPOL) were integrated into the IMAGINE environment for this purpose. Plate 64 shows the regional distributions of the various input parameters.

Results and Discussion

Mean NDVI for degraded/non-degraded steppe types were combined with the digital maps to generate IMAGINE SML models from Graphical Modeler prototypes. These were used to predict the locations of degraded pasture for each year for the two major grassland types. The analysis was based on the assumption that the mean NDVI for July and August—at the peak of the growing season—would be the best discriminator of degraded steppe, whereas the mean NDVI for May would provide an additional discriminator for lowland steppe.

MODEL TO MAP DEGRADED PASTURE (HULUNBUIR MIXED USAGE STEPPE)

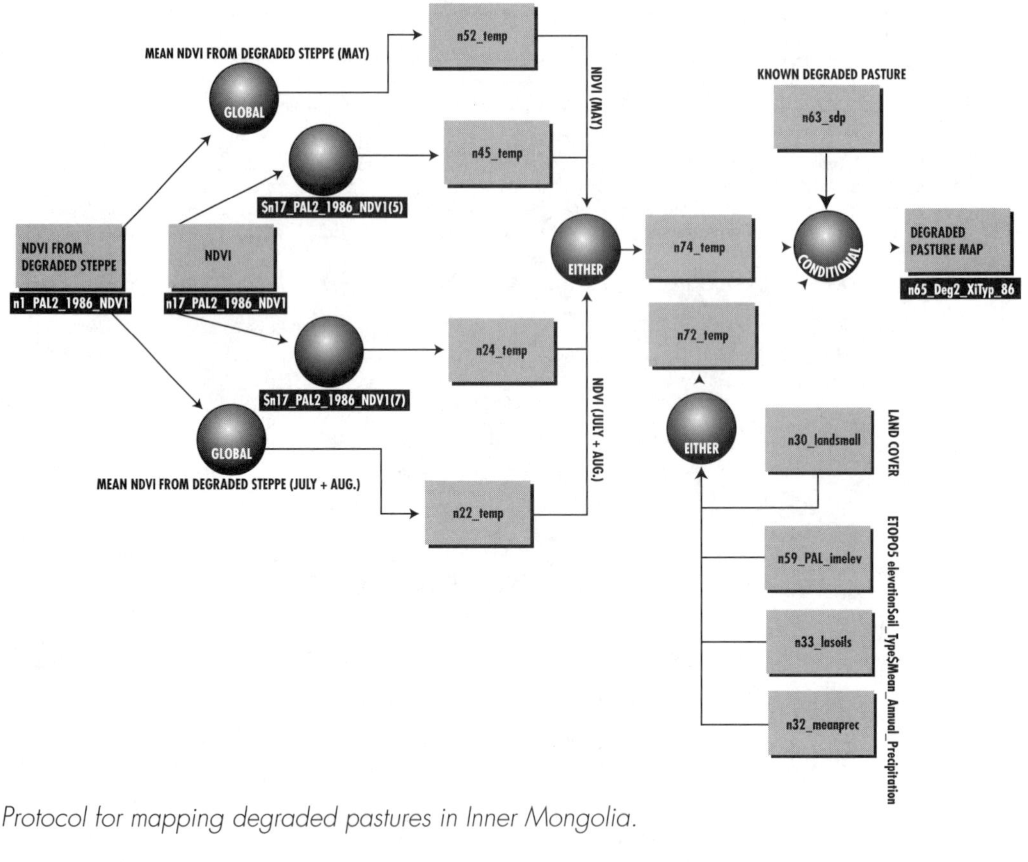

Protocol for mapping degraded pastures in Inner Mongolia.

The SML models follow a simple procedure for mapping predicted degraded steppe (PDS) for each year:

- The sum of the mean NDVI for selected SDP AOIs in the "signal" months is calculated, and an initial threshold for degraded steppe is determined.
- The threshold is adjusted so that the spatial distribution of predicted degraded steppe in 1986 matches as closely as possible the distribution of seriously degraded pasture shown by the SDP vector coverage.
- The value of each NDVI pixel for the corresponding "signal" months is examined to see if it falls below the adjusted threshold.
- If the value falls below the adjusted threshold and meets a secondary criterion designed to constrain the analysis to the same eco-climatic zone, the value is further tested against the rasterized SDP coverage on a cell-by-cell basis.

The thematic output images contain the following class codes:

- 0 - The area excluded from the analysis.
- 1 - "Hits," or where the model predicts steppe degradation in locations which coincide with the 1986 mapped SDP.
- 2 - False positives (where the model predicts steppe degradation in locations which do not coincide with the 1986 mapped SDP).
- 3 - False negatives (where the model fails to predict steppe degradation at the locations mapped as SDP on the vegetation map).
- 4 - The remaining bounded zone for the current model (cells where there is neither PDS nor SDP but which are not excluded from the analysis).

Although the three models were designed to operate on two specific steppe types within eco-climatic zones defined by mean annual precipitation, elevation, land cover, and soil type, they were not constrained geographically in any other way; both the typical steppe and mixed usage steppe models therefore flag pixels in other parts of IMAR as PDS. The results for all areas in 1986 and 1990 (see Plate 65) agree with local assessments of degradation, livestock density maps, and the Global Assessment of Human-Induced Soil Degradation.

Acknowledgments

The author wishes to thank the NOAA/NASA Pathfinder AVHRR Project Team at NASA Goddard Space Flight Center, Greenbelt, Maryland. He also thanks Professors Yong Shi Peng of Inner Mongolia University and Cui Haiting of Peking University for their compilation of the vegetation map for the series resource maps of Inner Mongolia. Finally, the author is grateful for timely and valuable suggestions from the staff of ERDAS, especially Xinghe Yang.

Change Detection of Pacific Coast Estuaries

Ruth E. Spell and Richard G. Kempka, Pacific Meridian Resources, Sacramento, California
Jon K. Graves, Columbia River Estuary Study Taskforce, Astoria, Oregon
Patrick T. Cagney, U.S. Army Corps of Engineers, Seattle, Washington

Challenge

The goal of the NOAA Coastwatch Change Analysis Project (C-CAP) is to develop a nationally standardized database of land cover in the coastal regions and to monitor land cover changes over time. To date, a pilot project has been completed for the Chesapeake Bay region where many of the C-CAP protocols were developed and tested.

The estuarine wetlands of the pacific northwest are relatively small and discontinuous compared to the extensive coastal wetlands along the Atlantic and Gulf coasts, yet they play an important role in the ecology and commerce of the region. In particular, estuarine wetlands support a highly productive fisheries industry that includes salmon, dungeness crab, and oysters. Much of the region's historic wetlands have already been lost, primarily due to the expansion via diking of agricultural land in the fertile floodplain. An estimated 50 percent of the original wetlands in Willapa Bay has been converted to other uses. Although currently protected from further conversion, the remaining wetlands are still affected by upland land management practices, such as logging, runoff from fertilized agricultural fields, and expanding development. Thus, not only the wetlands require inventorying and monitoring, but also the land cover and land use of the adjacent uplands.

The objectives of this project are to develop and test C-CAP protocols and methods for classifying and quantifying land cover change in both a large estuary and a bay on the west coast of the United States; produce a baseline land cover inventory based on recent satellite imagery using cost-effective, repeatable techniques; and, develop, refine, and commercialize techniques for assessing land cover changes for two different dates of satellite imagery covering coastal environments. The resulting spatial databases, when combined with other critical information concerning wetlands (e.g., fisheries habitat, water quality, marine resources, etc.), will be used to develop statistical and ecological models of coastal habitat change.

Data and Methodology

The project area shown in the following illustration is defined by 51 USGS 1:24,000 quads and covers approximately 1,624,572 acres (657,464 ha). An additional eight-quadrangle area (231,615 acres or 93,735 ha) for Grays Harbor was added to the land cover analysis.

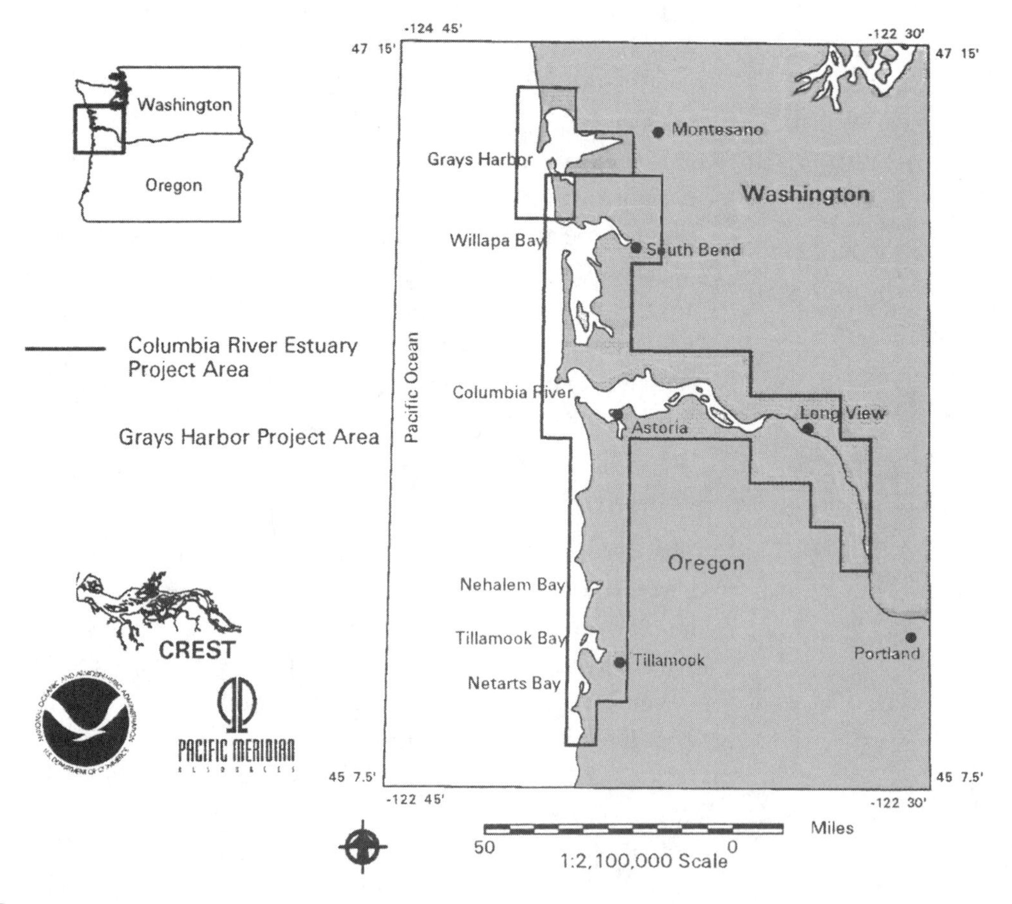

Project area map.

The classification scheme was adapted from NOAA C-CAP Version 1.0 protocols. This scheme was developed for use with satellite image data and was designed for cross-referencing with existing commonly used classification schemes. The classes are arranged hierarchically in three levels to facilitate analysis at different degrees of detail and to allow disaggregation of classes when field data or finer resolution imagery permit.

C-CAP classes are listed below. The boldface Level II classes are required by C-CAP; further subdivision to Level III is determined on a project-specific basis. Land cover was mapped to the required Level II classes, with the exception of the developed, woody, and palustrine woody classes, which were further defined to Level III.

CREST classification scheme based on NOAA C-CAP classification scheme

1.0 Upland	2.0 Wetland	3.0 Water and submerged lands
1.1 **Developed land**	2.2 **Unconsolidated shore**	3.1 **Water**
1.11 High intensity	2.3 **Estuarine emergent**	3.2 **Aquatic beds**
1.12 Low intensity	2.8 **Palustrine emergent**	
1.2 **Cultivated**	2.9 **Palustrine woody**	
1.3 **Grassland**	2.91 Scrub/shrub	
1.4 **Woody land**	2.92 Forest	
1.41 Scrub/shrub		
1.5 **Bare land**		

In addition to C-CAP, data from the National Wetlands Inventory (NWI) were used. The NWI wetlands are delineated on color infrared (CIR) photography and transferred to USGS quadrangles. Deepwater and wetland types are categorized into five major systems: marine, estuarine (tidal), palustrine (freshwater), riverine (rivers), and lacustrine (lakes). These are further divided into subclasses based on hydrography and vegetation. These data were used to assist in selection of training sites and stratifying the satellite imagery to reduce spectral confusion between upland and wetland classes.

Ancillary data sets used to assist in the classification included USGS 1:250,000 scale digital line graphs (DLGs); U.S. Bureau of Census 1:100,000 scale TIGER files; aerial photography collected from 1983 to 1992; and data from the Columbia River Estuary Data Development Program (CREDDP).

Two Landsat Thematic Mapper (TM) scenes were acquired for the project: September 18, 1992 for the baseline classification; and September 10, 1989 for change detection of land cover.

Field maps were produced from a preliminary unsupervised classification of the 1992 satellite image. These initial classes were used along with further clustering to select spectrally homogeneous areas as field sites. More than 100 field sites were identified and digitized. These digitized sites were overlaid on the clustered image and plotted as 1:24,000 scale field maps. Field data were collected from 255 separate sites located throughout the project area.

The 1992 image was reclassified using supervised signatures generated from the field data and the 150-class unsupervised classification. The supervised and unsupervised classes were clustered and assigned labels. This hybrid method combines the advantages of both the information-based supervised approach and the spectral-based unsupervised approach.

The results revealed confusion between wetland and upland classes. To improve the situation, diverse methods were evaluated and a new five-step classification methodology was adopted. The steps are summarized below.

1. Stratify the image into eco-regions. To reduce the effects of spectral variability across the study area and to facilitate processing, the project area image was

stratified into four eco-regions: Willapa Bay, the lower Columbia River Estuary, the upper Columbia River Estuary, and the Oregon coastal bays.

2. Stratify the NWI data. To reduce confusion between upland and wetland classes, each eco-region image was stratified into uplands and wetlands using available digital NWI data and digital CREDDUP.

3. Classify the strata. The upland and wetland strata for each eco-region were classified separately using a combined supervised and unsupervised approach. The wetland strata were classified into eight general classes, and the upland strata into six.

Initial Classes

Wetland Classes	Upland Classes
1. Emergent wetlands	1. Bright bare developed
2. Palustrine forest	2. Dark bare developed
3. Palustrine scrub/shrub	3. Grassland
4. Water	4. Upland forest
5. Unconsolidated shore	5. Upland scrub/shrub
6. Aquatic beds	6. Water
7. Bare	
8. Grassland	

4. Perform GIS modeling. Two GIS modeling procedures were used to refine the spectral classification. First, digital NWI data were used to subdivide the emergent wetland class into estuarine emergent and palustrine emergent. Second, in the upland areas, an urban mask was heads-up digitized using the 1:100,000 USGS DLG roads and overlaid ancillary map data. The bright and dark bare classes within the urban mask were visually analyzed to identify the

high- and low-intensity developed classes. The remaining non-urban bright and dark bare classes were then combined into one bare/exposed class.

5. Mosaic the strata. The upland and wetland strata for each eco-region were stitched together and the four eco-regions mosaicked to produce the final classified image. Manual GIS editing and localized cluster busting were used to remove seam lines and correct obvious misclassifications.

After the 1992 image was classified, change detection was performed on the 1989 image using the following equation:

$$Dijk = BVijk(T_1) - BVijk(T_2) + c$$

where

D = difference value

k = a single band

BV = single pixel

T_1 = Time 1 (1989)

i = pixel row

T_2 = Time 2 (1992)

j = pixel column

c = a constant

Experimentation with several change detection methods indicated that an image differencing approach would provide the best results. Image differencing involves the subtraction of pixel values in one band of a Time 1 image from the values of corresponding pixels in the same band of a registered Time 2 image. Accurate image registration is critically important to ensure that differences are due to land cover change rather than misregistration. The image subtraction results in an output image with a generally

Gaussian distribution of brightness values (BVs). A threshold is selected to identify the boundaries of change values in the tails of the distribution.

To determine the thresholds of change, a multitemporal image was first displayed with Band 5 from 1989 in the red and blue image planes and Band 5 from 1992 in the green image plane. With this band combination, areas of no change (i.e., pixels with relatively little change in value from 1989 to 1992) appeared as gray tones, whereas areas of change appeared in shades of purple (a decrease in value from 1989 to 1992) and green (an increase in value from 1989 to 1992). The difference image was overlaid, and image values were highlighted and visually compared to the purple and green shaded areas of the displayed multitemporal image to determine the thresholds of change. This procedure was performed with Band 5 to determine non-tidal change and with Band 3 to determine change in the tidal areas.

The Band 5 and Band 3 difference images were recoded to binary images of change (thresholded values) and no change. An NWI tidal mask was used to stratify the Band 5 change map to a non-tidal change mask and the Band 3 change map to a tidal change mask. The two change masks were combined and used to mask out the change areas of the raw 1989 image.

The 1989 change image was stratified into wetland and non-wetland images with an NWI mask and classified with the same techniques used for the 1992 classification. Because of limited field and ancillary data from 1989 and limited identification of aquatic beds due to the higher tidal stage in the 1989 image, the classes to be identified were aggregated into woody, grassland, bare/developed, unconsolidated shore, water/aquatic beds, and emergent wetlands. As in the 1992 classification, the emergent wetlands class was further subdivided into estuarine emergent and palustrine emergent through GIS modeling with the digital NWI data.

Results and Discussion

The following illustrations show the acreage totals for 1989 and 1992 and the acreage changes by class that occurred between the two dates. Overall, a change occurred in 101,827 acres (41,225 ha), or 6.27 percent of the project area. In the uplands, the greatest change was due to the effects of forestry clearcutting and replanting practices. A total of 42,313 acres (17,124 ha) in the 1989 woody class was lost, most of which was cleared and classed as bare/developed in 1992. Conversely, a gain of 20,104 acres (8,136 ha) took place in woody land from 1989 to 1992, with over half coming from regrowth of the clearcuts. This resulted in an aggregate loss of 22,209 acres (8,988 ha) of woody and an aggregate gain of 21,334 acres (8,634 ha) of bare/developed.

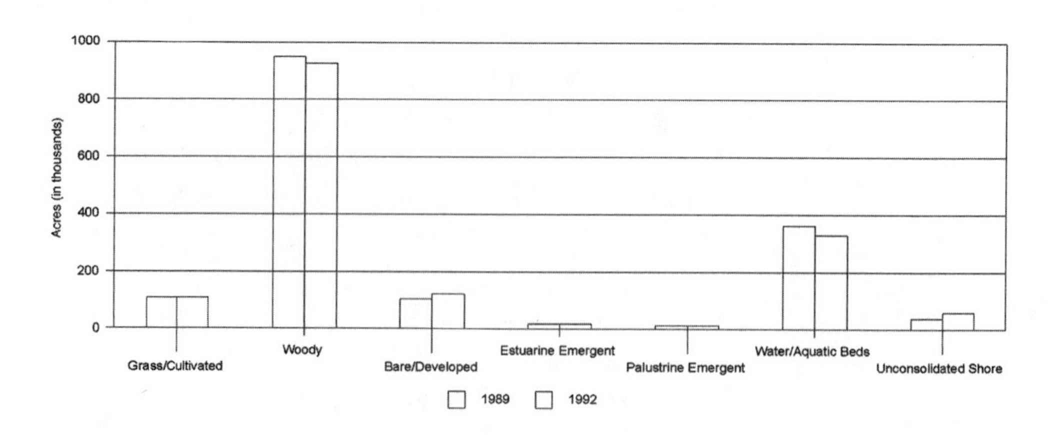

Land cover acreages for 1989 and 1992.

Matrix of land cover change from 1989 to 1992

1989 \ 1992 Landcover Class	Grass/ Cultivated	Woody	Bare/ Developed	Estuarine Emergent	Palustrine Emergent	Water/ Aquatic Beds	Shore	Total Losses	% of Project Area
Grass/ Cultivated	12,916	7,222	3,450	19	568	12	14	11,285	0.69%
Woody	7,439	40,172	34,502	22	274	35	41	42,313	2.60%
Bare/ Developed	5,165	11,867	16,792	163	191	434	741	18,561	1.14
Estuarine Emergent	49	143	70	2,664	9	9	129	409	0.03%
Palustrine Emergent	250	549	69	0	591	46	105	1,019	0.06%
Water	7	104	175	7	10	32,418	21,565	21,868	1.35%
Shore	48	219	1,629	1,546	365	2,565	21,030	6,372	0.39%
Total Gains	12,958	20,104	39,895	1,757	1,417	3,101	22,595	101,827	6.27%
% of Project Area	0.80%	1.24%	2.46%	0.11%	0.09%	0.19%	1.39%	6.27%	

In the wetland classes, the effects of the tidal stage difference between the two dates of imagery made it impossible to accurately identify true land cover change. A total of 22,595 acres (9,144 ha) of unconsolidated shore identified in 1992 could not be detected in 1989 due to the higher tidal stage in the 1989 image (monthly mean low = 146.1 cm). Some areas, however, experienced a loss of unconsolidated shore that was evident in spite of the lower tidal stage in 1992 (monthly mean low = 65.5 cm). The most noticeable of these losses occurred at the mouth of the Columbia River, where beach erosion is evident. Corresponding beach accretion is visible on the inland side, although this may be due to tidal stage differences.

Change in the wetland emergent classes was relatively small, with an aggregate increase of 398 acres (161 ha) of palustrine emergent and 1,348 acres (546 ha) of estuarine emergent. Some of the increase in the estuarine emergent class may be due to the exposure of more emergent vegetation by the lower 1992 tidal stage.

An accuracy assessment was performed by collecting approximately 30 "reference" sites for each of seven classes. Overall accuracy was 85 percent.

Benefits

The C-CAP protocols were tested in a Pacific Northwest coastal environment, and methods were developed to overcome difficulties in spectrally distinguishing certain regional land cover classes. These methods were used to classify the 1992 project area image and produce a baseline inventory of land cover. In addition, image differencing techniques were developed and applied to assess land cover changes in the project area between two dates of imagery (1989 and 1992). The resulting GIS database will assist local and regional land managers and planners in evaluating and managing the resources of the lower Columbia River Estuary and adjacent bays.

Acknowledgments

Cooperative partners in this project include NASA Earth Observation Commercialization Application Program (EOCAP), Columbia River Estuary Study Taskforce (CREST), Seattle District Army Corps of Engineers (ACOE), NOAA C-CAP, Pacific Meridian Resources, and Oregon Department of Land Conservation and Development. Primary funding was provided by the NASA EOCAP and NOAA C-CAP programs.

We would like to acknowledge the following organizations and people for their contributions to field work and project coordination: the NASA EOCAP program; Jerry

Dobson and Ed Bright of Oak Ridge Laboratory; Don Field and Bud Cross of NOAA, Beaufort Laboratory; Peter Britz of CREST; Bob Emmett of NOAA/NMFS, Hammond; Jeff Weber of the Oregon Department of Land Conservation and Development; Paul Hertzog, Allison Bailey, and Betty Bookheim of the Washington Department of Natural Resources; Kathleen Sayce of Shoal Water Botanical; Don Williamson of the Willapa Bay National Wildlife Refuge; and Merri Martz of the Seattle District of the ACOE.

A Method for Identifying Wetlands Mitigation Sites

Floyd Stayner, Water Resources Division, South Carolina Department of Natural Resources

Modified from Journal of the Urban and Regional Information Systems Association, 7(1):38-47, July 1995.

Challenge

Permitted land use, development pressures, and illegal fill activity continue to threaten the viability of the nation's wetlands. Although regulatory safeguards have been established to avoid or minimize the impacts of such activities, compensatory mitigation is sometimes required to replace the ecological loss of wetland destruction or fragmentation. In this study, potential mitigation sites on the South Carolina Coastal Plain are identified using a GIS and 1:24,000 scale information sources.

Mitigation involves replacing lost wetlands with wetlands of "comparable value." This requires mitigation sites to be evaluated for hydrologic function as well as location in the landscape. Potential mitigation sites must be considered as integrated components of the landscape, hydrologically linked to all other land uses/land covers within the watershed.

GIS can be used by regulators and managers to help identify and evaluate these landscape characteristics. The method proposed in this study identifies complexes of wetlands within a hydrologic unit that are physically amenable to restoration, enhancement, or protection. Sites determined to be physically suitable for wetlands mitigation are segregated into community type and further evaluated to determine potential "opportunity," or social/ecological benefits and to assess threats that may affect the site utility.

The South Carolina Department of Natural Resources (SCDNR) has been building a natural resources GIS for the last seven years. The primary sponsor of this project has been the National Oceanic and Atmospheric Administration working in partnership with SCDNR.

The Four Hole Swamp sub-basin in South Carolina is the study area for applying this model (see illustration). The Four Hole Swamp headwaters originate in the central Coastal Plain in South Carolina and drain about 650 sq mi (1,684 sq km). The area has been characterized as Atlantic Coast Flatwoods and Southern Coastal Plain, and contains some of the most productive agricultural land in South Carolina. In addition, the area contains one of the largest old-growth stands of tupelo and cypress in the United States and supports many rare plants and several federally threatened or endangered animal species. The habitats of these plant and animal species are threatened by hydrologic alterations.

Study area.

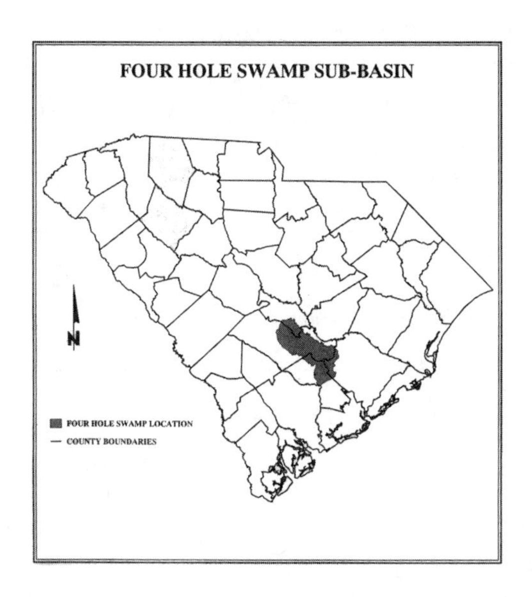

Data and Methodology

Data were developed by the SCDNR as part of the Natural Resources Decision Support System (NRDS) project that began in 1988. One of the objectives of the project was to develop a GIS specifically for addressing natural resource issues. The intent was to develop a system that could be used proactively to make natural resource management decisions that balance economic development with the need for resource stewardship. The following table describes available data coverages.

Available Data Coverages

Coverage	Source	Spatial Data Types
Mining and reclamation	SCDNR	polygon
Hazardous waste treatment, storage and disposal	S.C. Dept. of Health and Environmental Control (SCDHEC)	point
All landfills	SCDHEC	point
Archaeology	S.C. Dept. of Archaeology and Anthropology	polygon

Available Data Coverages

Coverage	Source	Spatial Data Types
National Register of Historic Places	S.C. Dept. of Archives and History, U.S. Dept. of the Interior	polygon/point
Protected areas (government parks, forests, refuges)	U.S. Geological Survey (USGS) topographic quadrangle maps	polygon
Sensitive species and communities of concern	SCDNR	point
Digital line graphs (separate coverages for roads and hydrography)	USGS	line
Soils	U.S. Natural Resources Conservation Service	polygon
Land use	1989 NAPP 1:40,000 photography, SCDNR	polygon
Wetlands	National Wetlands Inventory, U.S. Fish and Wildlife Service	polygon
Natural areas inventory	SCDNR	polygon

The data layers of primary importance are wetlands, land use, soils, roads, hydrography, and significant natural areas. These data sets adhere to national data classification systems and mapping standards as established by various federal programs. All data are based on the 1:24,000 scale USGS topographic map series. Several attributes pertinent to this study were added to the DLG data by SCDNR, including drainage order. All streams in the hydrography data layer were ordered by using the Strahler method of stream ordering.

SCDNR employed several quality control procedures to correct problems inherent in the original digital data. These procedures included edge matching and attribute correction where possible. However, some problems such as the age of the 1:24,000 maps available as source data for the DLG could not be corrected. Maps for the study area date from 1960, with the most recent map dating to 1989.

Other data layers obtained from various agencies include domestic waste permits, industrial waste permits, hazardous waste sites, archaeology sites, historic sites, sensitive species and communities of concern sites, and mining and reclamation sites.

The original intent of implementing this model was to use the ARC module with data in a traditional vector format, but data complexity made this impossible. The Four Hole Swamp sub-basin consists of portions of 17 topographic maps containing over 402,000 acres (162,000 ha). These individual maps were merged and clipped to the hydrounit boundary, creating large polygons that exceeded the maximum number of arcs per polygon (i.e., 10,000) permitted in ARC/INFO version 6.1.1. Another reason for using a raster-based approach was that required proximal analyses needed to be carried out in a timely manner. Vector processing would have taken large amounts of temporary storage and weeks of processing time. As a result of these issues, the model was implemented using the GRID module in ARC/INFO.

The first step in applying the model was defining mitigation classes. The three classes identified were restoration, enhancement, and protection. Restoration sites are characterized by soils of low productivity under cultivation classified as having hydric characteristics by NRDS. These lands, which typically occur on the margins of floodplains where the hydrologic regime is unpredictable, may have the highest potential for restoration to wetland vegetation.

Protection and enhancement sites were selected from the National Wetlands Inventory database. Types of wetlands alteration include dikes, impoundments, excavations, drains, or ditches. Altered wetlands, areas where the hydrology has been altered but hydrophytes have survived, may also provide high mitigation potential. These areas are called *enhancement sites.*

The remaining unaltered wetlands were identified as *protection sites.* These sites are assumed to be fully functional wetlands. Ecologically, these sites are extremely important, and their preservation is a desirable component of a mitigation plan. This methodology requires that connected mitigation sites be identified according to the status and position of currently existing, functional wetlands. Identifying these unmodified wetlands is necessary in order to perform proximal analyses.

Potential mitigation sites contained within a protected area were eliminated because these areas theoretically are already protected and not viable mitigation alternatives. The only protected area in the Four Hole Swamp is the Francis Beidler Forest, a private preserve managed by the National Audubon Society.

The wildlife habitat opportunity analysis evaluated the potential mitigation sites with regard to their ability to serve a wildlife habitat function. Many native species are in decline in South Carolina. Although population decline is attributable to a variety of causes, habitat loss and fragmentation are perhaps the most significant. Reduction in biological diversity and species quantity is directly related to the reduction in total area available for wildlife habitat.

The second step in this study was to assemble core habitat sites comprised of all the protection sites selected above, all upland forests (excluding pine plantation sites) from the uplands data layer, all protected areas (Francis Beidler

Forest), and all significant natural areas. On the basis of the model definition, these sites represented prime wildlife habitat areas in the sub-basin. Restoration and enhancement sites were added if adjacent to the core habitat sites.

The model defines multilane roads as significant barriers to wildlife movement. These roads were buffered to a distance of 131 ft (approximately 40m) to represent the highway right-of-way. The buffered roads were overlaid with the original core habitat sites. All coincident areas were eliminated, which resulted in division or fragmentation of the combined habitat sites.

Edge habitat is common across the Four Holes landscape, suggesting that existing landscape conditions meet the habitat needs for edge species. Thus, the habitat complexes with the above characteristics are further analyzed to determine adequate habitat interior. Each complex boundary is buffered inward by a distance of 328 ft (approximately 100m) to eliminate all edge habitat and determine interior habitat. The remaining habitat complexes represent areas containing interior habitat adequate for wildlife use. These complexes were restored to their original habitat boundaries.

The selected complexes were evaluated for size. The model defines habitat complexes of 40 acres (18 ha) or greater as containing adequate space for wildlife habitat. Given this criterion, only those containing at least 40 acres were selected. Finally, the original potential wetland mitigation sites were overlaid on the optimal wildlife habitat complexes. The coincident areas represent the potential mitigation sites meeting the wildlife opportunity analysis criteria.

The floodwater storage and water quality opportunity analyses evaluated the capability of the potential mitigation sites to perform hydrologic functions. Because wetland systems are extremely variable in structural characteristics, it is difficult to identify unifying concepts that would allow a convenient breakdown of wetland type on the basis of water quality or hydrologic function. However, watershed level characteristics are recognized as influencing a site's potential to serve these functions.

Wetland location within a watershed is an important determinant of its contribution to floodwater storage and water quality maintenance. Overbank flow is assumed to dominate the hydrologic function of high order streams, and adjacent wetlands are assumed to provide an opportunity to attenuate water flow during flood events by storing water. Wetlands associated with low order streams, the first to encounter pollutants from agricultural or urban areas, also provide an opportunity to enhance or maintain water quality.

High order streams were defined as all fourth, fifth, and sixth order streams, whereas low orders were defined as first, second, and third order streams. Because of the braided nature of parts of the Four Hole drainage system, some streams of the higher order network had to be added.

Potential flood storage sites were defined as any wetland site adjacent to a high order stream. In this study, wetland sites were used as a surrogate for elevation data. If elevation data had been available, the landscape topography could have been defined and used to assess hydrologic flow. No elevation data exist in a digital format for the study area; thus, this model assumes that wetland sites receive overbank flow from adjacent streams. All wetlands adjacent to high order streams were identified. Wetlands adjacent to wetland sites that are themselves adjacent to high order streams were identified as secondary high order wetlands.

Water quality sites were defined as wetland sites adjacent to low order streams. The model assumes that any adjacent wetland site receives overland flow before emptying into its related stream. All wetlands adjacent to low order streams were identified. Wetlands bordering wetlands adjacent to low order streams were identified as secondary low order wetlands.

Flood storage and water quality sites were combined with potential mitigation sites to identify sites providing single or dual hydrologic functions.

Results and Discussion

A composite overlay of the opportunity analyses was created depicting each potential mitigation site by class and by opportunity analysis criteria.

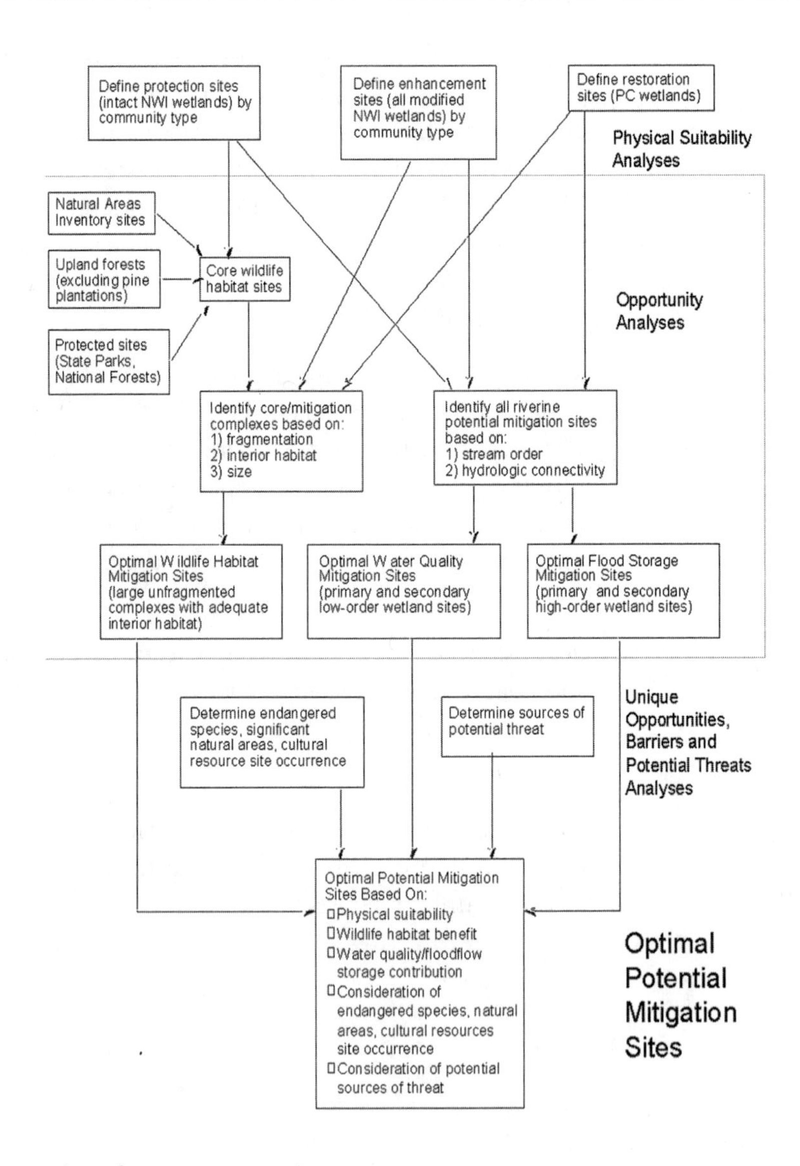

Composite overlay of opportunity analyses.

The composite map (Plate 66) illustrates the importance of the mainstem area in the Four Hole Swamp sub-basin. The area serves multiple opportunity functions in the sub-basin.

The potential mitigation sites identified in the overlay composite were evaluated with respect to the unique opportunities existing in the sub-basin. Unique opportunities were defined as the occurrence of sensitive species or communities of concern, archaeological sites, significant natural areas, or historic sites. These sites, in combination with the identified sites, present unique opportunities for mutual protection of important sub-basin resources.

Identified mitigation sites were evaluated with respect to potential threats existing in the sub-basin. Potential sources of threat were defined as hazardous waste sites (including generating, disposal, treatment, or storage sites), mining sites, and industrial and domestic waste sites. These sites present additional opportunities to ameliorate the impacts of respective activities.

The methodology was evaluated by field verification from a sample of identified mitigation sites. The methodology was successful in thoroughly inventorying the landscape for potential protection and enhancement sites. Restoration sites were not always identified because of an overly restrictive list of hydric soils.

Results from field verifications indicated that although the model successfully identified enhancement sites, a larger number of potential enhancement sites are in the field than are identified by the methodology. This is because a large number of sites identified as protection sites in the National Wetlands Inventory database have been modified in some way. The NWI database is dated, and therefore recent changes to the landscape may account for most of these unidentified sites.

The wildlife habitat component successfully identified potential mitigation sites that might serve as optimal habitat according to model definitions. Most restoration sites were eliminated from these results because of relatively small size and the model's requirements for interior habitat.

The floodwater storage and water quality analyses were not entirely successful for identifying distinct low and high order wetland sites. High order wetlands were consistently identified along the mainstem of the drainage system, but identifying low order wetlands was problematic because headwater areas as well as mainstem areas were also identified. The problem occurred because many small feeder streams empty into the mainstem area. Further refinement of stream orders might contribute to better definition of these areas. Elevation data would also improve characterization of hydrologic conditions.

In conclusion, this model provides a foundation to begin identifying potential wetland mitigation sites according to physical factors and various ecological functions. The methodology can be used effectively to better direct mitigation decisions in the regulatory arena. It does not replace field verification of sites but helps preliminary analyses narrow the number of candidate mitigation sites before field visits. In an era of diminishing budgets the method could save valuable resources.

Monitoring and Mapping the Belchatow Mining Complex in Poland

Stanislaw Mularz, Department of Photogrammetry and Remote Sensing Informatics, University of Mining and Metallurgy, Poland

Challenge

The Belchatow Mining Energy Complex (BMEC) has been in operation for more than 20 years. About ten years ago, concerns arose over the environmental degradation caused by BMEC mining and electricity generation. An affordable method had to be devised to detect, assess, and measure the environmental destruction related to BMEC and to track the success of environmental remediation programs.

BMEC is a major energy source for central Poland. Located about 6.5km (4 mi) south of the Lodz, BMEC is a large open pit lignite mine with a 350m (nearly 1,200ft) high dump body and an electric power plant.

The mine produces about one-third of all lignite mined in Poland, and the power plant supplies roughly 10 percent of the nation's electrical energy. The lignite deposit is 55m (180ft) thick and lies 150 to 250m (495 to 825ft) below the surface. The formation covers a narrow strip 25km (15 mi) long by 3km (1.8 mi) wide. The study region for this project included the immediate area surrounding the BMEC.

The mine has been in operation for two decades, and some environmental impacts are more obvious than others. Coverage of vegetation by the waste heap is readily visible, but the lingering effects of wastewater runoff still must be measured. The electricity plant, with its gas and dust emissions, is also believed to be causing environmental stress to vegetation in the area.

Combining SPOT panchromatic and multispectral imagery was the most cost-effective and efficient way to monitor the BMEC and its environment. Both SPOT 10m panchromatic and 20m multispectral imagery proved valuable because environmental degradation ranges from small-scale physical damage to broad, yet subtle, stress to vegetation and streams.

Results of the pilot study indicate the SPOT imagery is extremely useful and cost-effective for monitoring environmental degradation related to lignite mining and electricity generation. The processed SPOT images were viewed and visually analyzed in a GIS environment. This project primarily focused on preparing the imagery for analysis in the GIS. A continuation of the project will focus more on GIS analysis itself.

Data and Methodology

The Department of Photogrammetry and Remote Sensing Informatics at the University of Mining and Metallurgy spearheaded the BMEC study. The researchers obtained an August 1990 SPOTView image and a May 1986 SPOT multispectral image.

The SPOTView product is a 15′x15′ panchromatic image with 10m spatial resolution. This product and the SPOT XS image were GIS-ready, having been stretched, sharpened, and georeferenced by SPOT IMAGE in France before delivery.

The SPOT XS image was a level 1B product, which means it had been radiometrically corrected and geometrically processed to remove earth rotation effects. Level 1B products have also been resampled in the across-track direction to remove off-nadir imaging effects. The spatial resolution is 20m.

For GIS comparison with the satellite imagery, university researchers also obtained cartographic data, including 1:200,000 and 1:5,000 scale topographic maps. A 1:25,000 scale forest degradation inventory map was created from a color infrared aerial survey project flown in August 1985. A vegetative cover map of the dump site was also made from a black and white aerial survey conducted in July 1987.

Some reclamation activity has already begun at the site, and maps of these areas were also supplied to the researchers for the study.

For analysis and interpretation, the cartographic data sets were digitized and merged with the SPOT imagery for content enhancement and spatial comparison. Because the two SPOT images were acquired on different dates, they could be compared to detect changes in land cover.

A variety of standard image processing routines were run on the imagery. Two of the most important for this study were unsupervised and supervised classifications. Features from the cartographic maps and ground truthing were used as training data for the classification. The best results were achieved using a maximum likelihood supervised classification with the option of all pixels classified or 1 to 5 percent of pixels unclassified.

Results and Discussion

The complete extent of the open pit mine and its dump body were detected and measured with the satellite imagery, using both panchromatic and multispectral information. The artificial land forms of the mine and dump site displayed the tell-tale regular geometric shapes of artificial bodies, and the bare soil of the dump site had a distinctly different spectral signature than the surrounding forest and crop lands.

Magnification of the SPOT panchromatic image made it easy to distinguish the major components of the electric power plant: stack and vents, cooling tower, turbogenerator blocks, buildings, high voltage lines, and slag containers. The high resolution of the imagery allowed the researchers to visually identify and map these features.

All road surfaces could also be differentiated due to their unnatural linear shape. The entire transportation system around BMEC was mapped from the imagery.

The researchers located and quantified numerous land use changes by overlaying the pre-mine topographic maps with the more recent SPOT imagery. The results were displayed and visually examined in the GIS.

The following concerns received the most attention from the researchers:

- Dumping of waste products and construction material, as well as land clearing for construction of the plant, were conducted at the expense of the conifer forest. A large forest area was lost due to mine and plant development.
- The dump site has grown so large that ten small villages have been covered or consumed by the slag.
- Two large fish ponds have been destroyed in the area of the mine.
- Many acres of valuable cropland, grassland, and rangeland have been destroyed.
- Underground flow from the mine has apparently changed the natural flow regimes in the area and adversely impacted local water bodies.

The merged SPOT pan/multispectral image was compared with the 1985 forest map to determine the extent of damage to vegetation. The multispectral imagery clearly showed a variation in spectral values for coniferous trees located to the north/northwest of the electric power plant. This spectral variation could relate to vegetative stress. The area is being checked by field research to determine if the plant emissions are being carried by winds and damaging the conifer forest.

The news is not all bad at BMEC. Reclamation activities have been underway for almost a decade, and impacts can be measured with the satellite imagery. These reclamation projects have focused on revegetating the dump site with deciduous trees, brushwood, and grass.

A classification of the imagery showed that reclamation is complete in 4.2 percent of the dump site, advanced in 10.3 percent of the area, and in its initial stages in 28.9 percent.

Benefits

The results of the study show that SPOT imagery is very useful and economically attractive for conducting environmental inventories, monitoring, and mapping in an open-pit mine area.

The single most valuable application of the imagery was merging the panchromatic and multispectral data. This allowed the researchers to obtain very important thematic information about land and vegetative conditions as well as spatial information about physical aspects of the mine site. The imagery made it possible for the first time to assess the progress of the reclamation activity in a precise and cost-effective manner.

Vector and Raster Analysis of the Solitario Dome and Terlingua Uplift, West Texas

Michael Clark, Texas Legislative Council

Challenge

The Solitario Dome and Terlingua Uplift geologic areas are located west of Big Bend National Park. Located within Big Bend Ranch State Park on the northwestern edge of the Terlingua Uplift, Solitario Dome is a symmetric, structural dome about 12 km in diameter at the surface. Small, intrusive bodies of magmatic suites surrounded by circular ring sills in the Cretaceous sediments are found inside the dome. The Terlingua Uplift is about 29 km long and 13 km wide and is comprised of an asymmetrical anticline with a northwest trending axis that merges with the Solitario Dome to the northwest. General rock types include marine limestones, rhyolites, and tuffs. Collapsed breccia pipes, domes, grabens, fractures, and faults are structural features common in the Terlingua Uplift. These two areas represent distinct challenges for developing raster and vector data for GIS. The specific aims of the study were to (1) analyze GIS and remote sensing data for geological, structural, and lithologic features; (2) establish criteria and develop a methodology for GIS in a regional setting; and (3) analyze vector and raster data in a statistical modeling environment.

Analysis protocol.

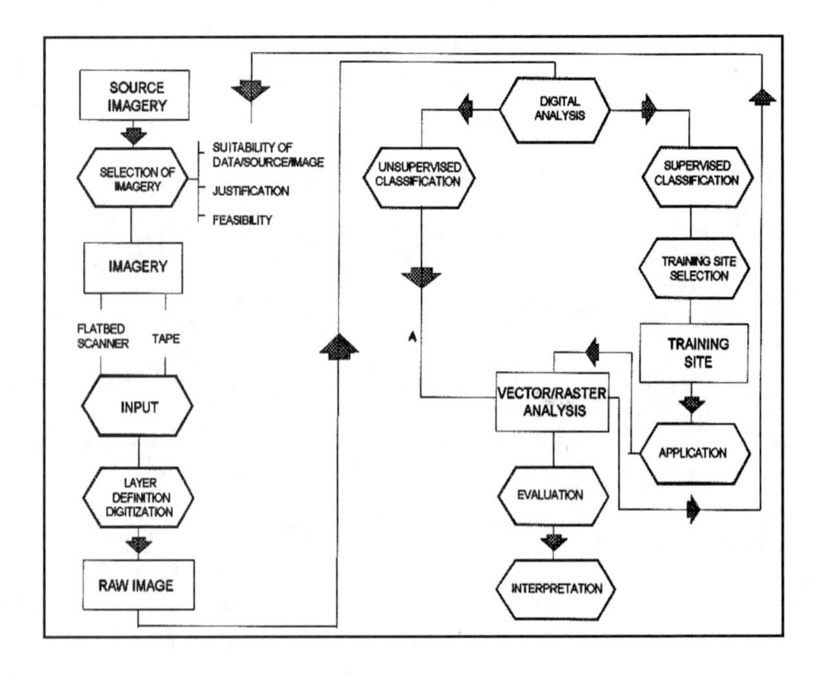

Data and Methodology

A data dictionary was built to define spatial features for data input and analysis. Among the considerations were lithology and morphology, and how to represent these features (i.e., polygons, lines, points, or raster layers). Once the data sets were defined, all spatial features were categorized. Hardware parameters, such as hard disk storage and computer time, determined the volume of data to be entered.

Data sources and equipment were selected based on the availability of laboratory resources at Stephen F. Austin State University. Raster data were derived from a 9″x9″ (225mm x 225mm) color aerial photograph for each study site. For the Solitario Dome, a published geologic map was used as a source for geologic ground data. An unpublished map was used for the northwest area of the Terlingua Uplift.

ARC/INFO (version 6.1.1) and ArcView (version 1.1) were used for vector analysis and data merging functions. ERDAS IMAGINE (version 8.1) was used for raster manipulations. The hardware included a network of Hewlett Packard 9000 series 720 model workstations. Each workstation had a 25″ 1280x1024 RGB monitor. Two stations were equipped with 24-bit accelerator cards used for high-quality graphic and image displays. An external 1.2 gigabyte hard drive was used for data storage along with 4 mm and 8 mm tape drives for data archiving. Calcomp 9100 E-size 36″x24″ digitizing tablets were used to digitize geologic maps. Photographs were scanned at 300 dpi with a high resolution 9″x11″ flatbed scanner.

Vector layers were defined on the basis of relative age of lithologic units and general rock types. Layers for the Solitario Dome included quaternary deposits, all sedimentary rocks, and Solitario magmatic suites and tuffs. Terlingua Uplift layers included quaternary deposits, all sedimentary rocks, pyroclastics rocks and intrusive magmatites, and breccia pipes.

Image classification was performed using unsupervised classification methods in ERDAS/IMAGINE. Five classes were determined to be of use based on features that could be seen in the image and experimentation with various numbers of fixed class settings on the imagery (Plates 67 and 68).

Results were obtained by integrating the vector and raster themes in ARC/INFO and comparing the two data sets using two methods. First, vector layers were overlaid on raster layers and visually compared. Second, both themes were converted to an ARC/INFO grid format for statistical analysis. Once in grid format, an arithmetic function was performed to calculate differences between the vector and raster layers to produce a statistical raster layer (Plates 69 and 70).

Results and Discussion

Unsupervised classification analysis of the Solitario Dome image showed the overall structural domal uplift pattern and the spatial distribution of the predominant, outward dipping Mesozoic sedimentary rock beds. Observed rock distribution patterns correlated well with the geologic map for the area. Steep cliffs appeared as red mimicking the relatively high topographic relief.

Grid analysis displayed pattern differences and similarities between the geologic base map and the results from unsupervised classification (see Plate 67). The yellow/blue and hash mark feature patterns correlated well with the base map features. Red feature patterns were not well correlated to the base map, indicating little, if any, relation between image classification and geologic base map.

The overall concentric ring pattern and outward dipping beds indicate a structural dome. Perpendicular to these concentric rings are numerous fracture sets interpreted as part of the hinge zones that formed during the doming episode. Smaller fractures superimposed on the dome may be related to the cooling of underlying intrusive bodies and the collapse of the central dome. Observed linear and morphologic structural patterns indicate extension as a direct result of locally strong vulcano-tectonic structural influence.

Unsupervised classification of the NW Terlingua Uplift image illustrates that overall spatial patterns correlate well with the base map (Plate 68). Lowes Valley appears (in gray) in the central portion of the image. In the right central portion of the image, a domal concentric pattern (Black Mesa) is detectable. The wide spatial distribution of the cretaceous limestones indicates the strong influence of at least two features on the overall image. The image is dominated by yellow and gray on the left and blue on the right, indicating an increase of topographical relief from left to right. Throughout the image, dark blue linear features indicate fractures, joints, and steep cliffs.

Grid analysis displays differences and similarities between the geologic base map and results from unsupervised classification (Plate 69). The blue and violet features in the central portion of the grid correlate well with the quaternary deposits of Lowes Valley. The yellow features of the right half correlate generally with surface cretaceous rocks. The red and orange features indicate no relation between image classification and the geologic base map. The color change of cretaceous limestones from east to west in the unsupervised classification coincides with the change in topography going from level to steep terrain. Throughout the unsupervised classification, dominant northwest and northeast trends overlap the terrain and influence drainage patterns. Given the extensive amount of vulcanic activity, the northeast trending fractures and grabens might indicate local extension stresses resulting from the intrusion of underlying magma.

From these observations, four geologic events can be summarized for each study area. Predomal activity, hinging of surface rocks, vulcanic activity, and post doming events occur in the Solitario Dome. Predomal fracturing and graben formation of the northeast joints/fractures, doming followed by vulcanic activity and formation of the northwest lineations, and formation of drainage patterns along fractures occur in the Terlingua Uplift.

Forest Management

Forest management is a topic of great interest to both the raster and vector communities. Early satellite scanners provided data on the extent of forest cover, and image-based applications quickly became popular as a tool for making global forest assessments. These tools progressed from single- to multi-stage statistical designs for calculating board footage of commercial stands, to more sophisticated attempts to merge rasters with economic data for complete *forest inventories.* In the 1980s, however, it became clear that raster data *alone* only supplied a small fraction of the overall data requirement for forest management solutions. Vector GIS added much of the rest.

In today's applications, combined raster and vector data processing is a standard approach toward more robust modeling of forest resources by public agencies and commercial interests at all levels. More importantly, perhaps, the case studies in this chapter are convincing evidence that GIS technology permits forest parameters to be modeled not only for what they reveal about the physical attributes of forests, but also for regulatory, licensing, and economic surveys. Spatial analysis based on changes recorded by time series data combined with relational attributes in vector format (and eventually with object attributes) holds vast potential for balancing forest industries with sustainable resources for the future.

Developing a Multi-scale Forests Database for Australia

Philip Tickle, National Forest Inventory, Bureau of Resource Sciences, ACT, Australia

Challenge

Since the advent of GIS technology, there has been a general recommendation from practitioners and software developers that data of different scales should be stored separately, and that most databases should consist of data of common scales and levels of attributing. This has meant that most natural resource management agencies have maintained independent local (e.g., 1:10,000-1:25,000), regional (e.g., 1:100,000) and state level (e.g., 1:500,000-1:1,000,000) databases for different planning purposes. As a consequence, the same area may have been mapped several times.

Reasons for the maintenance of this philosophy have centered around limitations in computer technology, and the use of traditional cartographic techniques focused on map production rather than development and maintenance of digital databases. With the cost of disk space declining exponentially, the performance of computers doubling, if not quadrupling every year, and the functionality of both raster and vector GIS improving at an equivalent pace, there are few reasons why an organization needs to maintain independent databases of different scales and varying levels of attributing.

In 1988 the Prime Minister of Australia launched the National Forest Inventory (NFI) as one element in a series of new initiatives to assist in the resolution of competing demands on Australia's native forests. The NFI is a partnership of commonwealth, state, and territory governments, and is supported by a small central project team.

The first task of the NFI was to produce a comprehensive overview of Australia's native forests. The ARC/INFO vector database used for this purposes consisted of a compilation of existing state level mapping at a nominal scale of 1:1,000,000. The statewide data sets were created using traditional cartographic techniques which ultimately meant standardizing the data, and in most cases, bringing the data back to the lowest common denominator. This expedient approach downplayed or ignored the finer scale data sets available for many areas. However, the approach revealed how little was known about the extent of Australian forests at the national level.

Since 1992, the NFI project team and its state partners have been undertaking regional and local scale inventories on a range of forest values with the aim of integrating these data sets to produce a new continental inventory capable of supporting regional and national level planning.

The aim of this study is to outline some of the methods used by the NFI Project Team to integrate data from a wide variety of sources and scales into a single, multiscale database that includes feature level metadata information and assessments of accuracy across the Australian continent. The advent of ARC/INFO GRID has substantially supported some new approaches to multiscale spatial database development and integration with vector data. A total of over 100 data sets are being used to develop a national database on Australian forests. An example using information on the state of Queensland (Qld) is given to illustrate some of the methods used. The Qld work involved integrating ten raster and vector data sets ranging in scale from 25m resolution TM to 1:5,000,000 scale vector-based mapping.

Data and Methodology

The National Forest Inventory Project Team is serviced by the most advanced GIS facility in Australia. It is maintained by the National Resource Information Centre (NRIC) within the Bureau of Resource Sciences. The Centre is equipped with multiple SUN SparcCenter 1000 and 4/690 multiprocessors running all modules of ARC/INFO; ERDAS IMAGINE and ERMAPPER image processing software; Oracle relational database management systems; and S-Plus for statistical analysis. A Thinking Machine CM2 Supercomputer is used to undertake classifications of remotely sensed and other raster data sets at approximately 100 times the speed of the SUN computers.

Approximately 100 gigabytes of on-line storage are integrated with a 50 gigabyte optical jukebox storage system and a terrabyte tape storage system for an on-line data archive.

In developing an inventory of Australia's forests the NFI Project Team has undertaken a bottom-up and auditing approach to digital database management. This approach works on the following principles:

- The lowest common denominator should not be used in developing a national database.
- The most current, finest scale and most highly attributed information should be used in any one area.

- One definitive data set should be stored in a corporate database, rather than independent data sets of different scales and currencies for any one area.
- Feature level metadata should be used to attribute differences in logical consistency.
- A corporate database should play an auditing role to identify gaps in information as well as providing the user with quantitative information on reliability.
- Both raster and vector data structures should be used in the single corporate database.

The NFI's state level databases consist of over 50 core attributes covering information on floristics (vegetation species), height, tree cover, tree age and growth stage, stand history, wood resources, land and ownership, and metadata.

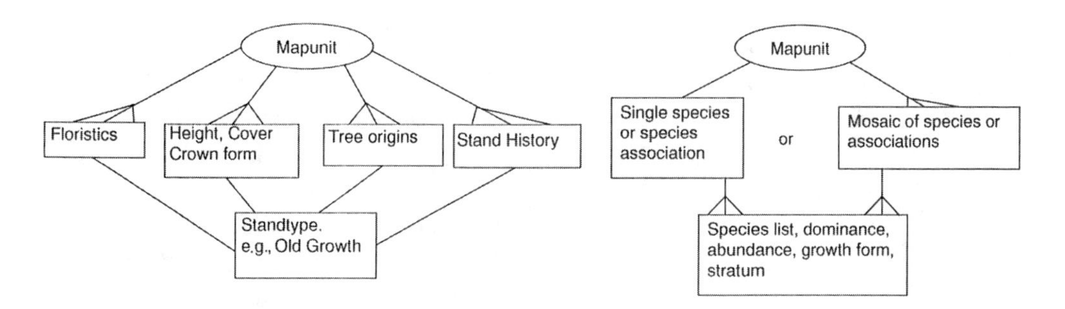

State level core attributes.

ARC/INFO's FREQUENCY command is used against all original attributes and a unique code (*mapunit*) is generated for each unique combination of attributes. These attributes, along with corresponding mapunit numbers are placed in a lookup table which has also been given the NFI attributes. The NFI attributes are populated with data (in many cases transformed) from the original attributes. Where attributes require that codes be generated, central (definitive) lookup tables are used. Every vegetation data set incorporated into the database (raster or vector) is populated with the same level of attributing and every attribute with missing data is filled with a null value (e.g., -9999 or NODATA).

The original data are given a unique number (referenced to a central table) and placed into a date-stamped archive area. The final data set (coverage or grid) is then placed into a reference area with only the mapunit attached to the polygon attribute table (PAT) or value attribute table (VAT) and a relational table for storage of the NFI attributes. The resultant coverages or grids refer to the same centralized relational tables stored in INFO or ORACLE (one-to-many relates).

Users can immediately integrate data of any scale and with any level of attributing. They can also undertake a gap analysis by querying fields for NODATA values, and they can relate back to the archive and track down the primary data sets used in the final data set if an update is to occur.

A great deal of effort has been spent internationally on metadata standards for use in meta-databases (directories) and to aid in data transfer (e.g., Spatial Data Transfer Standard, SDTS). However, most of the information contained within such metadata relates to the data set as a whole and any variations in a data set are only recorded as logical consistency text fields. The problem is that most of this information starts to disappear when users begin to integrate and overlay information to produce "value added" data sets.

In order to overcome this problem the NFI has implemented metadata management at both the data set and feature level. At the data set level, the SDTS-compatible information is stored using the NRIC Facility for Interrogating the National Directory of Australian Resources (FINDAR). But how are 50 different data sets appended together to produce a statewide data set? How are these data used to overlay a particular theme that consists of several other data sets?

The only way to keep track of this information is to store metadata at the feature level. A feature can be a single polygon or group of polygons with the same value for all attributes. As outlined previously, a unique code is given to every unique combination of attribute values within a data set. This unique code also records 18 metadata attributes that include information on everything from scale and sources of linework, to methods of wood volume derivation and data expiry dates (see following illustration). This ultimately means that the source of attribute values for each of the core attributes can immediately be identified, and that at any time an entire data set can be extracted from an integrated data set.

Metadata scheme.

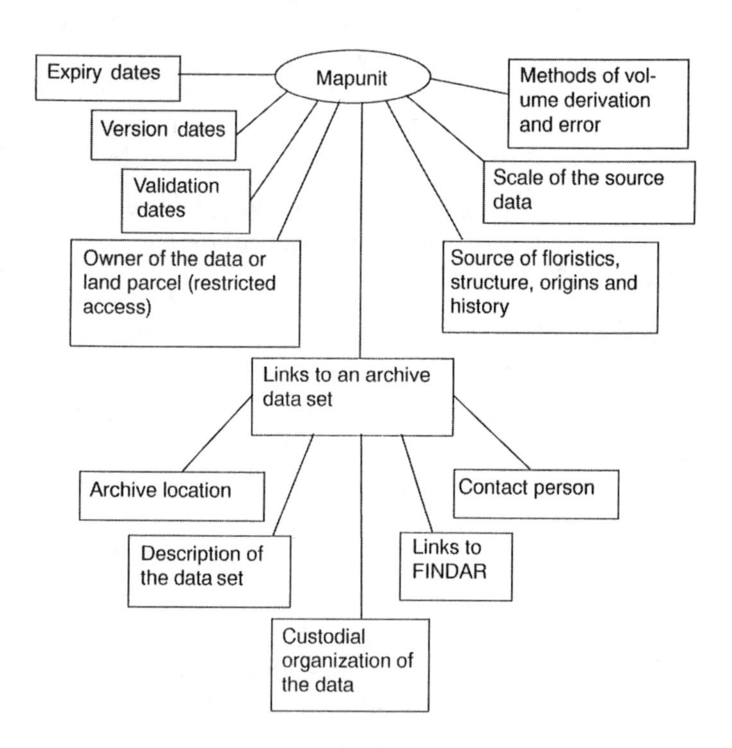

Results and Discussion

The area of dense and closed forest in the Qld in 1993 was reported at 6.814 million ha (16.83 million acres). In 1995, following statewide mapping at 1km resolution using Advanced Very High Resolution Radiometer (AVHRR) data, the figures were revised to 8.054 million ha (19.89 million acres) of dense and closed forest and 60.1 million ha (148.4 million acres) of sparse forest. However, very little was known about the reliability of these estimates.

To improve the reliability of estimates of forest area in Qld, the NFI Project Team, in collaboration with Qld agencies, compiled the most up-to-date digital data sets using the methods outlined above. Almost 65 percent of Qld's 174 million ha (429.78 million acres) had no validated vegetation data at a scale better than 1:5,000,000 in a readily accessible form. The remaining 35 percent of the state is covered by data sets at around 1:100,000 - 1:250,000 scale. Examples of the integrated forest cover and data set areas are shown in Plates 71 and 72, and the next illustration.

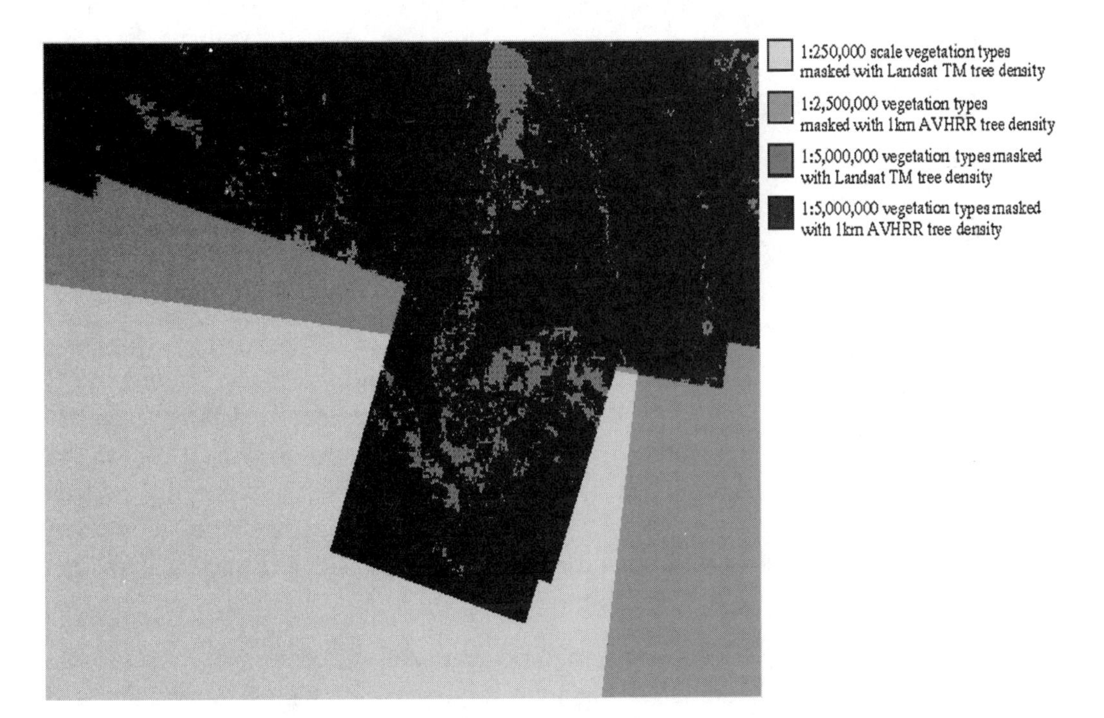

Integrated forest data sets.

A classification was undertaken using the supercomputing facility to produce 250 spectral classes from a compilation of 12 monthly mean AVHRR normalized difference vegetation indexes (NDVI). These classes were labeled using training data from 11,700 1km-square training sites randomly located within classified TM areas using GRID's RAND facility. Each site was given a mean tree cover density value using the STATISTICS function by weighting the mean by the number of TM cells in each tree cover class.

As the TM data were most reliable at detecting patch sizes greater than one hectare, GRID's BLOCKMAJORITY functions were used to generate a 1-ha minimum patch size (100x100m), before using GRID's ZONAL functions to effectively resample the data from 25m to 100m. This procedure was used instead of the RESAMPLE functions which can seriously degrade categorical data. The classified AVHRR imagery was then oversampled to 100m resolution to allow the ten grids to be merged.

Two methods were used to assess the accuracy of the AVHRR classification. The first was a simple error matrix on the overall classification accuracy, and the second was an analysis of forest area estimates.

The error matrix approach indicated that 85.6 percent of the 1km-square training sample sites had been correctly classified as forest/non-forest, and 78.5 percent of the sample sites were correctly classified according to tree cover density (sparse or dense). In most remote sensing applications these would be considered quite acceptable, but what do they mean in terms of area statements?

The second method analyzed the potential error in area estimates that may be associated with extrapolating the AVHRR classification across more than half the state. Forest area statements for both the TM- and AVHRR-derived data sets were calculated for 147 map sheets (see next illustration). In some areas there were enormous differences. For instance, in areas where the forest cover was sparse and comprised of small patches, the AVHRR underestimated the forest extent by up to 96.5 percent (e.g., arid regions). At the other end of the scale, where the vegetation was more dense but also comprised of isolated patches, the AVHRR overestimated the forest extent by up to 125.5 percent due to pixel saturation.

AVHRR area estimates in relation to "truth" data sets.

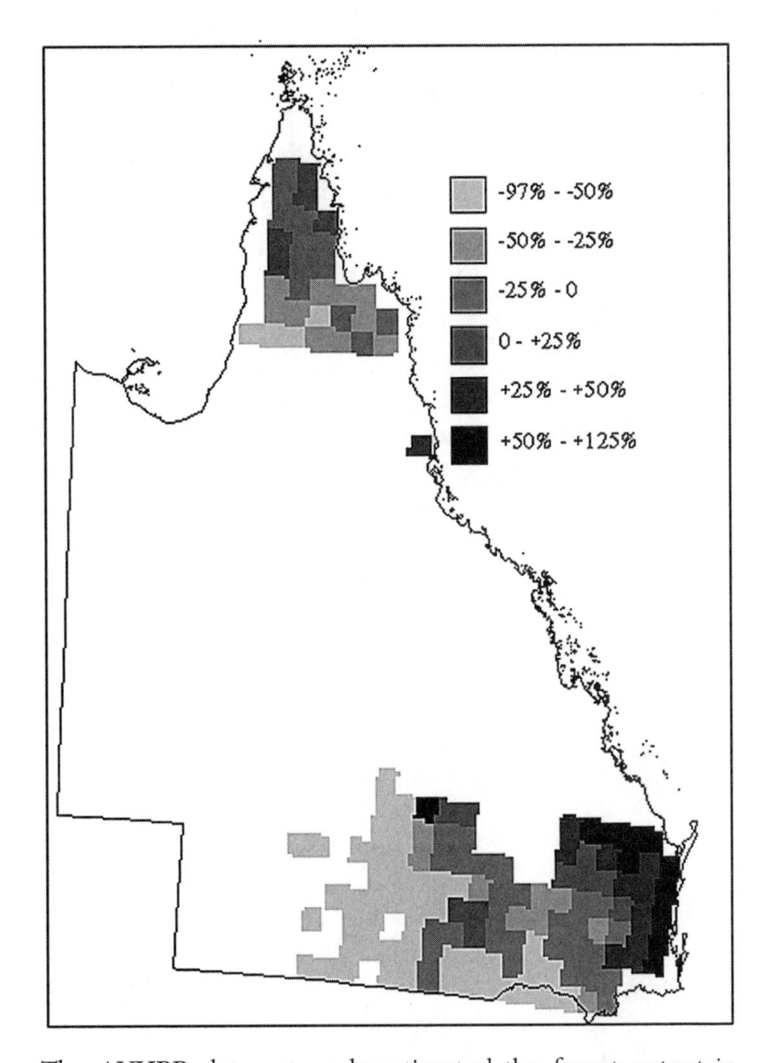

The AVHRR data set underestimated the forest extent in 110 of the 147 sample areas, with a mean underestimate of 26 percent plus or minus 7.84 percent at the 95 percent confidence interval. It should be noted that error bounds cannot be put on the existing higher resolution data because analyses such as these have not been previously carried out. Therefore, in areas where only AVHRR classifications are available, a range in area of forest extent can be given. The NFI team can now say that the area of sparse forest in Qld is somewhere between 40.28 and 43.48 million ha (99.49 million acres and 107.39 million acres), and the area of dense and closed forest is somewhere between 13.63 and 13.95 million ha (33.66 million acres and 34.45 million acres). This information is now also available at the national level.

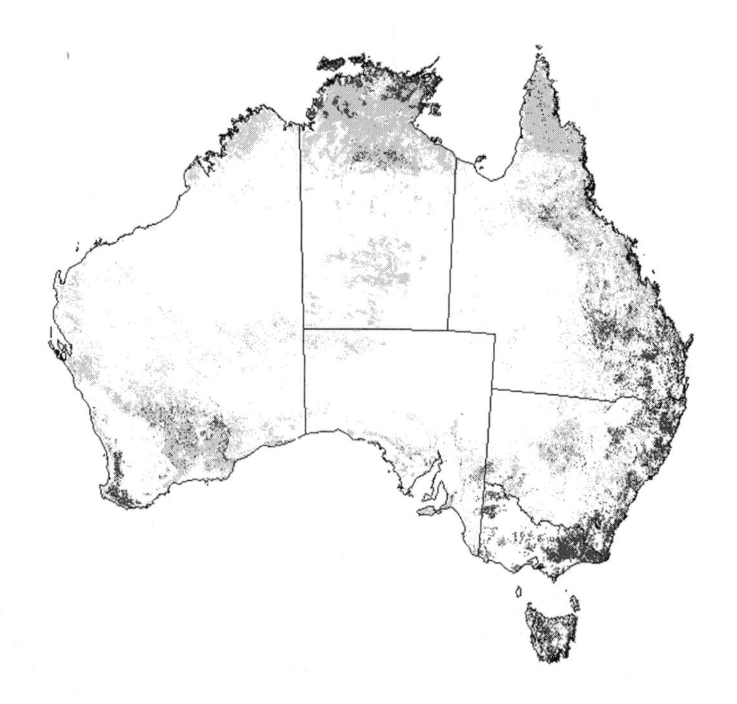

Distribution of sparse and dense forest throughout Australia.

Developing a multiscale national database describing Australia's forests could not occur without integrating both raster and vector technologies. ARC/INFO's vector and raster modules allow for flexible, definitive, multiscale databases. These are used to produce data-reduced raster or vector products for various purposes, while also maintaining the integrity of area statements by using imbedded metadata. By taking advantage of GRID's run-length encoded compression, and integrating feature level metadata, a user is no longer limited by the lowest common denominator in terms of scale and level of attribution.

Mapping Needleleaf Forests in Great Smoky Mountains National Park

John B. Rehder, Department of Geography, University of Tennessee

Challenge

This study highlights the uses of raster data for mapping needleleaf forests in a complex biosphere. In the Great Smoky Mountains, the necessity for forest cover inventory and assessment remains critical when one considers the variety of biota and the numerous natural and man-made threats to the biologic integrity of the Park. The Great Smoky Mountains National Park (GSMNP) contains 208,000 ha (513,968 acres) of land with thousands of botanical species located on terrain elevations ranging from 305 to 2,025m (1,000 to 6,700ft) with slopes up to 63°. The Park's ecological and topographic variety presents a unique challenge for analysis by remote sensing and GIS methods. Prior to the project, little was known of contemporary spatial patterns of all forest species on a park-wide scale. Scientific botanical studies had been made but most were at one-meter and one-fifth acre plot scales. Collectively, these studies led neither to a timely comprehensive inventory of the vegetative resources nor to a broad spatial mosaic representing the complex ecosystems of forest communities.

The subject of mapping the Park's forest cover resources has not been ignored. In 1941, a forest cover map of the entire Park was produced from field observations made on horseback and with help from the Civilian Conservation Corps (CCC). This map became the standard for the Park's forest cover until the mid-1980s. In the early 1970s, environmental concerns and an escalation of scientific research on the biota of the Park stimulated a modern comprehensive inventory. The Great Smoky Mountains National Park has been designated an International Biosphere Reserve, and the United Nations looks upon it as a world museum for biota. However, scientists and managers have been working at various spatial scales without a uniform spatial database.

The contract project from the National Park Service was the first of its kind to accomplish a park-wide survey of forest cover in the Great Smoky Mountains National Park using remote sensing with digital raster data acquired from an orbiting satellite. In the GIS Lab at the University of Tennessee, various remote sensing data were used to address the problem. Sources of remote sensing data came in image and digital formats from both aircraft and satellite platforms and were applied to mapping forest communities of red spruce, fraser fir, pine, and hemlock. Hardware and software consisted of an ERDAS digital image analysis system running on a Compaq desktop 386 computer, Cipher nine-track tape drive, Tektronix 4696 Ink Jet Printer, and Calcomp 9136 digitizer. The final maps were plotted on a Hewlett Packard Plotter 7580B.

Data and Methodology

The project used many different types of remote sensing data: Landsat digital data, NASA aircraft airborne digital scanner data and color infrared imagery, NAPP high altitude color infrared imagery, and low altitude aircraft imagery. Landsat digital data provided the base raster data for the park-wide maps.

MSS near infrared data (0.8 - 1.1mm) were used to detect and delineate needleleaf evergreen forests. A false color composite of green, red, and near infrared spectral bands provided the spectral separation of needleleaf species from broadleaf species. The base raster data from bands 4, 5, and 7 were used to produce the map of spruce, fir, pine, and hemlock coverages (Plate 73).

Another remote sensing raster data source was Landsat TM data. TM data are spectrally superior to those of the MSS. The best TM bands for needleleaf evergreen forest mapping are band 4 (.76–.90 μm) for detecting the presence of needleleaf forests and band 5 (1.55–1.75 μm) for needleleaf forest detection and moisture in broadleaf species.

The project goal was to create a series of forest cover maps that could be used by personnel at the Great Smoky Mountains National Park. As cartographic products, the maps were required to be in both paper and digital form with the latter capable of inputting into a GIS.

Signature extraction is the fundamental step in any spectral analysis of remote sensing data. Signatures are simply distinct characteristics seen on the imagery or discovered within the data that represent identity traits of objects on the ground. On visually interpreted imagery, signatures can include color, tone, texture, pattern, shadow, shape, size, height, and locational site characteristics within the context of the geographic setting. In digital analysis, however, signatures are largely limited to spectral signatures of color or tone that are derived from the numerical brightness levels of pixels. For example, a unique dark-toned signature representing needleleaf vegetation is produced by low levels of spectral reflectivity in the near infrared portion of the electromagnetic spectrum. In needleleaf vegetation, the narrow leaf structure limits the number of spongy mesophyll cells in the needles, thereby creating low reflectances of infrared radiation.

Conversely, healthy broadleaf vegetation in full foliage is highly reflective of near infrared radiation. This is because the larger, turgid cells of spongy mesophyll inside the leaf are geometrically shaped in such a way that a higher percentage of incoming near infrared radiation is reflected off the broadleaf surface back into the atmosphere and ultimately to the sensor. The needles in spruce, fir, pine, and hemlock forests cannot reflect as much electromagnetic radiation as broadleaf plants because of such differences in cell structure. In short, healthy broadleaf plants have bright signatures with high reflectance values; needleleaf plants have quite dark signatures resulting from low reflectance values in the near infrared spectrum.

The spruce-fir file was generated out of six rounds of supervised classification using eight samples and four classes of data. In the unsupervised classification, two classes out of 27 were declared spruce-fir and were deemed to correctly correlate with the supervised classification. The pine forests file was based on two rounds of unsupervised classification where all pines were in the same spectral signature class. In contrast, four rounds of supervised classification did not satisfactorily complete the pine pattern. In the hemlock forests, the hemlock GIS file was based on two rounds of an unsupervised classification in which three of 27 classes represented hemlock. In addition, four rounds of a supervised classification were made but none provided a complete portrayal of the distribution. The primary problems in classifying all needleleaf forests were errors of commission where shadows, unconsolidated stands, understory species, and mixed species tended to create more classified pixels than expected.

Results and Discussion

The spruce-fir forest zone at elevations above 1500m (about 5,000 ft) is a unique relic forest from the late Quaternary period dating to about 8,000 years ago. The zone is believed to reflect the effects of acid rain on the Red Spruce (*Picea rubens*). However, stands of Fraser Fir (*Abes fraseri*) are experiencing widespread mortality from insect infestations of the Balsam Wooly Adelgid (*Adelges piceae ratz*). Such threats to this rare biotic museum were part of the reason for the park-wide inventory of forest cover.

From a remote sensing perspective, the spruce-fir zone is the most identifiable group among needleleaf evergreen forest types in the Park. Low level, dark spectral signatures from the spruce-fir zone are used for signature extraction. On near infrared data, such as from MSS band 4 (0.8μm–1.1μm) or from TM band 4 (.76μm–.90μm), spruce-fir exhibit digital numbers (DNs) ranging 20 to 66 out of a DN range from 0 to 127. By comparison, broadleaf species have DNs ranging from 57 to 100. The differences are so clear that spruce-fir are unlikely to be confused with broadleaf communities.

An important attribute of the spruce-fir zone is high elevation. This parameter prevents spruce-fir signatures from being confused with needleleaf pine species at lower elevations. However, hemlock forests cover a broader range of elevations from 300m to 2,025m (1,000 to 6,700ft) so it is possible to confuse hemlock with spruce-fir using elevation alone. Fortunately, hemlock forests have a slightly brighter spectral signature and a much more scattered spatial distribution than spruce-fir. The map in Plate 73 shows the spruce-fir zone in the blue pattern along the higher mountain peaks and ridges in the northern half of the Park. The zone is accurately portrayed; the map was verified through on-site ground truth reconnaissance and through consultation with expert field biologists.

The pine forests of the Great Smoky Mountains National Park are largely located at elevations below 1,500m (red pixels on Plate 73). Spectrally, pine forests have low DNs in the near infrared spectrum. Digital numbers range from 33 to 71 out of a total DN range of 0 to 255. Pine signatures are not as dark as the signatures in the spruce-fir zone; however, they are much darker than all other broadleaf species and are just slightly darker than hemlock forests. In addition to spectral reflectance and elevation parameters, pines favor south-facing slopes and disturbed land. These habitat preferences combine to give pines a unique pattern on the landscape spectrally and spatially.

Hemlock forests (green pixels on Plate 73) occupy the widest range of elevations from 300 to 2025m (1000 to 6700ft) and of DNs than either pine or spruce-fir. Still, through the process of elimination, it was possible to produce a map pattern of hemlock forests by first creating files of spruce-fir and pine. Out of 27 DN classes, hemlock patterns represented three. Spectrally, hemlock DNs are in the range 53 to 72. Sample site verification confirmed the hemlock pattern on Plate 73.

The needleleaf forests in the Great Smoky Mountains National Park were mapped from raster-based data, verified through field reconnaissance, and approved by National Park Service personnel. It is estimated that the map for spruce-fir is 95+ percent correct, the pine pattern is 90 percent correct, and the hemlock pattern is 80 perent correct.

Information Technologies to Support the Licensing of Forestry Activities

G. W. Turner, R. M. C. Ruffio, and M. W. Roberts, Environment Protection Authority, New South Wales, Australia

Challenge

Since mid-1992, the New South Wales Environment Protection Authority has issued pollution control licenses to State Forests of NSW, the state agency responsible for forest resource management. The licenses take a best management practices approach, and require compliance with a range of operational conditions outlined in a timber harvesting plan.

It is critical that the licensing program incorporate the best available technologies to be effective. GIS and associated information technologies, such as remote sensing and GPS, offer substantial benefits through support to the auditing, assessment, and monitoring of forestry activities and the environmental condition of forestry and surrounding lands. The technologies are being evaluated in the EPA with a view to adopting them in the licensing program as operational procedures.

Several features described in the harvesting plan, including filter strips, protection strips, log dumps, snig tracks, and coupe boundaries, are potentially amenable to detection and monitoring by remote sensing. It is possible to make comparisons of field conditions during or after logging (as interpreted from imagery) with digital versions of licensed harvesting plans approved prior to the commencement of activities.

Data and Methodology

Plate 74 shows spectral enhancement highlighting features associated with logging activities. IMAGINE software automatically applies a two-standard deviation linear stretch on displayed images, but it is often useful to generate a "manual" stretch. In the example, bias toward the visible red and green bands was introduced in preference to the near infrared band. This improves the ability to discriminate features such as log dumps and tracks within the logged area.

Digital harvesting plans created in vector format and approved prior to logging can be overlaid on post-logging enhanced satellite imagery as a preliminary desk audit to check for general compliance. At the very least, this raster/vector merge in the preliminary desk audit step may eliminate compartments that are compliant and provide prioritization criteria for further auditing of non-compliant compartments. Features such as log dumps or access roads with the greatest discrepancy in the comparison, for example, can be earmarked for further detailed investigation.

Next, by using environmental information in the GIS, discrepancies located within or near high degradation risk areas, such as highly erodible soils, or adjacent to sensitive areas such as wildlife corridors or buffer strips, may also be flagged for further scrutiny. Maps showing GIS buffers facilitate interpretation of disturbed sites with respect to stream proximity. Therefore, the combination of raster/vector merges and buffering technology not only allows discrepancies to be flagged for possible non-compliance, but also provides an indication of possible environmental implications, such as log dumps sited close to drainage lines.

Another element of the technology is raster-based digital elevation models (DEMs). These models provide a potential for monitoring forestry activity in relation to slope. For example, logging carried out on slopes greater than a prescribed threshold, say 300, can be identified by the GIS and thus assist in targeting compartments for specialized scrutiny of proposed harvesting operations. In the case of post-harvest auditing, detailed information on slope can be overlaid with both satellite imagery and harvesting plans to provide some indication of the possible impact of logging features, particularly where possible license breaches may have occurred. In the example shown in Plate 74, a generalized slope classification, generated from a 125m grid DEM, is displayed with SPOT imagery showing the location and intensity of logging activity.

Plate 75 illustrates the assistance provided by terrain models in interpreting the catchment setting of proposed or completed activities. In this example, the surface model was generated using ARC TIN, and the log dumps and compartment boundary extracted from the harvesting plan have been overlaid. In the following figure, a DEM was used to derive a creek longitudinal profile draining from a nest of proposed log dumps. This information enables one to flag gradient break points, sediment traps, or other potential impact sites.

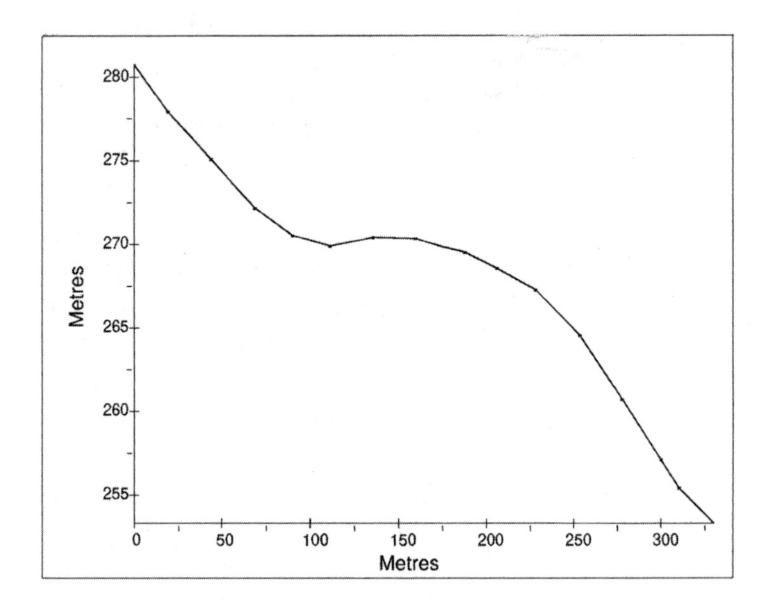

Longitudinal profile of a creek segment derived from a surface model.

Benefits

These visualization and interrogation techniques demonstrate the advantages to licensers of appreciating a three-dimensional landscape perspective of activities (or proposed activities) in the absence of field inspection opportunities. Furthermore, image processing and GIS analytical tools are available at the auditing stage for evaluating factors critical to non-point source pollution, including slope angle, length or shape, and the siting of features such as log dumps in relation to the terrain.

Satellite or other remotely sensed data representing field conditions can be combined with management information, harvesting plans, and other resource information, such as terrain or environmentally sensitive areas. The technology also allows plans, cross-sections, and three-dimensional visualizations to assist in both approving or setting conditions to planned activities, or auditing during and after operations.

Timber Harvest Scheduling with Adjacency Constraints

Bruce Carroll, Boise Cascade Corporation
Vaughan Landrum and Lisa Pious, Pacific Meridian Resources

Challenge

The U.S. Forest Service originally developed FORPLAN to help each National Forest develop management plans as required by law. Boise Cascade Corporation uses a customized version of FORPLAN on its corporate timberlands to predict potential benefits of management. FORPLAN is used by the Forest Service and private industry to schedule forest land management activities over time. However, FORPLAN is non-spatial and cannot model Oregon, Washington, and California regulations that constrain timber harvesting based on whether adjacent stands have recently been harvested. Therefore, FORPLAN may produce solutions that cannot be implemented in the field. To resolve this problem, Boise Cascade Corporation retained Pacific Meridian Resources to link ARC/INFO with FORPLAN to produce an application called the Spatial Feasibility Test (SFT) that tests the feasibility of FORPLAN solutions against adjacency constraints.

Data and Methodology

FORPLAN is a matrix generator for linear programming specifically designed for managing large areas of forest land. FORPLAN accepts tabular data on the land base (i.e., acres of each stand type), potential silvicultural and management prescriptions for each stand type, revenue and cost streams associated with each prescription, and any number of constraints on the land base (e.g., harvest levels must be non-declining over time). FORPLAN then submits the matrix to a linear program that attempts to maximize an objective (e.g., maximize lumber production) subject to the given constraints. The result is a FORPLAN allocation of the prescriptions that must be applied to a given acreage of each stand type in each time period to produce the maximum objective.

The primary drawback to FORPLAN is that it is non-spatial. There is no guarantee that a FORPLAN solution will be implementable because FORPLAN cannot analyze site-specific cases, such as seasonal inaccessibility or logging practicalities. When adjacency constraints are in effect, as they are in Washington, Oregon, and California, FORPLAN alone may vastly overestimate the potential benefits of management.

A typical regulation for an Oregon site, for example, might require a maximum clear-cut size of 120 acres (48 ha) and a four-year lag between adjoining harvests. This implies the following adjacency constraint: The combined area of adjoining polygons allocated to the clear-cut prescription must not exceed 120 acres in any four-year span. Because of this constraint, there may be no way to distribute harvests across the land base and meet the FORPLAN allocation and still comply with the adjacency constraint.

To solve this problem, Boise Cascade retained Pacific Meridian Resources to produce a Spatial Feasibility Test. SFT is an ARC/INFO based application that disaggregates a FORPLAN solution by applying it to an actual coverage of stand polygons. SFT can apply adjacency constraints to any silvicultural treatment code (such as clearcut) that the user specifies, with an associated maximum acreage (such as 120).

SFT is not an optimizer. It progresses through the list of stand polygons, and allocates the current polygon according to the FORPLAN solution. SFT checks the polygon's neighbors for adjacency violations for all time periods (e.g., there may be 20 five-year periods, for a simulation total of 100 years). If none are found, the polygon is assigned the prescription specified by FORPLAN, and SFT moves to the next polygon.

This method is deterministic, based solely on the order in which the polygons are checked. To counter this weakness, SFT allows the user to re-sort the polygon list on any database item, including a random number, and submit numerous runs to SFT for the same adjacency constraints. SFT automatically compares the set of runs, and ranks them according to a user-specified criterion. The user can then apply the result of a chosen run to the original stand coverage, and produce maps and analyses using standard ARC/INFO and ArcView.

Results and Discussion

The following two figures demonstrate SFT output. The first illustration shows the effect of adjacency constraints on harvest scheduling. Shaded areas are scheduled for harvest at some time in the 20-period run. Unshaded areas are never harvested. The unconstrained solution meets the FORPLAN allocation for the area. The constrained solution falls far short of the FORPLAN goal. The second and third illustrations show the harvest pattern chosen by SFT for the first four periods.

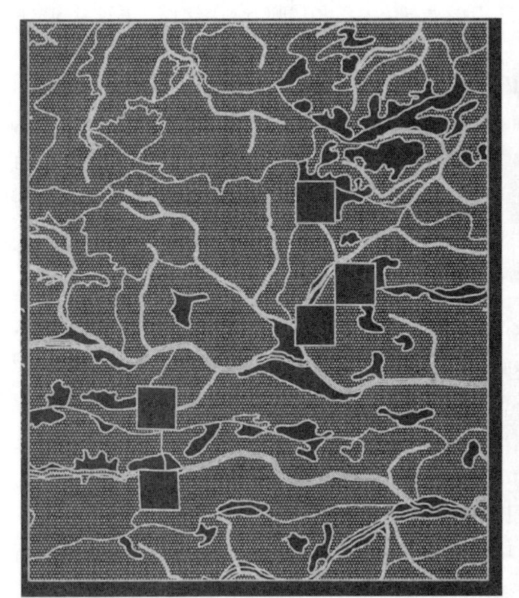

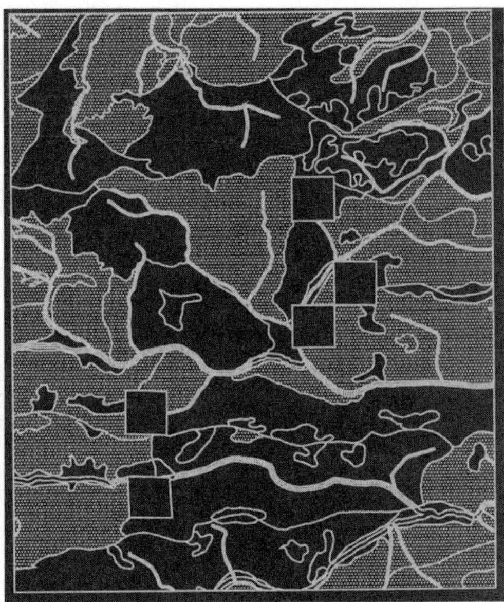

Effect of adjacency constraints on harvest scheduling. The unconstrained solution appears at the left and the constrained at the right.

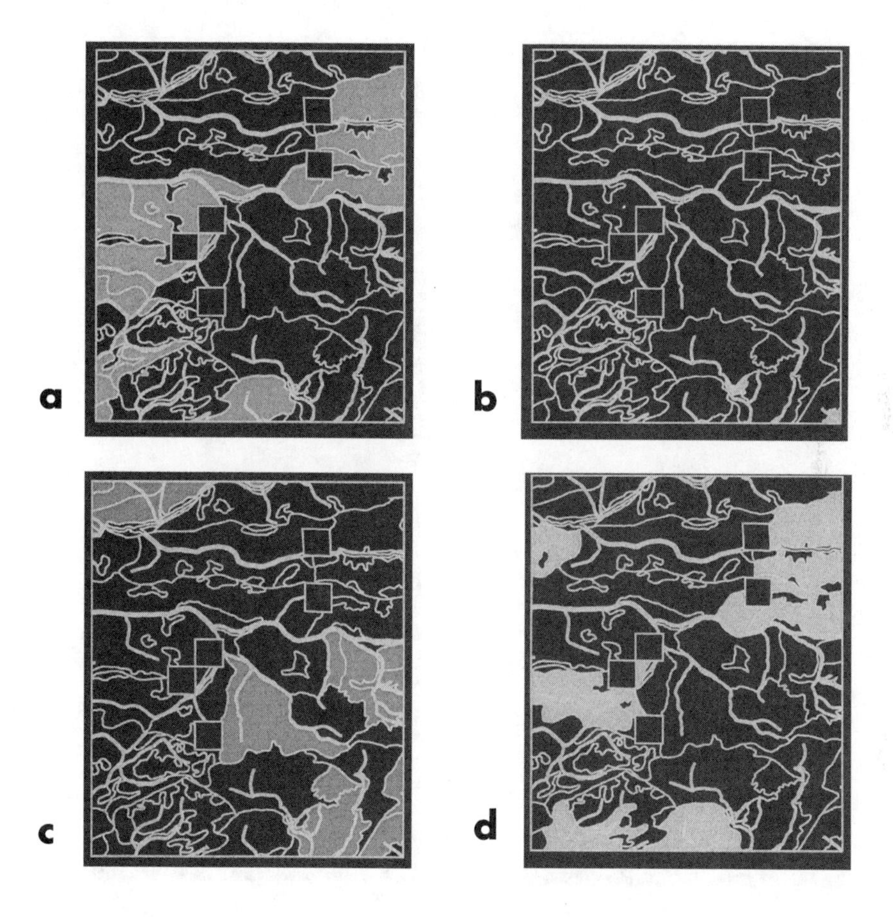

Constrained solution, first four periods. Stands allocated for harvesting are shaded.

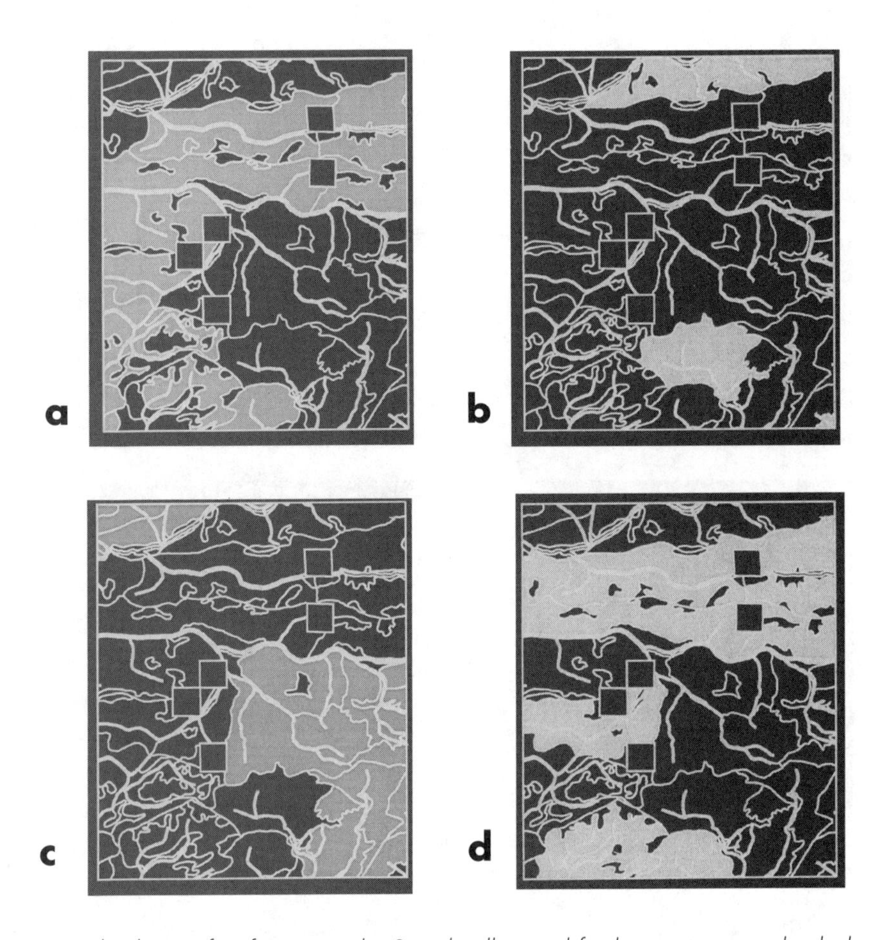

Unconstrained solution, first four periods. Stands allocated for harvesting are shaded.

Boise Cascade is using SFT with FORPLAN in an iterative cycle. FORPLAN is run first, then numerous SFT runs are performed using applicable adjacency constraints. If SFT is able to allocate the full FORPLAN solution, no further cycling is needed, since SFT has proven that the FORPLAN solution is attainable given current adjacency constraints.

If SFT cannot allocate the full FORPLAN run, then FORPLAN will be rerun with tighter constraints. For example, if the best SFT run was able to allocate only 75 percent of the FORPLAN solution, then FORPLAN may be rerun with an objective that is 75 percent of the previous FORPLAN run. SFT is then run on the FORPLAN output, and the iterative cycle continues. When SFT is able to allocate the entire FORPLAN solution, then Boise Cascade is assured that the FORPLAN goals are attainable.

SFT is not intended to be an intelligent harvest scheduler because it does not create a logical harvesting pattern across the landscape based on transportation networks or working circles. SFT's main purpose is to test the feasibility of Boise Cascade Corporation's FORPLAN runs under adjacency constraints. With SFT, Boise Cascade Corporation will use FORPLAN with increased confidence that FORPLAN's results are realistic and implementable.

Benefits

SFT applies a FORPLAN solution to a coverage of timber stands. The user specifies a maximum adjoining size for any set of silvicultural treatments (e.g., maximum of 120 acres for clearcuts) in any period. SFT can also apply a regulation in which stands that are adjoining for less than a given percent (e.g., ten percent in Washington) of their perimeter are not considered adjacent. SFT produces reports and coverages that itemize how well SFT was able to apply the FORPLAN solution to the ground.

Boise Cascade will use SFT in an iterative manner: run FORPLAN; run SFT to apply adjacency constraints; reformulate FORPLAN to more closely match SFT outputs; rerun FORPLAN and SFT; and continue until SFT is able to fully allocate the FORPLAN solution. The resulting FORPLAN solution will then be implementable under adjacency constraints.

Fire!™ Using GIS to Predict Fire Behavior

Kass Green, Mark Finney, Jeff Campbell, David Weinstein, and Vaughan Landrum, Pacific Meridian Resources

(Reprinted from the Journal of Forestry, 93 (5):21–25, published by the Society of American Foresters, 5400 Grosvenor Lane, Bethesda, MD 20814-2198. Not for further reproduction.)

Challenge

Resource managers have long sought spatial models of wildland fire growth. Non-spatial models based on the Rothermel fire spread equation, such as BEHAVE, provide a means to predict fire behavior based on inputs of fuels, weather, and topography for a specific location. However, short-term fire growth predictions over complex landscapes require repeated calculations, becoming impractical for long-term projections of large fires over highly heterogeneous landscapes; hence, the need for spatial computer models of fire growth.

In Fall 1993, the U.S. Marine Corps at Camp Lejeune, North Carolina, retained Pacific Meridian Resources to create a GIS layer of fuels for a 15-million ha area including and surrounding Camp Lejeune, and to use this layer to develop a GIS-based fire behavior model. The result is FIRE!™, a computer model that integrates fire behavior modeling into the ARC/INFO GIS environment.

Data and Methodology

FIRE! is an ArcTools-based GIS application that integrates spatial fuels and topographic data with temporal weather, wind settings, and initial fuel moistures to predict forest fire behavior across both time and space. The process allows a user to model fire behavior by defining a fire "scenario." The following figure presents a flow diagram. The user specifies the appropriate fuels, canopy cover, slope, elevation, and aspect layers required for the simulation. In addition, non-spatial data sets including weather, wind, initial fuel moistures, and fuel model adjustment factors can be created, specified, and edited. Finally, scenario parameters of ignition location, ignition shape, run time, and resolution are designated.

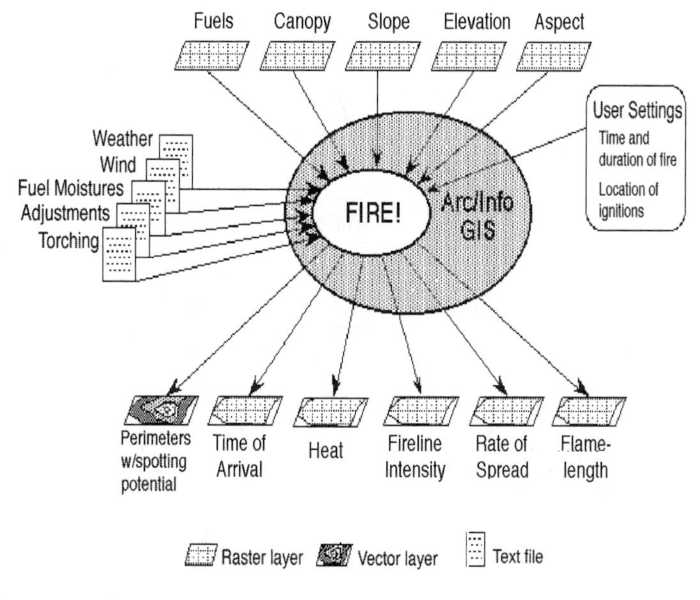

FIRE! simulation flow diagram.

As the simulation progresses, vector representations of fire perimeters are graphically displayed (see next illustration). Maximum spotting distance is computed from user inputs and displayed. At the conclusion of the simulated burn, FIRE! also displays raster representations of time of arrival, heat, fireline intensity, rate of spread, and flame length for the burned area. Plots of the simulation results may be generated using built-in plotting templates or customized by the user with the full suite of plotting tools available in ARC/INFO. All of the output data are preserved as ARC/INFO coverages and grids, which may be further analyzed.

FIRE! simulation results.

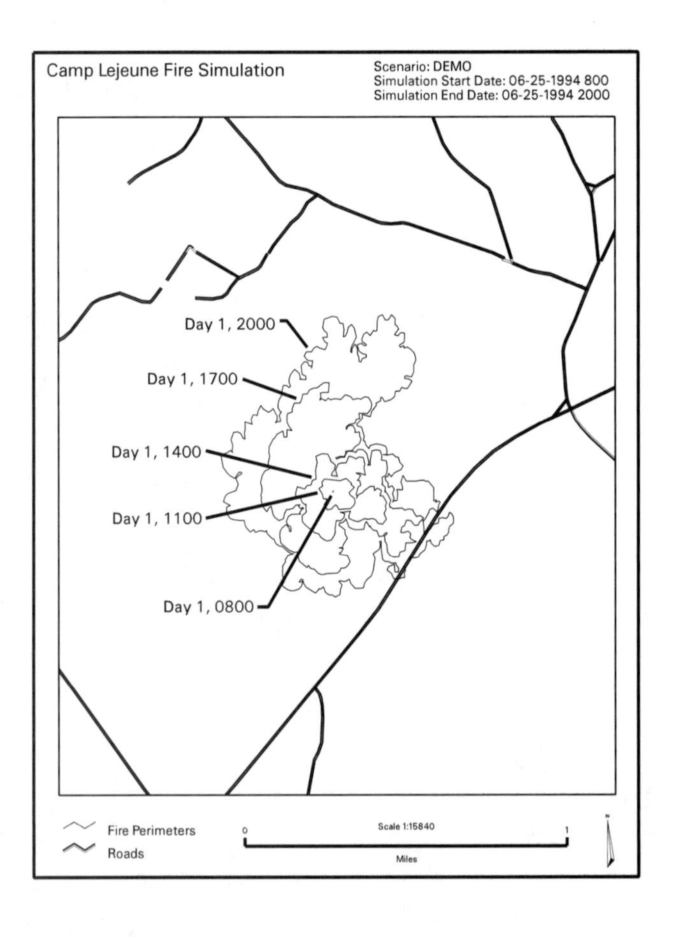

The engine of the FIRE! application, responsible for all complex computations necessary for simulating fire behavior, is FARSITE (Fire Area Simulator), a C++ program developed by Systems for Environmental Management.

FARSITE interacts seamlessly within the ARC/INFO environment as a component of FIRE!, enhancing the spatial display and query capabilities of the GIS for fire modeling and analysis.

Vector or wave type models have been shown to closely simulate fire growth with varying winds. The vector approach in FARSITE and FIRE! propagates the fire front in a fashion similar to a wave, shifting and moving continuously in time and space. Vector models solve for the position of the fire front at specified times. FARSIGHT uses a technique for vector propagation, known as Huygen's principle, to expand surface fire fronts in two dimensions. While rasters are still used to represent the underlying landscape and to record fire characteristics during the simulation, the fire perimeters are processed and stored as continuous vectors.

The fuels layer was created from TM imagery using a combined supervised/unsupervised approach. Image classification was enhanced with ancillary GIS data layers, including various past and present land use and land cover characteristics, and aerial photography. The classification was further refined with other GIS coverages for overstory vegetation types, soil types, and recent forest management activities. Extensive use was also made of field collected data. Ground data for fuels, overstory and understory vegetation cover, and tree crown cover were collected for sample locations throughout the project area.

The classifications resulted in raster data layers depicting the continuous variation of fire fuels and tree crown cover present. The spatial detail provided by the pixel classification provides a realistic prediction of fire behavior when used in the GIS-based model because the pixels contain the complexity and composition of land cover characteristics.

A raster forest tree cover layer was also developed by classifying the TM data. Initially, water, bare/non-flammable, and other non-forest fuel classes were masked from the imagery. These areas were assigned a crown cover class of 0 percent. For the remaining areas, unsupervised classifications were conducted to label crown cover in broad categories.

FIRE! also accepts raster coverages for topography in a variety of units (e.g., degrees or percent for slope). For this study, topographic coverages were created in ARC GRID from raster contour layers provided by Camp Lejeune.

Given the required GIS modeling parameters, various fire scenarios can be simulated. User-specified parameters include the burn simulation start and end dates and times, and the spatial and temporal resolution of calculations performed during the simulation. A spatial resolution of greater than 25m may be specified for scenarios covering very large areas in which only a gross estimation of fire behavior over a long time period is desired. Specifying a greater spatial resolution reduces the computational requirements of the model resulting in a faster simulation. However, scenarios requiring detailed information regarding fire behavior throughout a simulation area should utilize a spatial resolution at least as small as the input data sets provide.

Finally, a data set identifying the location and configuration of a fire ignition must be specified. Fire ignitions may be established as points, lines, and/or polygons and are entered interactively by clicking on the screen at the desired ignition locations. The user may also adjust the predicted rate of spread to match actual observed rate of spread by fuel type.

Field data regarding past prescribed burns and wild fires collected by Camp Lejeune Forestry Division personnel were extensively used in the fire behavior model calibration process. This involved fuel-specific adjustments to rate of spread similar to those used for the BEHAVE system. Base personnel consistently collect detailed data regarding fire behavior (spread rates, flame lengths, burn perimeter) for fires burning on the base. Global positioning system (GPS) units are used to map fire perimeters. The fire behavior model was calibrated using these detailed data from past fires to ensure the most accurate and reliable fire behavior predictions under local conditions. Incorporating these data during the model calibration was among the most important tasks of the entire project.

Results and Discussion

As the model performs the necessary fire behavior calculations, vectors are displayed indicating the fire's perimeter at a user-specified time interval. The vectors may be displayed over the fuels raster data layer or the original TM image. At the completion of the simulation, raster data layers are produced providing the flame length, fireline intensity, time of arrival, heat per unit area, and rate of spread of the fire for every pixel within the burned perimeter (Plate 76).

FIRE! is an example of a GIS model that goes beyond inventory, monitoring, and display to allow ecosystem managers to simulate the spatial outcomes of future management and policy decisions. By making the ability to vary critical model assumptions readily accessible to the manager, FIRE! tests the sensitivity of decisions to assumptions of weather, fuels, and topography. Future applications will allow the economic costs and benefits of these decisions to be considered by incorporating ignition risk, suppression costs and land value into the model.

Analyzing the Cumulative Effects of Forest Practices

Kass Green, Pacific Meridian Resources

(Reprinted, reformatted, and condensed with permission from Geo Info Systems, February 1993. Copyright by Advanstar Communications, Inc.)

Challenge

One of the most compelling challenges forest managers face is identifying, measuring, and monitoring cumulative effects of forest practices across both time and space. Cumulative effects are the changes to the environment caused by the spatial and temporal interaction of natural ecosystem processes with two or more forest practices. Attention has focused primarily on fragmentation of wildlife habitat, accelerated erosion, increased flooding, deterioration of water quality, and changes in stream morphology affecting fish habitat.

A basic premise of cumulative effects analysis is that landscapes differ in their sensitivities to forestry activities. For example, impacts of timber harvesting on water quality vary spatially with differences in climate, geologic materials, vegetation cover, and terrain. Consequently, the risks to water quality, and fish and wildlife habitat associated with the range for forest management activities vary from place to place. Because all watersheds cannot be examined simultaneously, a method is needed to determine the order in which basins will be visited for cumulative effects assessment.

In Spring 1991, the Washington State Department of Natural Resources (DNR) retained Pacific Meridian Resources to create a land cover GIS layer for the forested areas of Washington and use it to perform the DNR-designed wildlife and hydrologic screens. Four types of screens have been identified for key categories of hazard and risk: slope instability, fisheries, wildlife, and hydrology. Fisheries and slope instability screens were prepared by DNR staff.

Data and Methodology

To use remotely sensed data to map vegetation, correlations must be established between the variation found in the remotely sensed data (aerial photography or satellite imagery) and the variation that occurs in vegetation on the ground. To establish such links, we used a multistage approach that relied on integrating image processing; interpreted high-altitude, 1:67,600-scale, black-and-white aerial photography; field reconnaissance; and review of draft maps by DNR regional foresters.

First, the seven bands of the TM data were reduced to bands 3, 4, 5, and a ratio of band 3 to band 4. The first three bands were used because they are most indicative of vegetation. The ratio compensated for confusion caused by shadows on north slopes. Next, each scene was cropped to show only the area being classified. The relevant area of each scene was determined by the following:

- Relationship with adjacent overlapping scenes.
- Major ecological regions (so that areas in different ecological regions of the state were classified separately).
- USFS and National Park Service lands already classified.
- Extent of the study area.

An ISODATA unsupervised clustering algorithm was then run on the imagery for each area. Unsupervised classes (or spectral clusters) were identified using photo interpretation and minimal field reconnaissance. Spectral clusters that obviously correspond to one of the five classes in the classification system were colored in a dull color that symbolized the class represented. Unknown or confused spectral clusters were given bright, garish colors.

The colored spectral clusters were then plotted as hardcopy maps at the same scale as the high-altitude photography to allow direct comparison. The maps were reviewed for misidentification and to identify the brightly colored unknown spectral classses. Following review by Pacific Meridian staff, the plots were reviewed by experienced DNR reginal foresters. DNR foresters were asked to delineate misidentified areas and answer questions about confused or mislabeled areas which were then corrected.

For analysis, the raster coverages were then transformed into polygons for each water resouce inventory area (WRIA). Polygons simplify the vegetation coverage, imitate the photo interpreter's delineation of vegetation types, and eliminate the noise associated with the image pixel classification. The minimum size for polygons was ten acres for non-contrasting land cover types and five acres for contrasting types (e.g., non-forested areas next to closed canopy stands).

Accuracy assessment of the new forest cover GIS was performed by DNR's Forest Practices Division staff. DNR divided the state into 15 ecoregions, and each scene was divided into the ecoregions within it. Polygons of at least five acres were delineated on every twentieth 1:12,000 aerial photo for four selected townships within each scene/ecoregion combination. Photo interpretation of the polygons was compared with the correlating classification. An error matrix with user's and producer's accuracy was created for each ecoregion/scene combination.

The priorities for cumulative effects analysis may have to be changed over the long term as forest cover changes in each basin. Another step in cumulative effects analysis will be keeping the forest cover GIS up to date. Because the forest cover GIS coverage is derived from digital satellite imagery, changes can be monitored by directly comparing past images with newly acquired, up-to-date images. Multidate TM imagery provides an efficient means to identify and measure change in land cover over time, particularly change from loss in vegetation cover. As part of a contract funded by the National Aeronautics and Space Administration (NASA), Pacific Meridian Resources has developed production methods for using satellite imagery to detect land cover and land use change. DNR staff were trained to use these methods during a two-day, on-site training session.

The project was completed in two phases. Phase one involved building the GIS coverages required to prepare the screens. The second phase involved the actual implementation of the screens. All data and results were delivered to DNR as ARC/INFO files. The following GIS coverages were required to prioritize the watersheds for further examination:

- Watershed basins
- Forest type
- Current forest cover (for both the wildlife and hydrology screens and to be used as baseline vegetation data for future projects including long-term monitoring of cumulative effects)
- Precipitation (for the hydrology screen)
- Rain-on-snow (also for the hydrology screen)
- Soil hazard rating
- Slope
- Fish presence or absence and hatchery locations

WRIAs selected for screening analysis were WRIA subdivisions and are fourth- to fifth-order in size. Over 200 basins were designated. Approximately 30 were later excluded because they did not contain significant tracts of commercial forest. The basin and WRIA GIS coverage was provided to Pacific Meridian by DNR.

The first step in developing the forest cover layer was to specify cover classes. A simple classification system was agreed upon that would both identify vegetation in critical stages of maturity for wildlife and hydrology, and make it possible to use existing Forest Service GIS forest cover data. The following table describes the classes. The classes were derived from size and crown closure classifications developed for Region 6 of the Forest Service by Pacific Meridian Resources. This existing data covered 15.2 million acres of forested lands. Data for the remaining 16.6 million acres of private, tribal, state, and other government lands were captured by classifying parts of 15 terrain corrected and georeferenced TM scenes.

Description of Land Cover Classes

Class	Definition	Wildlife Equivalency	Hydrologic Equivalency
1	> 10% tree crown closure in trees ≥ 21″ DBH with > 70% total crown closure and < 75% of the crown in hardwoods or shrubs	Late seral stage	Hydrologically mature (100% mature)
2	<10% tree crown closure in trees ≥ 21″ DBH with > 70% total crown closure and <75% of the crown in hardwoods or shrubs	Mid seral stage	Hydrologically mature (100% mature)
3	10 - 70% total crown closure and < 75% of the crown in hardwoods or shrubs	Early seral stage	Hydrologically immature (50% mature)

Description of Land Cover Classes

Class	Definition	Wildlife Equivalency	Hydrologic Equivalency
4	< 10% crown closure and/or > 75% of the crown in hardwoods or shrubs	Early seral stage	Cleared forest (0% mature)
5	Water	Open water	Open water (100% mature)
6	Non-forested	Non-forest land	Non-forested (100% mature)

Two of the GIS coverages are "precipitation" and "rain-on-snow." The precipitation coverage was digitized by DNR using NOAA 10-year 24-hour precipitation isohyetal data. The rain-on-snow coverage was created by DNR as a map of five precipitation zones likely to have various amounts of snow on the ground in early January. National Weather Service and cooperative snow survey and elevation data were used to delineate the rain-on-snow zones. Both coverages were reprojected onto UTM zones and clipped to each WRIA for analysis.

Soil hazard ratings for state and private commercial forest lands are available from DNR's existing GIS. Because these data are not available for federal lands, an average soil hazard rating was derived using slope classes from USGS 3-arc second digital elevation data. The following five classes were used: 0-5%, >5-27%, >27-49%, >49-70%, and >70%.

The Washington Rivers Information System (WRIS), a GIS database maintained by the Washington Department of Wildlife, contains site-specific information on state, tribal, and federal hatchery facilities and the presence and absence of fish species. An updated version of this database is maintained by DNR and was used as input to the fisheries screen.

Having assembled the eight required data sets, phase two of the project (analysis) could be undertaken. The hydrology screens were prepared using ERDAS's raster GIS module on individual WRIAs. The slope instability screens were prepared in ARC GRID and the fisheries were computed using ARC/INFO.

Wildlife screen scores for each WRIA were based on the following six factors:

- ❐ Number of acres of forest land within the WRIA.
- ❐ Proportion of the forested type within the WRIA in late seral stage habitat.
- ❐ Proportion of the forested type within the WRIA in mid-successional forest cover.
- ❐ Proportion of the forested type within the WRIA in functional late seral stage habitat (defined as areas with more than 60 continuous acres in the late seral cover type).
- ❐ Proportion of the forested type within the WRIA in large patches of late seral stage habitat (defined as areas with more than 640 continuous acres in the late seral cover type).
- ❐ Distribution of late seral stage habitat by basin within the WRIAs.

The hydrology screen is based on the change in available water during a hypothetical 10-year, 24-hour rain on snow storm from an assumed fully forested condition to the current forested condition. The GIS model calculates the area weighted averages of storm precipitation plus snow melt for each basin. The difference between these calculated values for a given basin represent the change in water available for runoff that could occur in a storm as a result of timber harvest. The greater the difference, the greater the potential for increased peak flows and damaging effects downstream. The data produced from this screen were normalized to values between 0 and 100.

The slope instability screen was derived from a combination of the soil hazard, slope, and rain-on-snow data sets. First, the soil hazard and slope GIS layers were combined to produce a layer that indicated where steep and unstable areas were located. This layer was then combined with the rain-on-snow layer to show where storms would increase the probability of slides. The combined data were area weighted for each basin and normalized to result in basin values between 0 and 100.

The fisheries screen used the WRIS data to calculate the proportion of fish to total mileage of the river systems in each basin. Additional weight was given to basins with one or more fish hatcheries. Most values for this screen were fractions of one. A few basins had values greater than one when a high proportion of utilized fish habitat and hatcheries was present.

Results and Discussion

The hydrologic, slope instability, and fisheries screens were combined to prioritize the forest basins of the state for watershed analysis. The following equation was used:

$$C = \frac{H + S}{2(F)}$$

where

C = the combined ranking

H = the value from the hydrologic screen

S = the value from the slope instability screen

F = the value from the fisheries screen

Responding to directives from policymaking authorities, Washington's Department of Natural Resources staff along with cooperators in Washington's timber/fish/wildlife agreement have begun to analyze and regulate forest practices with regard to potential for causing significant cumulative effects. GIS and remote sensing are critical tools in this process. Performing the analysis in a GIS provides methods and results that are easily tested for sensitivity to assumptions, repeatable, and accessible to interested parties. Use of satellite imagery to create the forest cover GIS allowed for both cost- and time-saving efficiencies, and resulted in a coverage which is easily updated and assessed for change.

Acknowledgments

The authors would like to thank the staffs of both the Department of Natural Resources and Pacific Meridian Resources for the hard work that made this project possible. We are particularly indebted to Deanna Harper, Paul Hardwick, and the DNR regional foresters who so patiently reviewed the forest cover classifications.

Using GIS to Produce Fire Control Maps

Sonny Parafina, Cook-Hurlbert, Inc., Austin, Texas

Challenge

Maps play a central role in fighting forest fires. From locating sites for staging areas to providing information about vegetation type, topography, and cultural features, maps are a primary source of information for coordinating fire fighting efforts. However, for that information to be effective, map data must be maintained and updated on a regular basis. Rapid changes in vegetation cover and cultural features can make maps outdated before they are printed. The Texas Forest Service (TFS) faced this problem in the piney woods of east Texas. Most TFS maps were produced in the early 1970s, and the last updates took place in the early 1980s. In 1992, the TFS selected the Mapping Sciences Laboratory at Texas A&M University to digitally reproduce their 15′ quad Fire Control Map Series.

The Fire Control Map Series encompasses an area from the counties along the Texas-Louisiana border to the hill country of central Texas, and the counties along the Texas-Oklahoma border to the coastal plains along the Gulf of Mexico. Updating such a large area using manual cartography and aerial photography was prohibitively expensive. The Mapping Sciences Laboratory proposed developing a combined vector and raster GIS database in ARC/INFO with a custom cartographic interface to produce publication quality maps.

Data and Methodology

Creating seamless ARC/INFO coverages of roads, hydrography, and cultural features for the project by digitizing USGS 7.5′ quad sheets would have take several years. The Census Bureau TIGER92 files were proposed, but positional and attribute table errors were of some concern to the TFS. An alternative source of road network data was available from the Texas Department of Transportation (TxDOT) in the form of highway files digitized at 1:24,000 scale in MicroStation. While positionally accurate to ± 40ft (12m), the TxDOT files had the disadvantage of lacking attribute tables. The solution that the Mapping Sciences Laboratory and the TFS reached was to use the TIGER92 files as the base for the road network and the TxDOT design files as source data for both positional and data conflation. In addition, a cooperative arrangement of quality control and error checking was established between the two agencies. The hydrography coverage was created from the 1:24,000 USGS Digital Line Graphs, and the cultural features were also derived from TIGER92. Additional cultural features, such as churches, schools, and cemeteries not present in the TIGER92 or the TxDOT highway files were digitized from the previous 15′ quad series annotated by foresters in the field offices.

The original fire control maps were printed in black and green, with the green areas representing vegetation. Vegetation cover was compiled from aerial photographs. The TFS Fire Control mapping project used TM imagery. Six full TM scenes were required for full coverage of the project area. The scenes were purchased with geometric corrections to reduce time spent in processing the imagery. TFS originally proposed classifying the vegetation coverage into eight categories, but after reviewing sample maps, field foresters and forestry technicians in the district offices asked to reduce vegetation coverage to four categories to make the maps easier to use. The classified imagery was delivered with an 85 percent overall accuracy. Ground verification was accomplished through aerial photographs, site visits, and aerial videography.

The TFS purchased a Sun SparcStation LX dedicated to processing data and producing maps for the fire control mapping project using ARC/INFO software. The TFS project also used an Intergraph 6880 workstation for processing the TM imagery using Intergraph's ISI-2 remote sensing software. Various PCs were also used for manipulating the TxDOT highway design files in MicroStation and digitizing. Storage of the processed data was difficult because of the size of both the vector and raster data sets. The solution was to write the data on multiple magneto-optical disks and provide the user with a menu system in ARC/INFO for extracting the data.

Creation of the vector database began with processing the TIGER92 files for all counties in the project area into ARC/INFO coverages. Similarly, to prepare the TxDOT highway files for translation to ARC/INFO, the counties were joined together in MicroStation and saved as a large design file. An AML program was written to translate the TxDOT highway file into coverages that corresponded with the CAD file levels. Additional cultural features not present in TIGER92 or the TxDOT highway files were digitized.

Check plots of each quad sheet were inspected on screen and the TIGER92 derived road network was checked for positional accuracy against the TxDOT road network. When discrepancies between the two coverages were detected, the TIGER92 road network was adjusted. Both agencies also agreed to use the Roads of Texas atlas as the definitive reference. Hardcopy check plots were printed for quality control and visually inspected by both Mapping Sciences Laboratory and TFS personnel before changes were committed to the database. Each Thematic Mapper scene was processed individually and the resulting classified scenes provided the input for ARC/INFO's GRID module. Intergraph's ISI-2 remote sensing package was used to process the TM scenes.

Several enhancement techniques were applied to each scene to improve the classification accuracy. First, the brightness of each scene was increased using a histogram equalization. A Kauth-Thomas tasseled cap transformation was performed on each scene to yield a soil brightness index (SBI), green vegetation index (GVI), yellow stuff index (YSI), and non-such index (NSI). The green vegetation index or greenness was retained and the other indices were discarded. The greenness index is highly correlated with leaf area index and biomass. Greenness was useful in distinguishing between the different vegetation types represented on the fire control maps.

A supervised classification was performed on TM band 7, tasseled-cap greenness index, and TM band 1. Training sites were manually selected for each of the land use categories (pine, bottom land hardwoods, and brush) specified by TFS along with other categories (such as water, urban, or pastures) that would be helpful in differentiating the land use categories. The training set histograms were iteratively examined until the land use signature was normally distributed because a multimodal distribution indicates possible overlap between land use categories. The classification was performed using the maximum likelihood decision rule. The classified scenery was exported in a TIFF format from ISI-2. The classified TIFF image was georeferenced in ARC/INFO by a program that reads the header of the TM scene and writes a corresponding ARC/INFO world file. The world file is an ASCII text file that contains six parameters that ARC/INFO uses to georeference imagery.

Verification of the accuracy of the classified image was accomplished through sampling recent aerial photographs and aerial videography flown for another multistage sampling, remote sensing project. An ARC/INFO AML was written to generate a random sampling frame of coordinates that fell within areas covered by aerial photography or videography. The sample points were visually checked and tabulated. The resulting statistics provide a benchmark for the overall accuracy of the classification. Each TM scene underwent several iterations of this process until a minimum criteria of 85 percent overall accuracy was achieved. Once certified as meeting the accuracy criteria, the classified TIFF image was converted to the ARC/INFO GRID data structure. The land use classifications were written to the *GRID.vat* table using an AML that populated the land use field based on the values in the grid color table. Once this translation process was completed, users could query the land use grids.

Results and Discussion

GIS professionals often criticize using the technology to produce paper maps. However, paper maps are still crucial to day-to-day operations of field foresters, as well as fire fighting operations. As one forester put it, "You can't fold up a computer and stick it in your back pocket." Once the initial investment in hardware and software has been made, the cost of producing and maintaining paper maps decreases in comparison to manual methods. An added bonus is the ability to rapidly produce specialty maps that focus on a particular area or attribute of a region.

A graphical cartographic interface comprised of a set of AML routines was written to extract the data from both the vector layers and the raster layer for producing publication quality maps. The data were subsetted from the GIS into 15′ quads, reprojected, annotated, and placed into a map template generated by the cartographic interface. A test plot was printed for visual inspection, and the final copy was written as color separates to Encapsulated PostScript (eps) files and sent to an offset printer.

A few processes were applied to the subsetted data to improve the overall appearance of the map. For example, a 3x3 low pass filter was run over the land use grid to reduce the speckled appearance of the map. Although such alterations reduce the accuracy of the land use classification, TFS felt that such enhancements made the maps more serviceable in the field. Given that these changes were made on subsetted data, the original raster database remained unchanged.

In general, the TFS mapping project was a study in work flow automation. The end products were a vector/raster GIS and a cartographic subsystem, which in themselves are neither glamorous nor cutting edge examples of GIS technology. Most of the geographic data processing was accomplished through batch processes that ran at night or on weekends, making computing time available for other projects, and for students and researchers. A project manager, two full-time GIS technicians, and three part-time student workers staffed the two-year project. Approximately 65 percent of staff members' time was absorbed in quality control and quality assurance.

List of Contributors

CONTRIBUTORS	Organization Name
Ahl, Douglas	Lockheed Martin Stennis Operations
Baumgartner, David	Kansas Applied Remote Sensing Program, University of Kansas
Bennett, Teri Brotman	Earth Data Analysis Center, University of New Mexico
Benson, Jan	Facility for Information Management, Analysis, and Display, Los Alamos National Laboratory
Bowyer, Julie K.	National Resource Information Centre
Brock, John	EROS Data Center
Budge, Amelia	Earth Data Analysis Center, University of New Mexico
Byrne, Michael	Roads and Traffic Authority of New South Wales
Cagney, Patrick T.	U.S. Army Corps of Engineers
Campbell, Jeff	Pacific Meridian Resources
Carlson, George R.	Breedlove, Dennis, and Associates, Inc.
Carnevale, Sue	San Diego Association of Governments
Carroll, Bruce	Boise Cascade Corporation
Carroll, Crista S.	Albuquerque District Office, USDI Bureau of Land Management
Carver, Steve	School of Geography, University of Leeds

CONTRIBUTORS	Organization Name
Chen, Dong Mei	Department of Geography, San Diego State University
Chernin, Philip R.	Camp, Dresser, and McKee, Inc
Cheshire, Heather	Computer Graphics Center, North Carolina State University
Chopping, Mark	Department of Geography, University of Nottingham
Clandillon, Stephen	SERTIT
Clark, Michael	Texas Legislative Council
Cornford, Dan	School of Geography, University of Birmingham
Craven, David	Department of Geography, George Mason University
Cullen, Bradley	Department of Geography, University of New Mexico
Curtiss, Richard	Department of Geography, University of New Mexico
Driver, Julie	Department of Geography, University of Regina
Egbert, Stephen L.	Kansas Applied Remote Sensing Program, University of Kansas
Elder, Tom	Arizona Department of Water Resources
Esnault, Dominique	SERTIT
Estes, Jack	Department of Geography, University of California Santa Barbara
Feldman, Sandra C.	Bechtel Corporation
Finney, Mark	Pacific Meridian Resources
Flint, Scott	California Department of Fish and Game
Foresman, Timothy	Department of Geography, University of Maryland
Fraipont, Paul de	SERTIT
Georgieff, Didier	DDAF du Bas-Rhin
González Murgía, René G.	Centro de Calidad Ambiental, Instituto Tecnológico y de Estudios Superiores de Monterrey
Graves, Derek	Department of Geography, University of South Carolina
Graves, Jon K.	Columbia River Estuary Study Taskforce
Green, Kass	Pacific Meridian Resources
Greene, Bob	Facility for Information Management, Analysis and Display, Los Alamos National Laboratory
Guzzetti, Fausto	Institute for Hydrogeological Protection, Consiglio Nacionale delle Ricerche (CNR-IRPI)
Haack, Barry	Department of Geography, George Mason University
Hanning, Richard	Fluor Daniel Siting and Consulting

CONTRIBUTORS	Organization Name
Hennemann, Jean-Daniel	DDAF du Bas-Rhin
Henry, Robert J.	Formerly Breedlove, Dennis, and Associates, Inc.; currently Houston Advanced Research Center
Huang, Xuequiao	Geography Department, University of South Carolina
Huffman, Rodney L.	Department of Biological and Agricultural Engineering, North Carolina State University
Isacks, Bryan L.	Department of Geological Sciences and Institute for the Study of the Continents, Cornell University
Jakubauskas, Mark E.	Department of Geography, University of Oklahoma
Jensen, John R.	Geography Department, University of South Carolina
Jordan, Thomas	Center for Remote Sensing and Mapping Science (CRMS), University of Georgia
Kasouf, D.	Pacific Meridian Resources
Kempka, Richard G.	Pacific Meridian Resources
Khorram, Siamak	Computer Graphics Center, North Carolina State University and International Space University
Klaver, Robert	EROS Data Center
Klein, Andrew G.	Department of Geological Sciences and Institute for the Study of the Continents, Cornell University
Kramber, William J.	Idaho Department of Water Resources
Landrum, Vaughan	Pacific Meridian Resources
Lee, Re-Yang	Kansas Applied Remote Sensing Program, University of Kansas
Lefeuvre, Alain	DDAF du Bas-Rhin
Lewis, Andrew T.	Pacific Meridian Resources
Lewis, John E.	McGill University
Lewis, Kari	California Department of Fish and Game
Lo, Thomas H.C.	South Florida Water Management District
López Baros, Shirley	Earth Data Analysis Center, University of New Mexico
Lozano García, Diego Fabián	Centro de Calidad Ambiental, Instituto Tecnológico y de Estudios Superiores de Monterrey
Ma, Zhenkui	Montana Cooperative Wildlife Research Unit, University of Montana
Martinko, Edward A.	Kansas Applied Remote Sensing Program, University of Kansas
McAndrew, Dewayne	U.S. Bureau of Reclamation

CONTRIBUTORS	Organization Name
McDonald Jampoler, Susan	Department of Geography, George Mason University
Messina, Joe	SPOT Image Corporation
Miller, Keith E.	Camp, Dresser, and McKee, Inc.
Moore, Donald	EROS Data Center
Morain, Stan	Earth Data Analysis Center, University of New Mexico
Morisette, Jeffrey T.	Computer Graphics Center, North Carolina State University
Morse, Anthony	Idaho Department of Water Resources
Mularz, Stanislaw	Department of Photogrammetry and Remote Sensing Informatics, University of Mining and Metallurgy
Muldavin, Esteban	New Mexico Natural Heritage Program, University of New Mexico
Nadeau, Andrew	EROS Data Center
Neville, Paul R. H.	Earth Data Analysis Center, University of New Mexico
Nizam, Yousef	Faculty of Engineering and Architecture, American University of Beirut
Nuñez Brown, Dana	Parsons Brinckerhoff
O'Leary, John F.	Department of Geography, San Diego State Universtiy
Palmer, David	Idaho Department of Water Resources
Parafina, Sonny	Cook-Hurlbert, Inc.
Parrott, Robert	San Diego Association of Governments
Paulson, Douglas J.	Earth Data Analysis Center, University of New Mexico
Pelletier, Ramona E.	NASA Stennis Space Center
Pious, Lisa	Pacific Meridian Resources
Pugh, James A.	Breedlove, Dennis, and Associates, Inc.
Redmond, Roland L.	Montana Cooperative Wildlife Research Unit, University of Montana
Rehder, John B.	Department of Geography, University of Tennessee
Reichenbach, Paola	Institute for Hydrogeological Protection, Consiglio Nacionale delle Ricerche (CNR-IRPI)
Reid, Frederic A.	Ducks Unlimited, Inc.
Renaud, Anthony	Maricopa County Department of Transportation
Ribeiro da Costa, João	Univerisdade Nova de Lisboa
Ritman, Kim	National Forest Inventory, Bureau of Resource Sciences
Roberts, M.W.	Environmental Protection Authority

CONTRIBUTORS	Organization Name
Roberts, Mindy	Camp, Dresser, and McKee, Inc.
Rowland, James	EROS Data Center
Ruffio, R.M.C.	Environmental Protection Authority
Seeparr, Joseph	Remote Sensing Research Unit, University of California-Santa Barbara
Smoot, James. C.	Lockheed Martin Stennis Operations
Spell, Ruth E.	Pacific Meridian Resources
Stallings, Casson	Computer Graphics Center, North Carolina State University
Stayner, Floyd	Water Resources Division, South Carolina Department of Natural Resources
Stow, Douglas	Department of Geography, San Diego State University
Tickle, Philip	National Forest Inventory, Bureau of Resource Sciences
Turner, G.W.	Waters and Catchments Branch, Environmental Protection Authority
Veitch, Simon M.	National Resource Information Centre
Walser, Ed	Lockheed Martin Stennis Operations
Watkins, Russell	Department of Geography, University of Auckland
Weinstein, David	Pacific Meridian Resources
Welch, Roy	Center for Remote Sensing and Mapping Science (CRMS), University of Georgia
Whistler, Jerry L.	Kansas Applied Remote Sensing Program, University of Kansas
White, Robert	Bonneville Power Administration
Wright, Richard D.	CESAR, Department of Geography, San Diego State University

Contact Points

Teri Brotman Bennett
Earth Data Analysis Center
University of New Mexico
2500 Yale Blvd. SE, Suite 100
Albuquerque, New Mexico 87131
tbennett@spock.unm.edu

Jan Benson
Los Alamos National Laboratory
P.O. Box 1663, Mailstop D452
Los Alamos, New Mexico 87545
benson@fimad6.lanl.gov

Michael Byrne
Roads and Traffic Authority of New South Wales
52 Rothschild Avenue
Rosebery, New South Wales 2018 Australia
byrnem@rta.nsw.gov.au

Bruce Carroll
Boise Cascade Corporation
One Jefferson Square
Boise, Idaho 83728

Crista S. Carroll
USDI Bureau of Land Management
435 Montano NE
Albuquerque, New Mexico 87107

Steve Carver
University of Leeds
LS2 9JT
Leeds, United Kingdom
steve@geography.leeds.ac.uk

Philip R. Chernin
Camp, Dresser, and McKee, Inc.
10 Cambridge Center
Cambridge, Massachusetts 02132

Mark Chopping
Geography Department
University of Nottingham
University Park NG7 1DQ
Nottingham, United Kingdom
chopping@geography.nottingham.ac.uk

Michael Clark
Texas Legislative Council
621 Russet Valley
Cedar Park, Texas 78613
70641.3012@compuserve.com

Dan Cornford
School of Geography
University of Birmingham
B15 2TT Edgbaston
Birmingham, United Kindom
d.cornford@bham.ac.uk

Bradley Cullen
University of New Mexico
Bandelier West, Room 219
Albuquerque, New Mexico 87131
bcullen@unm.edu

Julie Driver
Department of Geography
University of Regina
Regina, Saskatchewan, Canada
driver@cas.uregina.ca

Tom Elder
Arizona Department of Water Resources
500 N. 3rd Street
Phoenix, Arizona 85004

Jack Estes
University of California Santa Barbara
1629 Ellison Hall
Santa Barbara, California 93106
estes@geog.ucsb.edu

Sandra C. Feldman
Bechtel Corporation
P.O. Box 193965
San Francisco, California 94119-3965
scfeldma@bechtel.com

Didier Georgieff
DDAF du Bas-Rhin
2 Rue des Mineurs
Strasbourg Cedex, 67070 France
dgieff@mail.sdv.fr

Kass Green
Pacific Meridian Resources
5915 Hollis Street, Building 8
Emeryville, California 94608

Fausto Guzzetti
Consiglio Nacionale delle Ricerche
Via della Madonna Alta
126 - 06100, PERUGIA, Italy
fausto@kenoby.irpi.unipg.it

Barry Haack
Department of Geography
George Mason University, MS 1E2
Fairfax, Virginia 22030
bhaack@gmu.edu

Robert J. Henry
Houston Advanced Research Center
4800 Research Forest Drive
The Woodlands, Texas 77381
rjh@harc.edu

Mark E. Jakubauskas
University of Oklahoma
634 Energy Center
Norman, Oklahoma 73019
jakubaus@uoknor.edu

John R. Jensen
University of South Carolina
S. Bull Callcott Building, Room 114
Columbia, South Carolina 29208

D. Kasouf
Pacific Meridian Resources
149 Forest Avenue Suite 615
Marietta, Georgia 30060

Richard G. Kempka
Pacific Meridian Resources
9823 Old Winery Place, Suite 16
Sacramento, California 95827

Robert Klaver
EROS Data Center
Science and Applications Branch
Sioux Falls, South Dakota 57198
bklaver@edcsnw27.cr.usgs.gov

Andrew G. Klein
Cornell University
Ithaca, New York 14850
klein@geology.cornell.edu

William J. Kramber
Idaho Department of Water Resources
1301 N. Orchard St.
Boise, Idaho 83706
bkramber@idwr.state.id.us

Thomas H.C. Lo
South Florida Water Management District
P.O. Box 24680, 3301 Gun Club Road
West Palm Beach, Florida 33406
thomas.lo@sfwmd.gov

Shirley López Baros
Earth Data Analyis Center
University of New Mexico
2500 Yale Blvd. SE, Suite 100
Albuquerque, New Mexico 87131
slopez@spock.unm.edu

Diego Fabián Lozano García
Centro de Calidad Ambiental
Sucursal de Correos "J," C.P. 64849
Monterrey, Nuevo Leon, México
dlozano@campus.mty.itesm.mx

Joe Messina
SPOT Image Corporation
1897 Preston White Dr.
Reston, VA 22091-4368
messina@spot.com

Stan Morain
University of New Mexico
2500 Yale Boulevard SE Suite 100
Albuquerque, New Mexico 87131
smorain@spock.unm.edu

Jeffrey T. Morisette
North Carolina State University
Box 7106
Raleigh, North Carolina 27695-7106
jtmorise@unity.ncsu.edu

Stanislaw Mularz
University of Mining and Metallurgy
Department of Photogrammetry and
Remote Sensing Informatics
Al. Mickiewicza 30
30-059 Krakow, Poland

Paul R. H. Neville
University of New Mexico
2500 Yale Blvd. SE, Suite 100
Albuquerque, New Mexico 87131
pneville@spock.unm.edu

Yousef Nizam
Faculty of Engineering and Architecture
American University of Beirut
Beirut, Lebanon

Dana Nuñez Brown
Parsons Brinckerhoff
505 South Main Street, Suite 900
Orange, California 92668

Sonny Parafina
Cook-Hurlbert, Inc.
1911 Jollyville Road
Austin, Texas 78759
sparafina@cook-hurlbert.com

Robert Parrott
San Diego Association of Governments
Suite 800, First Interstate Plaza, 41 B
San Diego, California 92101
bpa@sandag.cog.ca.us

Douglas J. Paulson
Earth Data Analysis Center
University of New Mexico
2500 Yale Blvd SE, Suite 100
Albuquerque, New Mexico 87131
dpaulson@spock.unm.edu

Roland L. Redmond
University of Montana
Botany 205
Missoula, Montana 59812
red@selway.umt.edu

John B. Rehder
University of Tennessee
408 Geography and Geology Building
Knoxville, Tennessee 37916
rehder@utkux.utcc.utk.edu

Anthony Renaud
Maricopa County DOT
2901 West Durango Street
Phoenix, Arizona 85009
trenaud@gis.mcdot.maricopa.gov

João Ribeiro da Costa
Universidade Nova de Lisboa
Quinta da Torre
2825 Monte Caparica, Portugal
jrc@uninova.pt

Kim Ritman
Bureau of Resource Sciences
P.O. Box E11, Queen Victoria Terrace
Parkes, ACT 2600 Australia
kritman@mailpc.brs.gov.au

Ruth E. Spell
Pacific Meridian Resources
9823 Old Winery Place, Suite 16
Sacramento, California 95827

Casson Stallings
North Carolina State University
Box 7106
Raleigh, North Carolina 27695-7106
casson@ncsu.edu

Floyd Stayner
South Carolina Department of Natural Resources
1201 Main Street, Suite 100
Columbia, South Carolina 29201
stayner@stayner.dnr.state.sc.us

Doug Stow
San Diego State University
College of Arts and Sciences
San Diego, California 92182-4493
stow@sdsu.edu

Philip Tickle
National Forest Inventory
Bureau of Resource Sciences
P.O. Box E11, Queen Victoria Terrace
Parkes, ACT Australia
philip@henric.nric.gov.au

G.W. Turner
Waters and Catchments Branch
Environmental Protection Authority
Locked Bag 1502
Bankstown 2200, New South Wales, Australia

Simon M. Veitch
National Resource Information Centre
P.O. Box E11, Queen Victoria Terrace
Parkes, ACT 2600 Australia
simon%yoric@amphoric.nric.gov.au

Russell Watkins
University of Auckland
Private Bag 92019
Auckland, New Zealand
rl.watkins@auckland.ac.nz

Roy Welch
University of Georgia
Department of Geography
Athens, Georgia 30602
rwelch@risc.crms.ega.edu

Robert White
Bonneville Power Administration
Box 3621
Portland, Oregon 92709
rgwhite@bpa.gov

Richard D. Wright
San Diego State University
5500 Campanile Drive
San Diego, California 92182-4493
wright@typhoon.sdsu.edu

Numerics

1990 inventory, updating vector land use
inventories 314

A

access, wilderness continuum mapping 286
accuracy
polygon mode filter, accuracy assessment 233
precision distinguished 6
Active Microwave Imager-Synthetic Aperture
Radar (AMI-SAR)
AMI-SAR 90
Advanced Very High Resolution Radiometer
(AVHRR) 85
drought patterns 248
raster characteristics 19
Advanced Visible/Infrared Imaging Spectrometer
(AVIRIS), raster characteristics 18
aerial photography
biodiversity mapping 374
agriculture
agro-environmental monitoring
ARC/INFO GRID module 240
benefits 247
data and methodology 240
normalized difference vegetation index 242
project described 238
Rhine corridor 238
SPOT XS georeferenced image 240
desert irrigation water use
crop calendars 255
CropWatch program 255
groundwater depletion 254
project described 254
SPOT XS Level 1B scenes 256
drought patterns
Advanced Very High Resolution
Radiometer 248
Global Inventory Monitoring and Modeling
Studies 249
Normalized Difference Vegetation
Indices 248
project described 247
long-term change in irrigated land
ARC/INFO AML 263
CONTROLPOINTS routine 260
land use/land cover classes 261
project described 259
scanned aerial photography 260

shift in land and water use patterns 260
velum plat maps 260
polygon mode filter
decreasing within-field heterogeneity 237
GLEAMS 232
image processing and accuracy assessment 233
Landsat Thematic Mapper 233
National Wetlands Inventory data 233
project described 231
speckle in land cover classifications 231
tasseled cap transformation 232
agro-environmental monitoring
Rhine corridor 238
SPOT XS georeferenced image 240
airborne remote sensors
balloons 4
electromagnetic spectrum 8
ground resolution and raster characteristics 19
passive instruments 12
raster characteristics 18
rocketry, airborne remote sensors 4
weather satellites 5
along-track stereo imaging, passive instruments 12
analog data
electromagnetic spectrum 8
raster characteristics, reflectance curve 15
apparent naturalness, wilderness continuum mapping 286
atmospheric corrections, preprocessing 52
Australian National Wilderness Inventory, wilderness continuum mapping 285
AVHRR-NDVI composites, grassland environments 380

B

banding, preprocessing 51
bandwidth, raster characteristics 17
Beirut urban growth mapping
change detection 361
growth saturation index calculation 359
IDRISI PC-based software 360
land use change identification 359
land value estimation 359
project described 358
spectral classes 361
Belchatow Energy Mining Complex
project described 408
biodiversity mapping
project described 373
airborne videography 374
Nature Conservancy's Biological Conservation Database 375
New Zealand landscape structure 282
biophysical land units
Boolean logic 290
project described 289
biophysical naturalness, wilderness continuum mapping 286
bit systems, raster characteristics 20
Bonneville Power Administration, lightning analysis 264
Boolean logic, biophysical land units 290
boundaries of topographic divisions, topographic divisions for Italy 299

C

calibration, raster characteristics 25
cattle grazing determination, Tamaulipas, Mexico project 354
cell-based GIS processing
ARC/INFO GRID module 35
ARC/INFO IMAGE INTEGRATOR 36
change detection
Beirut urban growth mapping 361
image differencing technique 48
land cover change enhancement. *See* Land cover change enhancement.
PCA technique 48
reference image technique 47
image display and processing 47
classification of categories, theme extraction 43
cluster analysis, topographic divisions for Italy 299
coastline polygon coverage, global atmospheric water vapor storage 274
collateral information processing, raster characteristics 27
commercial forest areas, Tierra del Fuego forestry and biodiversity 305
Compact Airborne Spectrographic Imager (CASI), raster characteristics 18
compliance monitoring, water resources management applications 216

CONTROLPOINTS routine, long-term change in irrigated land 260
cost of imagery, Murray Darling Basin database 342
Countryside Information System, wilderness continuum mapping 286
CREST classification, Pacific coast estuaries 389
crop calendars, desert irrigation water use 255
CropWatch program, desert irrigation water use 255
cross-track stereo imagery, passive instruments 12

D

dark pixel subtraction, earth corrections 53
data and information
  accuracy and precision distinguished 6
  data output verification 5
  GIS protocols for decision making 6
  goal of information systems 5
  metadata, described 7
  modeling of processes 5
  quality control 7
  spatial and non-spatial data in GIS 7
data collection systems
  Active Microwave Imager-Synthetic Aperture Radar (AMI-SAR) 90
  European Resource Satellite (ERS-1)
    data availability 90
    program objectives 88
    digital data products 106
  Indian Remote Sensing Satellite (IRS)
    digital data products 105
    Linear Imaging Self-Scanning System (LISS) 95
    program objectives 92
    system characteristics 93
  Landsat
    digital data products 104
    multispectral scanner 79
    program objectives 77
    thematic mapper 80
    primary satellite sensor systems 71
  RADARSAT sensor 100
  RADARSAT, program objectives 97
  RADARSAT, system characteristics 98
  raster storage formats and media, MSI data 108
  SPOT, described 72
  SPOT, system characteristics 73
  SPOT, digital data products 101
  storage formats and media, SPOT image data
    SPOT, storage formats and media 109
  TIROS
    Advanced Very High Resolution Radiometer (AVHRR) 85
    history 82
    system characteristics 83
data dictionary construction, Solitario Dome and Terlingua Uplift 414
data fusion
  image display and processing 49
decision rules, theme extraction 46
descriptive statistics, topographic divisions for Italy 299
desert irrigation water use
  crop calendars 255
  CropWatch program 255
  groundwater depletion 254
  project described 254
  SPOT XS Level 1B scenes 256
detector corrections, preprocessing 51
detector drift, raster characteristics 25
detector sensitivity, raster characteristics 25
digital cartography, image formation 2
digital data
  electromagnetic spectrum 8
  passive instruments 12
  raster characteristics 17
digital data products
  European Resource Satellite (ERS-1) 106
  Indian Remote Sensing Satellite (IRS) 105
  Landsat 104
  SPOT (Satellite Pour l'Observation de la Terre) 101
DN (digital number) value, pixel 30
DN values, radiometric resolution 31
DN values, thematic data 31
drought patterns
  project described 247
dynamic algorithm compilation, glacier mapping 223

E

earth corrections
  atmospheric corrections, dark pixel subtraction and linear regression 53
  atmospheric corrections, solar altitude problems 54
  rectification 34
  registration 34
  earth rotation problems 54
Earth Observing System program, glacier mapping 219
earth rotation problems, earth corrections 54
ecological boundaries, New Zealand landscape structure 281
edge detection filter, spatial enhancement 39
edge enhancement filters, spatial enhancement 39
edge habitat, wetlands mitigation sites 403
environment and mineral exploration
  uranium mine contamination, project described 366
effective spatial resolution, raster characteristics 22
electromagnetic spectrum
  active and passive remote sensor systems 9
  airborne remote sensors 8
  digital and analog data 8
  fluorescence studies 8
  image creation from sensor data 7
  laser technologies 10
  long wavelengths infrared region 9
  mapping through energy reflection 8
  microwave sensing 9
  mid-wavelength infrared region 9
  multispectral scanners 10
  radar 9
  radiometry 9
  reflective infrared radiation 9
  remote sensing described 8
  satellite remote sensing data 9
  short wavelength infrared region 9
  spectrum span 8
electromechanical imaging spectrometers, passive instruments 13
electromechanical scanners, passive instruments 12
endemism in species, New Zealand landscape structure 282
energetics model, wetlands restoration areas 209
enhancement sites, wetlands mitigation 401
environment and mineral contamination
  Belchatow Energy Mining Complex, project described 408
  biodiversity mapping
    airborne videography 374
    Nature Conservancy's Biological Conservation Database 375
    project described 373
  grassland environments
    Global Assessment of Human Induced Soil Degradation database 379
    Inner Mongolia Autonomous Region 378
    People's Republic of China 378
    predicted degraded steppe 385
    project described 378
  Pacific coast estuaries
    CREST classification 389
    field sites 391
    NOAA C-CAP Version 1.0 protocols 388
    project described 386
  Solitario Dome and Terlingua Uplift
    project described 413
  wetlands mitigation sites
    edge habitat 403
    field verifications 407
    flood storage sites 404
    floodwater storage and water quality opportunity analyses 404
    Four Hole Swamp sub-basin 398
    high order streams 404
    project described 397
    quality control 400
    restoration, enhancement, and protection 401
    South Carolina Coastal Plain 397
    water quality sites 405
    wetland location 403
    wildlife habitat opportunity analysis 402
environment and mineral exploration
  ARC/INFO hydrologic modeling software 370
  Grants Uranium Belt 366
  ISODATA unsupervised clustering algorithm 369
  Landsat Thematic Mapper 368

potential contamination pathways 370
principal component analysis 368
environmental noise, correcting 308
ER mapper, glacier mapping 222
ER Mapping Annotation Overlay, glacier mapping 224
ERS-1
data availability 90
program objectives 88
European Resource Satellite (ERS-1), digital data products 106
experimental hyperspectral sensors, raster characteristics 18
extraction of change areas, updating vector land use inventories 319

F

false color composites, RGB color composite 32
field of view, raster characteristics 16
field sites, Pacific coast estuaries 391
field verifications, wetlands mitigation sites 407
flash density, lightning analysis 266
flood storage sites,
wetlands mitigation sites 404
floodwater storage analysis, wetlands mitigation 404
fluorescence studies, electromagnetic spectrum 8
formation of raster 15
Four Hole Swamp sub-basin, wetlands mitigation sites 398

G

gain and offset corrections, preprocessing 51
geographic information systems. *See* GIS 2
Geolink, glacier mapping 223
GeoMet Data Services, Inc., lightning analysis 265
geometric corrections, preprocessing 52
geometric processing, land cover change enhancement 309
geostationary satellites, global atmospheric water vapor storage 273
GIS
data fusion 2
image display and processing. *See* image display and processing 29
image formation 1
protocols for decision making 6
quality control 7
remote sensing links 5
research applications 2
site-specific GISs 18
spatial and non-spatial data in 7
spatial data analysis 2
spectral data 17
temporal resolution 23
glacier mapping
dynamic algorithm compilation 223
Earth Observing System program 219
ER mapper 222
ER Mapping Annotation Overlay 224
Geolink 223
Landsat Thematic Mapper imagery 222
project described 219
SPOT data 221
GLEAMS, polygon mode filter 232
Global Assessment of Human Induced Soil Degradation database 379
global atmospheric water vapor storage
project described 272
coastline polygon coverage 274
geostationary satellites 273
International Satellite Cloud Climatology C2 data set 273
polar orbiting satellites 273
water column amount 273
Global Inventory Monitoring and Modeling Studies, drought patterns 249
Grants Uranium Belt 366
grassland environments
AVHRR-NDVI composites 380
Global Assessment of Human-Induced Soil Degradation database 379
Inner Mongolia Autonomous Region 378
People's Republic of China 378
predicted degraded steppe 385
project described 378
grayscale display, pixels 30
Great Britain, wilderness continuum mapping 283
ground control points 26
ground resolution
aerial photography 19
raster characteristics 19

ground sampling distance, raster characteristics 16
ground truthing, irrigated water use inventory 227
groundwater depletion, desert irrigation water use 254
growth saturation index calculation, Beirut urban growth mapping 359

H

hardware, image formation 2
heads-up digitizing
  irrigated water use inventory 229
  Tierra del Fuego forestry and biodiversity 306
high order streams, wetlands mitigation sites 404
high resolution visible sensor, raster characteristics 16
HSV transformation
  hue-saturation-value space 33
hydrology
  glacier mapping
    dynamic algorithm compilation 223
    Earth Observing System program 219
    ER mapper 222
    ER Mapping Annotation Overlay 224
    Geolink 223
    Landsat Thematic Mapper imagery 222
    project described 219
    SPOT data 221
  water resources management applications
    Landsat data 212
    compliance monitoring 216
    project described 211
    SPOT data 212
    water use permits 213
  wetlands restoration areas
    digital line graphs 207
    energetics model 209
    National Wetlands Inventory 206
    project described 205
    riparian category 208
    waterfowl habitats 209
Hyperspectral IMAGER (HSI), raster characteristics 18
hyperspectral scanners, raster characteristics 17

I

image differencing technique, change detection 48
image display and processing
  cell-based GIS processing
    ARC/INFO GRID module 35
    ARC/INFO's IMAGE INTEGRATOR 36
  change detection 47
  data fusion 49
  density slicing 32
  digital number (DN) value 30
  earth corrections
    atmospheric corrections
      dark pixel subtraction and linear regression 53
      earth rotation problems 54
      solar altitude problems 54
    rectification 34
    registration 34
  grayscale display 30
  HSV transformation, hue-saturation-value space 33
  orthorectification 34
  preprocessing
    atmospheric corrections 52
    banding or striping 51
    detector corrections 51
    gain and offset corrections 51
    geometric corrections 52
    line dropout 51
    line start errors 51
    radiometric correction 51
  radiometric resolution, DN values 31
  rectification, ground control points 34
  RGB color composite, false color composites 32
  spatial enhancement
    edge detection filter 39
    edge enhancement filters 39
    noise reduction 38
    smoothing filters 38
  spectral enhancement
    indices 40
    principal components analysis 42
    tasseled cap transformation 42
  thematic data, DN values 31

theme extraction
classification of categories 43
decision rules 46
iterative self-organizing data analysis technique 44
Mahalanobis distance rule 46
maximum likelihood rule 46
minimum distance rule 46
supervised clustering 45
unsupervised clustering 44
image formation
data and information
accuracy and precision distinguished 6
data output verification 5
GIS protocols for decision making 6
goal of information systems 5
metadata described 7
modeling of processes 5
quality control 7
spatial and non-spatial data in GIS 7
electromagnetic spectrum
active and passive remote sensor systems 9
airborne remote sensors 8
digital and analog data 8
image creation from sensor data 7
laser technologies 10
long wavelengths infrared region 9
mapping through energy reflection 8
microwave sensing 9
mid-wavelength infrared region 9
multispectral scanners 10
radar 9
radiometry 9
reflective infrared radiation 9
remote sensing described 8
satellite remote sensing data 9
short wavelength infrared region 9
spectrum span 8
visible region 9
GIS and remote sensing links 5
GIS research applications 2
image formation
airborne remote sensor 4
remote sensing data uses 3
modeling 2
passive instruments
along-track stereo imaging 12
cross-track stereo imagery 12
digital data 12
electromechanical imaging spectrometers 13
electromechanical scanners 12
off-nadir viewing 12
push-broom scanning 13
radiometers 11
scanners and imaging spectrometers 12
scatterometers 12
solid-state imaging spectrometers 14
visible and near infrared region 11
whisk-broom scanning 13
raster characteristics
pointing and accuracy of telescopes and mirrors 25
Advanced Very High Resolution Radiometer (AVHRR) 19
Advanced Visible/Infrared Imaging Spectrometer 18
aerial photography and ground resolution 19
airborne remote sensors 18
analog reflectance curve 15
bandwidth 17
bit systems 20
calibration and registration 25
collateral information processing 27
Compact Airborne Spectrographic Imager (CASI) 18
detector drift 25
detector sensitivity 25
effective spatial resolution 22
experimental hyperspectral sensors 18
field of view 16
formation of raster 15
ground control point selection 26
ground resolution distance 19
ground sampling distance 16
high resolution visible sensor 16
Hyperspectral Imager (HSI) 18
image-to-image registration 26
imaging spectrometers 15
instantaneous field of view 16
Landsat Thematic Mapper 16

Lewis spacecraft 18
minimum spectral resolution 20
multispectral scanners 17
noise-equivalent change in reflectance 25
non-photographic sensors 19
Nyquist Theorem 15
pixel-to-pixel registration 26
platform stability 27
quantization 21
radiometric resolution 19
rate of data collection equation 24
resolution 19
rubbersheeting 27
satellite remote sensing data 14
scanners 15
signal-to-noise ratio 25
site-specific GISs 18
spectral channel selection 22
spectral channels 15
spectral library 23
spectral properties 22
spectral resolution 19, 22
swath width 24
telemetry rates and carrier wave bandwidths 14
temporal resolution 23
voltage range 20
word lengths 21
remote sensing 1
remote sensing language 3
weather balloons 4
image processing, polygon mode filter 233
image-to-image registration, raster characteristics 26
IMAGINE SML models, grassland environments 384
imaging spectrometers
raster characteristics 15
indices, spectral enhancement 40
information science, image formation 2
Inner Mongolia Autonomous Region, grassland environments 378
instantaneous field of view, raster characteristics 16
International Satellite Cloud Climatology C2 data set 273
irrigated water use inventory
Landsat data
normalized difference vegetation index 228
manual digitizing 229
on-screen digitizing 229
IRS
digital data products 105
Linear Imaging Self-Scanning System (LISS) 95
program objectives 92
system characteristics 93
ISODATA clustering
theme extraction 44

K

Kansas Applied Remote Sensing Program, map development 328
Kansas Data Access and Support Center, map development 334
Kathmandu urban growth
future growth prediction 348
Kathmandu Valley 344
project described 343
road network 346
spectral confusion 346

L

land cover change enhancement
change detection processing
multitemporal overlay 309
geometric processing 309
multidate classification of transition sequences 312
multitemporal image differencing 310
multitemporal remote sensing 307
post-classification comparison approach 311
project described
radiometric processing 309
land use change identification, Beirut urban growth mapping 359
land use/land cover
long-term change in irrigated land 261
map development
Kansas Applied Remote Sensing Program 328

Kansas Data Access and Support Center 334
project described 328
Murray Darling Basin database
cost of imagery 342
metadata 340
project described 334
Tijuana River Watershed Project
heads-up digitizing 325
project described 323
U.S.-Mexico border GIS 322
updating vector land use inventories
1990 inventory 314
1995 inventory 318
extraction of change areas 319
multi-date satellite imagery 313
preprocessing 318
project described 314
San Diego Association of Governments 313
transportation modeling 320
land value estimation, Beirut urban growth mapping 359
Landsat
digital data products 104
irrigated water use inventory 229
program objectives 77
system characteristics
multispectral scanner 79
thematic mapper 80
water resources management applications 212
Landsat MSS, New Zealand landscape structure 280
Landsat thematic mapper
glacier mapping 222
polygon mode filter 233
raster characteristics 16
Tierra del Fuego forestry and biodiversity 304
uranium mine contamination 368
landscape analysis
biophysical land units
Boolean logic 290
potential plant communities 291
project described 289
satellite remote sensing data 290
New Zealand landscape structure
biodiversity 282
ecological boundaries 281
endemism in species 282
Landsat MSS 280
polygon decrease 281
predator introduction 282
project described 279
Regional Bureau of Evaluation Project 279
spatial heterogeneity of land cover types 280
SPOT XS 280
Tierra del Fuego forestry and biodiversity
commercial forest areas 305
GPS technologies 303
heads-up digitizing 306
Landsat TM 304
project described 301
wetland area buffering 305
topographic divisions for Italy
boundaries of topographic divisions 299
cluster analysis 299
descriptive statistics 299
digital elevation models 293
mean height values archive 294
morphometric criteria 295
morphometric parameters 300
project described 293
wilderness continuum mapping
apparent naturalness 286
Australian National Wilderness Inventory 285
biophysical naturalness 286
Countryside Information System 286
Great Britain 283
multiple criteria evaluation approach 284
project described 283
remoteness from access 286
remoteness from population 286
laser technologies, electromagnetic spectrum 10
Lewis spacecraft 18
lightning analysis
Bonneville Power Administration 264
flash density 266
GeoMet Data Services, Inc. 265
magnetic direction finding technology 265
National Lightning Detection Network 265
project described 264
line dropout, preprocessing 51
line start errors, preprocessing 51
Linear Imaging Self-Scanning System (LISS) 95
linear regression, earth corrections 53

long wavelengths infrared region, electromagnetic spectrum 9
long-term change in irrigated land
land use/land cover classes 261
project described 259
scanned aerial photography 260
shift in land and water use patterns 260
velum plat maps 260

M

magnetic direction finding technology, lightning analysis 265
Mahalanobis distance rule, theme extraction 46
manual digitizing, irrigated water use inventory 229
map development
Kansas Applied Remote Sensing Program 328
Kansas Data Access and Support Center 334
project described 328
maximum likelihood 355
maximum likelihood
theme extraction 46
mean height values archive, topographic divisions for Italy 294
metadata
described 7
Murray Darling Basin database 340
meteorology 265
global atmospheric water vapor storage
coastline polygon coverage 274
geostationary satellites 273
International Satellite Cloud Climatology C2 data set 273
polar orbiting satellites 273
project described 272
water column amount 273
lightning analysis
Bonneville Power Administration 264
flash density 266
magnetic direction finding technology 265
National Lightning Detection Network 265
project described 264
SPOT information 266
winter road maintenance
ARC/INFO GRID module 268
expenditure data 270
project described 267
weighted lane length 270
winter index 268
Microsoft Excel 5.0, Tierra del Fuego forestry and biodiversity 305
microwave sensing
active systems 9
electromagnetic spectrum 9
mid-wavelength infrared region, electromagnetic spectrum 9
mineral exploration. *See* environment and mineral exploration
minimum distance
Tamaulipas, Mexico project 355
theme extraction 46
minimum spectral resolution, raster characteristics 20
modeling
image formation 2
processes 5
morphometric parameters, topographic divisions for Italy 295
MSI data, raster storage formats and media 108
multidate classification of transition sequences, change detection enhancement 312
multiple criteria evaluation approach, wilderness continuum mapping 284
Multiple Species Conservation Planning, Tijuana River Watershed Project 325
multispectral scanners
electromagnetic spectrum 10
Landsat 79
raster characteristics 17
multitemporal image differencing 311
multitemporal overlay, change detection enhancement 309
multitemporal remote sensing. *See* Land cover change enhancement 307
Murray Darling Basin database
cost of imagery 342
metadata 340
project described 334

N

National Lightning Detection Network, lightning analysis 265
National Wetlands Inventory

polygon mode filter 233
wetlands restoration areas 206
Nature Conservancy's Biological Conservation Database, biodiversity 375
NDVI. *See* Normalized difference vegetation index 228
New Zealand landscape structure
biodiversity 282
ecological boundaries 281
endemism in species 282
Landsat MSS 280
polygon decrease 281
predator introduction 282
project described 279
Regional Bureau of Evaluation Project 279
spatial heterogeneity of land cover types 280
SPOT XS 280
NOAA C-CAP Version 1.0 protocols, Pacific coast estuaries 388
noise, correcting environmental and sensor noise 308
noise-equivalent change in reflectance, raster characteristics 25
non-photographic sensors, raster characteristics 19
normalized difference vegetation index
agro-environmental monitoring 242
drought patterns 248
grassland environments 379
irrigated water use inventory 228
Nyquist theorem, raster characteristics 15

O

off-nadir viewing, passive instruments 12

P

Pacific coast estuaries
CREST classification 389
field sites 391
NOAA C-CAP Version 1.0 protocols 388
project described 386
passive instruments
along-track stereo imaging 12
cross-track stereo imagery 12
digital data 12
electromechanical imaging spectrometers 13
electromechanical scanners 12
off-nadir viewing 12
push-broom scanning 13
radiometers 11
satellite and airborne instruments 12
scanners
passive instruments 12
scatterometers 12
solid-state imaging spectrometers 14
visible and near infrared region 11
whisk-broom scanning 13
PCA technique, change detection 48
People'4s Republic of China, grassland environments 378
pixels, DN (digital number) value 30
pixels, grayscale display 30
pixel-to-pixel registration, raster characteristics 26
platform stability, raster characteristics 27
pointing and accuracy of telescopes and mirrors, raster characteristics 25
polar orbiting satellites, global atmospheric water vapor storage 273
polygon decrease, New Zealand landscape structure 281
polygon mode filter
decreasing within-field heterogeneity 237
GLEAMS 232
image processing and accuracy assessment 233
Landsat Thematic Mapper 233
National Wetlands Inventory data 233
project described 231
speckle in land cover classifications 231
tasseled cap transformation 232
post-classification comparison approach, change detection enhancement 311
potential contamination pathways, uranium mine contamination 370
potential plant communities, biophysical land units 291
predator introduction, New Zealand landscape structure 282
predicted degraded steppe, grassland environments 385
preprocessing
detector corrections 51
updating vector land use inventories 318
atmospheric corrections 52

banding or striping 51
gain and offset corrections 51
geometric corrections 52
line dropout 51
line start errors 51
radiometric correction 51
primary satellite sensors systems, described 71
principal component analysis, uranium mine contamination 368
push-broom scanning, passive instruments 13

Q

quality control
data and information 7
wetlands mitigation sites 400
quantization, raster characteristics 21

R

radar, electromagnetic spectrum 9
RADARSAT
program objectives 97
RADARSAT sensor 100
system characteristics 98
radiometric correction, preprocessing 51
radiometry
electromagnetic spectrum 9
passive instruments 11
processing, land cover change enhancement 309
raster characteristics 19
resolution, raster characteristics 20
raster characteristics
Advanced Very High Resolution Radiometer (AVHRR) 19
Advanced Visible/Infrared Imaging Spectrometer (AVIRIS) 18
airborne remote sensors 18
analog reflectance curve 15
bandwidth 17
bit systems 20
calibration and registration 25
collateral information processing 27
Compact Airborne Spectrographic Imager (CASI) 18
detector drift 25
detector sensitivity 25
effective spatial resolution 22
experimental hyperspectral sensors 18
field of view 16
formation of raster, described 15
ground control point selection 26
ground resolution distance 19
ground sampling distance 16
high resolution visible sensor 16
Hyperspectral Imager (HSI) 18
image-to-image registration 26
imaging spectrometers 15
instantaneous field of view 16
Landsat Thematic Mapper 16
Lewis spacecraft 18
minimum spectral resolution 20
multispectral scanners 17
noise-equivalent change in reflectance 25
non-photographic sensors 19
Nyquist Theorem 15
pixel-to-pixel registration 26
platform stability 27
pointing and accuracy of telescopes and mirrors 25
quantization 21
radiometric resolution 19
rate of data collection equation 24
resolution 19
rubbersheeting 27
satellite remote sensing data 14
scanners 15
signal-to-noise ratio 25
site-specific GISs 18
spectral channel selection 22
spectral channels 15
spectral library 23
spectral properties 22
spectral resolution 19
swath width 24
telemetry rates and carrier wave bandwidths 14
temporal resolution 23
voltage range 20
word lengths 21
raster storage formats and media, MSI data 108
rate of data collection equation, raster characteristics 24
rectification
earth corrections 34
ground control points 34

reference image technique, change detection 47
reflective infrared radiation, electromagnetic spectrum 9
Regional Bureau of Evaluation Project, New Zealand landscape structure 279
registration
- earth correction 34
- raster characteristics 25

remote sensing
- active and passive sensor systems 9
- data uses 4
- electromagnetic spectrum 8
- GIS links 5
- image formation 1

remoteness, wilderness continuum mapping 286
resolution, raster characteristics 19
restoration sites, wetlands mitigation sites 401
RGB color composite, false color composites 32
Rhine corridor, agro-environmental monitoring 238
riparian category, wetlands restoration areas 208
road network, Kathmandu urban growth 346
rubbersheeting, raster characteristics 27

S

San Diego Association of Governments, updating vector land 313
satellite remote sensing data
- biophysical land units 290
- electromagnetic spectrum 9
- passive instruments 12
- raster characteristics 14
- temporal resolution 23

scanned aerial photography, long-term change in irrigated land 260
scanners
- raster characteristics 15

scatterometers, passive instruments 12
sensor noise, correcting 308
short wavelength infrared region, electromagnetic spectrum 9
signal-to-noise ratio, raster characteristics 25
smoothing filters, spatial enhancement 38
software, image formation 2
solar altitude problems, earth corrections 54
solid-state imaging spectrometers, passive instruments 14
Solitario Dome and Terlingua Uplift
- project described 413

South Carolina Coastal Plain, wetlands mitigation sites 397
spatial data analysis, GIS 2
spatial enhancement
- edge detection filter 39
- edge enhancement filters 39
- noise reduction 38
- smoothing filters 38

spatial heterogeneity of land cover types, New Zealand landscape structure 280
speckle in land cover classifications, polygon mode filter 231
spectral channels, raster characteristics 15
spectral classes, Beirut urban growth mapping 361
spectral confusion, Kathmandu urban growth 346
spectral enhancement
- indices 40
- principal components analysis 42
- tasseled cap transformation 42

spectral library, raster characteristics 23
spectral properties, raster characteristics 22
spectral resolution
- raster characteristics 22

spectral resolution, raster characteristics 19
SPOT
- described 72
- digital data products 101
- glacier mapping 221
- lightning analysis 266
- storage formats and media 109
- system characteristics 73
- water resources management applications 212
- XS Level 1B scenes, desert irrigation water use 256

SPOT XS
- georeferenced image, agro-environmental monitoring 240
- New Zealand landscape structure 280

storage formats and media, SPOT 109
striping, preprocessing 51
supervised clustering, theme extraction 45
swath width, raster characteristics 24
synoptic global coverage, raster characteristics 17

T

Tamaulipas, Mexico project
cattle grazing determination 354
data and methodology 351
minimum distance and maximum likelihood classifications 355
tasseled cap transformation
polygon mode filter 232
spectral enhancement 42
telemetry rates, raster characteristics
carrier wave bandwidths, raster characteristics 14
Television and Infrared Observation Satellite. *See* TIROS 82
temporal resolution, raster characteristics 23
temporally dynamic landscape analysis 307
thematic data, DN values 31
thematic mapper, Landsat 80
theme extraction
classification of categories 43
decision rules 46
iterative self-organizing data analysis technique 44
Mahalanobis distance rule 46
maximum likelihood rule 46
minimum distance rule 46
supervised clustering 45
unsupervised clustering 44
Tierra del Fuego forestry and biodiversity
commercial forest areas 305
GPS technologies 303
heads-up digitizing 306
Landsat TM 304
project described 301
wetland area buffering 305
Tijuana River Watershed Project 327
heads-up digitizing 325
Multiple Species Conservation Planning 325
project described 323
U.S.-Mexico border GIS 322
TIROS
Advanced Very High Resolution Radiometer (AVHRR) 85
history 82
system characteristics 83
topographic divisions for Italy
boundaries of topographic divisions 299
cluster analysis 299
descriptive statistics 299
mean height values archive 294
morphometric criteria 295
morphometric parameters 300
project described 293
transportation modeling, updating vector land use inventories 320

U

U.S.-Mexico border GIS, Tijuana River Watershed Project 322
ultraspectral scanners, raster characteristics 17
unsupervised clustering, theme extraction 44
updating vector land use inventories
extraction of change areas 319
extraction of change areas
1990 inventory 314
1995 inventory 318
multidate satellite imager 313
preprocessing 318
project described 314
San Diego Association of Governments 313
transportation modeling 320
uranium mine contamination
Grants Uranium Belt 366
ISODATA unsupervised clustering algorithm 369
Landsat Thematic Mapper 368
potential contamination pathways 370
principal component analysis 368
project described 366
urban and regional planning
Beirut urban growth mapping
change detection 361
growth saturation index calculation 359
land use change identification 359
land value estimation 359
project described 358
spectral classes 361
Kathmandu urban growth
future growth prediction 348
Kathmandu Valley 344
project described 343
road network 346
spectral confusion 346
Tamaulipas, Mexico project

cattle grazing determination 354
described 350
minimum distance and maximum likelihood classifications 355

V

vector land use inventories, multidate satellite imagery 313
velum plat maps, long-term change in irrigated land 260
verification of data output 5
visible and near infrared region, passive instruments 11
visible region, electromagnetic spectrum 9
voltage range, raster characteristics 20

W

water column amount, global atmospheric water vapor storage 273
water quality sites, wetlands mitigation sites 405
water resources management applications
compliance monitoring 216
project described 211
SPOT data 212
water use permits 213
water use permits, water resources management applications 213
waterfowl habitats, wetlands restoration areas 209
weather balloons, remote sensing data 4
weather satellites, airborne remote sensors 5
weighted lane length, winter road maintenance 270
wetland area buffering, Tierra del Fuego forestry and biodiversity 305
wetland location, mitigation sites 403
wetlands mitigation sites
edge habitat 403
field verifications 407
flood storage sites 404
floodwater storage and water quality opportunity analyses 404
Four Hole Swamp sub-basin 398
high order streams 404
project described 397
quality control 400
restoration, enhancement, and protection 401
South Carolina Coastal Plain 397
water quality sites 405
wetland location 403
wildlife habitat opportunity analysis 402
wetlands restoration areas
digital line graphs 207
energetics model 209
National Wetlands Inventory 206
project described 205
riparian category 208
waterfowl habitats 209
whisk-broom scanning, passive instruments 13
wilderness continuum mapping
apparent naturalness 286
Australian National Wilderness Inventory 285
biophysical naturalness 286
Countryside Information System 286
Great Britain 283
multiple criteria evaluation approach 284
project described 283
remoteness from access 286
remoteness from population 286
wildlife habitat opportunity analysis, wetlands mitigation sites 402
winter index, winter road maintenance 268
winter road maintenance
expenditure data 270
project described 267
weighted lane length 270
winter index 268
within-field heterogeneity, polygon mode filter 237
word lengths, raster characteristics 21

More OnWord Press Titles

Pro/ENGINEER and Pro/JR. Books

INSIDE Pro/ENGINEER, 2E
Book $49.95 Includes Disk

Pro/ENGINEER Quick Reference, 2E
Book $24.95

Pro/ENGINEER Exercise Book, 2E
Book $39.95 Includes Disk

Thinking Pro/ENGINEER
Book $49.95

Pro/ENGINEER Tips and Techniques
Book $59.95

INSIDE Pro/JR.
Book $49.95

MicroStation Books

INSIDE MicroStation 5X, 3d ed.
Book $34.95 Includes Disk

INSIDE MicroStation 95, 4E
Book $39.95 Includes Disk

MicroStation Reference Guide 5.X
Book $18.95

MicroStation 95 Quick Reference
Book $24.95

MicroStation 95 Productivity Book
Book $49.95

MicroStation Exercise Book 5.X
Book $34.95 Includes Disk
Optional Instructor's Guide $14.95

MicroStation 95 Exercise Book
Book $39.95 Includes Disk
Optional Instructor's Guide $14.95

MicroStation for AutoCAD Users, 2E
Book $34.95

Adventures in MicroStation 3D
Book $49.95 Includes CD-ROM

Build Cell for 5.X
Software $69.95

101 MDL Commands (5.X and 95)
Optional Executable Disk $101.00
Optional Source Disks (6) $259.95

Windows NT

Windows NT for the Technical Professional
Book $39.95

SunSoft Solaris Series

SunSoft Solaris 2. User's Guide*
Book $29.95 Includes Disk

SunSoft Solaris 2. for Managers and Administrators*
Book $34.95

SunSoft Solaris 2. Quick Reference*
Book $18.95

*Five Steps to SunSoft Solaris 2.**
Book $24.95 Includes Disk

SunSoft Solaris 2. for Windows Users*
Book $24.95

The Hewlett Packard HP-UX Series

HP-UX User's Guide
Book $29.95 Includes Disk

HP-UX Quick Reference
Book $18.95

Five Steps to HP-UX
Book $24.95 Includes Disk

Softdesk Books

Softdesk Architecture 1 Certified Courseware
Book $34.95 Includes CD-ROM

Softdesk Architecture 2 Certified Courseware
Book $34.95 Includes CD-ROM

Softdesk Civil 1 Certified Courseware
Book $34.95 Includes CD-ROM

Softdesk Civil 2 Certified Courseware
Book $34.95 Includes CD-ROM

INSIDE Softdesk Architectural
Book $49.95 Includes Disk

INSIDE Softdesk Civil
Book $49.95 Includes Disk

Other CAD

Manager's Guide to Computer-Aided Engineering
Book $49.95

Fallingwater in 3D Studio: A Case Study and Tutorial
Book $39.95 Includes Disk

Geographic Information Systems (GIS)

INSIDE ARC/INFO
Book $74.95 Includes CD-ROM

ARC/INFO Quick Reference
Book $24.95

INSIDE ArcView
Book $39.95 Includes CD-ROM

ArcView Developer's Guide
Book $49.95

ArcView/Avenue Programmer's Reference
Book $49.95

101 ArcView/Avenue Scripts: The Disk
Disk $101.00

ArcView Exercise Book
Book $49.95 Includes CD-ROM

INSIDE ArcCAD
Book $39.95 Includes Disk

The GIS Book, 3d ed.
Book $34.95

INSIDE MapInfo Professional
Book $49.95

Raster Imagery in Geographic Information Systems
Book $59.95

GIS: A Visual Approach
Book $39.95

Interleaf Books

INSIDE Interleaf (v. 6)
Book $49.95 Includes Disk

Adventurer's Guide to Interleaf Lisp
Book $49.95 Includes Disk

Interleaf Exercise Book
Book $39.95 Includes Disk

Interleaf Quick Reference (v. 6)
Book $24.95

Interleaf Tips and Tricks
Book $49.95 Includes Disk

OnWord Press Distribution

End Users/User Groups/Corporate Sales

OnWord Press books are available worldwide to end users, user groups, and corporate accounts from your local bookseller or computer/software dealer, or from Softstore/CADNEWS Bookstore: call 1-800-CADNEWS (1-800-223-6397) or 505-474-5120; fax 505-474-5020; write to CADNEWS Bookstore, 2530 Camino Entrada, Santa Fe, NM 87505-4835, or e-mail ORDERS@HMP.COM. CADNEWS Bookstore is a division of SoftStore, Inc., a High Mountain Press Company.

Wholesale, Including Overseas Distribution

High Mountain Press distributes OnWord Press books internationally. For terms call 1-800-4-ONWORD (1-800-466-9673) or 505-474-5130; fax to 505-474-5030; e-mail ORDERS@HMP.COM, or write to High Mountain Press, 2530 Camino Entrada, Santa Fe, NM 87505-4835, USA. Outside North America, call 505-474-5130.

On the Internet: http://www.hmp.com

OnWord Press 2530 Camino Entrada, Santa Fe, NM 87505-4835 USA